T0338119

DISTRIBUTED COOPERATIVE CONTROL OF MULTI-AGENT SYSTEMS

DISTRIBUTED COOPERATIVE CONTROL OF MULTI-AGENT SYSTEMS

Wenwu Yu
Southeast University, China

Guanghui Wen
Southeast University, China

Guanrong Chen
City University of Hong Kong, China

Jinde Cao
Southeast University, China

This edition first published 2016

Published by John Wiley & Sons Singapore Pte. Ltd., 1 Fusionopolis Walk, #07-01 Solaris South Tower, Singapore 138628, under exclusive license granted by Higher Education Press for all media and languages excluding Simplified and Traditional Chinese and throughout the world excluding Mainland China, and with non-exclusive license for electronic versions in Mainland China.

For details of our global editorial offices, for customer services and for information about how to apply for permission to reuse the copyright material in this book please see our website at www.wiley.com.

Library of Congress Cataloging-in-Publication Data

Names: Yu, Wenwu, 1982-
Title: Distributed cooperative control of multi-agent systems / Wenwu Yu [and four others].
Description: Singapore : John Wiley & Sons, Inc., [2016] | Includes bibliographical references and index.
Identifiers: LCCN 2016018240 (print) | LCCN 2016023245 (ebook) | ISBN 9781119246206 (cloth) | ISBN 9781119246237 (pdf) | ISBN 9781119246220 (epub)
Subjects: LCSH: Chaotic synchronization. | Synchronization. | System analysis.
Classification: LCC Q172.5.S96 D57 2017 (print) | LCC Q172.5.S96 (ebook) | DDC 003/.7–dc23
LC record available at https://lccn.loc.gov/2016018240

Typeset in 11/13pt TimesLTStd by SPi Global, Chennai, India
Printed and bound in Singapore by Markono Print Media Pte Ltd

10 9 8 7 6 5 4 3 2 1

Contents

Preface ix

1 Introduction 1
1.1 Background 1
1.1.1 Networked Multi-agent Systems 1
1.1.2 Collective Behaviors and Cooperative Control in Multi-agent Systems 2
1.1.3 Network Control in Multi-agent Systems 4
1.1.4 Distributed Consensus Filtering in Sensor Networks 5
1.2 Organization 6

2 Consensus in Multi-agent Systems 11
2.1 Consensus in Linear Multi-agent Systems 11
2.1.1 Preliminaries 11
2.1.2 Model Formulation and Results 13
2.2 Consensus in Nonlinear Multi-agent Systems 15
2.2.1 Preliminaries and Model Formulation 15
2.2.2 Local Consensus of Multi-agent Systems 16
2.2.3 Global Consensus of Multi-agent Systems in General Networks 19
2.2.4 Global Consensus of Multi-agent Systems in Virtual Networks 26
2.2.5 Simulation Examples 29
2.3 Notes 30

3 Second-Order Consensus in Multi-agent Systems 31
3.1 Second-Order Consensus in Linear Multi-agent Systems 32
3.1.1 Model Formulation 32
3.1.2 Second-Order Consensus in Directed Networks 33
3.1.3 Second-Order Consensus in Delayed Directed Networks 37
3.1.4 Simulation Examples 41

3.2 Second-Order Consensus in Nonlinear Multi-agent Systems 42
3.2.1 Preliminaries 42
3.2.2 Second-Order Consensus in Strongly Connected Networks 45
3.2.3 Second-Order Consensus in Rooted Networks 50
3.2.4 Simulation Examples 53
3.3 Notes 54

4 Higher-Order Consensus in Multi-agent Systems **56**
4.1 Preliminaries 56
4.2 Higher-Order Consensus in a General Form 58
4.2.1 Synchronization in Complex Networks 58
4.2.2 Higher-Order Consensus in a General Form 59
4.2.3 Consensus Region in Higher-Order Consensus 60
4.3 Leader-Follower Control in Multi-agent Systems 64
4.3.1 Leader-Follower Control in Multi-agent Systems with Full-State Feedback 65
4.3.2 Leader-Follower Control with Observers 67
4.4 Simulation Examples 69
4.4.1 Consensus Regions 69
4.4.2 Leader-Follower Control with Full-State Feedback 70
4.4.3 Leader-Follower Control with Observers 70
4.5 Notes 71

5 Stability Analysis of Swarming Behaviors **73**
5.1 Preliminaries 73
5.2 Analysis of Swarm Cohesion 76
5.3 Swarm Cohesion in a Noisy Environment 80
5.4 Cohesion in Swarms with Switched Topologies 82
5.5 Cohesion in Swarms with Changing Topologies 84
5.6 Simulation Examples 93
5.7 Notes 95

6 Distributed Leader-Follower Flocking Control **96**
6.1 Preliminaries 96
6.1.1 Model Formulation 97
6.1.2 Nonsmooth Analysis 99
6.2 Distributed Leader-Follower Control with Pinning Observers 103
6.3 Simulation Examples 110
6.4 Notes 114

7 Consensus of Multi-agent Systems with Sampled Data Information **115**
7.1 Problem Statement 116

7.2 Second-Order Consensus of Multi-agent Systems with Sampled Full Information 117
7.2.1 Second-Order Consensus of Multi-agent Systems with Sampled Full Information 119
7.2.2 Selection of Sampling Periods 122
7.2.3 Design of Coupling Gains 123
7.2.4 Consensus Region for the Network Spectrum 125
7.2.5 Second-Order Consensus in Delayed Undirected Networks with Sampled Position and Velocity Data 125
7.2.6 Simulation Examples 128
7.3 Second-Order Consensus of Multi-agent Systems with Sampled Position Information 132
7.3.1 Second-Order Consensus in Multi-agent Dynamical Systems with Sampled Position Data 132
7.3.2 Simulation Examples 139
7.4 Consensus of Multi-agent Systems with Nonlinear Dynamics and Sampled Information 142
7.4.1 The Case with a Fixed and Strongly Connected Topology 145
7.4.2 The Case with Topology Containing a Directed Spanning Tree 149
7.4.3 The Case with Topology Having no Directed Spanning Tree 155
7.5 Notes 158

8 Consensus of Second-Order Multi-agent Systems with Intermittent Communication 159
8.1 Problem Statement 159
8.2 The Case with a Strongly Connected Topology 161
8.3 The Case with a Topology Having a Directed Spanning Tree 165
8.4 Consensus of Second-Order Multi-agent Systems with Nonlinear Dynamics and Intermittent Communication 167
8.5 Notes 172

9 Distributed Adaptive Control of Multi-agent Systems 174
9.1 Distributed Adaptive Control in Complex Networks 175
9.1.1 Preliminaries 175
9.1.2 Distributed Adaptive Control in Complex Networks 176
9.1.3 Pinning Edges Control 178
9.1.4 Simulation Examples 181
9.2 Distributed Control Gains Design for Second-Order Consensus in Nonlinear Multi-agent Systems 183
9.2.1 Preliminaries 184
9.2.2 Distributed Control Gains Design: Leaderless Case 186
9.2.3 Distributed Control Gains Design: Leader-Follower Case 190
9.2.4 Simulation Examples 194
9.3 Notes 196

10 Distributed Consensus Filtering in Sensor Networks **198**
10.1 Preliminaries 199
10.2 Distributed Consensus Filters Design for Sensor Networks with Fully-Pinned Controllers 201
10.3 Distributed Consensus Filters Design for Sensor Networks with Pinning Controllers 205
10.4 Distributed Consensus Filters Design for Sensor Networks with Pinning Observers 207
10.5 Simulation Examples 210
10.6 Notes 213

11 Delay-Induced Consensus and Quasi-Consensus in Multi-agent Systems **214**
11.1 Problem Statement 214
11.2 Delay-Induced Consensus and Quasi-Consensus in Multi-agent Dynamical Systems 217
11.3 Motivation for Quasi-Consensus Analysis 223
11.4 Simulation Examples 224
11.5 Notes 228

12 Conclusions and Future Work **229**
12.1 Conclusions 229
12.2 Future Work 230

Bibliography **232**

Index **241**

Preface

In the natural, social and technological worlds, there are many large-scale complex networks with multiple agents for which centralized control is often difficult or even impossible to apply. Therefore, distributed cooperative control of multi-agent systems has been widely investigated for its easy implementation, strong robustness, and high self-organizability. In a multi-agent system under a network communication structure, to cooperative with other agents, everyone needs to share information with its adjacent peers so that all agents can agree on a common goal of interest. Recently, some progress has been made in analyzing collective behaviors in such dynamical networks for which closely related focal topics are synchronization, consensus, swarming and flocking. However, there are very few books focusing on distributed cooperative control of multi-agent systems addressing a broad spectrum of scientific interest. It is now clear that the impact of cooperative control of multiple autonomous agents in engineering and technology is prominent and will be far-reaching. Thus, an in-depth study with detailed analysis of this subject will benefit both theoretical research and engineering applications in the near-future development of related technologies.

The authors of this book have been working together on distributed cooperative control of multi-agent systems for about seven years with some relatively comprehensive results developed on the topic. This book summaries their main contributions in the field with general background knowledge and information, for a broad discipline, including particularly dynamics of general multi-agent systems, for example first-order, second-order and higher-order consensus, as well as swarming and flocking behaviors. Some technical issues about multi-agent systems with sampled data information transmission, missing control input, adaptive control and filters design are also investigated.

This book presents the basic knowledge along with a thorough review of the-state-of-the-art progress in the field. The contents of the book are summarized as follows: (1) first-order, second-order, and higher-order consensus are discussed for both linear and nonlinear multi-agent systems in Chapters 2-4; (2) stability analysis of a general swarming model with hybrid nonlinear profiles, stochastic noise, and switching topologies is investigated in Chapter 5; (3) distributed leader-follower flocking control for multi-agent dynamical systems with time-varying velocities is studied in Chapter 6; (4) hybrid control of multi-agent systems including sampled-data control and intermittent control is further discussed in Chapters 7 and 8; (5) fully adaptive control protocols for multi-agent systems are designed in Chapter 9; (6) some applications to distributed consensus filtering in sensor networks are presented in Chapter 10; (7) an interesting problem for delay-induced consensus in multi-agent systems with second-order dynamics is addressed in Chapter 11; (8) conclusions are drawn with future research outlook in Chapter 12.

This work is supported by the National Natural Science Foundation of China under Grant Nos. 61322302, 61304168, and by the Natural Science Foundation of Jiangsu Province of China under Grant No. BK20130595. The first author, Wenwu Yu, would like to express his deepest gratitude to his wife Lingling Yao and little daughter, Zhiyao Yu for their love and unconditional support; this academic book a good gift for them.

Wenwu Yu, Guanghui Wen, Guanrong Chen, and Jinde Cao
Winter, 2015

1

Introduction

1.1 Background

1.1.1 Networked Multi-agent Systems

Most large-scale systems in nature and human societies, such as biological neural networks, ecosystems, metabolic pathways, the Internet, the WWW, and electrical power grids can be described by networks with nodes representing individuals in the system and edges representing the connections between them. Recently, the study of various complex networks and systems has attracted increasing attention from researchers in various fields of physics, mathematics, engineering, biology, and sociology alike [8, 35, 62, 90, 117, 118, 119, 123, 142].

In the early 1960s, Erdös and Rényi (ER) proposed a random-graph model, which laid a solid foundation for modern network theory [35]. In a random network, each pair of nodes is connected with a certain probability. In order to describe a transition from a regular network to a random network, Watts and Strogatz (WS) proposed an interesting small-world network model [123]. Then, Newman and Watts (NW) modified the original WS model to generate another version of the small-world model [80]. Meanwhile, Barabási and Albert (BA) proposed a scale-free network model, in which the degree distribution of the nodes follows a power-law form [8]. Since then, small-world and scale-free networks have been extensively investigated worldwide.

Cooperative and collective behaviors in networks of multiple autonomous agents have also received considerable attention in recent years due to the growing interest in understanding the amazing animal group behaviors, such as flocking and swarming, and also due to their emerging broad applications in sensor networks, UAV (unmanned air vehicles) formations, and robotic teams. To coordinate with other agents in a network, every agent needs to share information with its adjacent peers so that all can agree on a common goal of interest, such as the value of some measurement in a sensor network, the heading in a UAV formation, or the target position of a robotic team.

Distributed Cooperative Control of Multi-agent Systems, First Edition.
Wenwu Yu, Guanghui Wen, Guanrong Chen, and Jinde Cao.

Recently, some progress has been made in analyzing cooperative control for collective behaviors in dynamical multi-agent systems, for which some closely related focal topics are synchronization [90, 117, 118, 142], consensus [15, 57, 77, 81, 98, 99, 101, 115], swarming [44, 45], and flocking [82]. More details can be found in survey papers [4, 17, 84, 120].

1.1.2 Collective Behaviors and Cooperative Control in Multi-agent Systems

Synchronization is a typical collective behavior in nature. Since the pioneering work of Pecora and Carroll [90], chaos control and synchronization have received a great deal of attention due to their potential applications in secure communications, chemical reactions, biological systems, and so on [143, 145]. Typically, there are large numbers of nodes in real-world complex networks. In recent years, a lot of work has been devoted to the study of synchronization in various large-scale complex networks [14, 70, 117, 118, 142]. In [117, 118], local synchronization was investigated by the transverse stability to the synchronization manifold, where synchronization was discussed on small-world and scale-free networks. In [132, 134], a distance from the collective states to the synchronization manifold was defined, and based on this, some results were obtained for global synchronization of coupled systems [14, 70]. A general criterion was derived in [142], where the network sizes can be extended to be much larger than those given in [14, 70]. However, it is still very difficult to ensure global synchronization in general large-scale networks due to the computational complexity. Recently, global pinning synchronization for a class of complex networks with switched topologies was addressed in [130] by using tools from stability analysis of switched systems.

The consensus problem has a long history in the field of computer science especially for distributed computing [74]. The idea of consensus was originated from statistical consensus theory by DeGroot [28], which was revisited two decades later for pattern recognition using multi-sensors [10]. Usually, it refers to the problem of how to reach agreement among a group of autonomous agents in a dynamically changing environment [99]. One of the main challenges in solving such a consensus problem is that an agreement has to be reached by all agents in the whole dynamic network while the information of each agent is shared only locally. Various models have been used to study the consensus problem. Vicsek et al. studied a discrete-time system that models a group of autonomous agents moving in the plane with the same speed but different headings [115]. It was shown, through simulation, that using a distributed averaging rule, agents could eventually move in the same direction without centralized coordination. Vicsek's model by nature is a simplified version of the model proposed earlier by Reynolds [101]. Analysis on Vicsek's model, or its continuous-time version, shows that the connectivity of the time-varying graph that describes the neighboring relationships within the group is key in achieving consensus [15, 57, 81, 77, 98]. In particular, in [81], Olfati-Saber and Murray established the relationship between the algebraic connectivity (also called the Fiedler eigenvalue [37]) and the speed of convergence when the underlying directed graph is balanced. A broader class of directed graphs that may lead to reaching consensus are those that contain spanning trees [98], which are also called rooted graphs [15].

It is interesting to observe that Vicsek's model is similar to a class of models discussed in synchronization of complex networks [14, 70, 117, 118, 134, 142]. In 1998, Pecora and Carroll made use of a master stability function to study the synchronization of coupled complex networks [90]. To date, stability and synchronization of small-world and scale-free networks have been investigated extensively using this master stability function method.

In the literature, most work on the consensus problem considered the case where agents are governed by first-order dynamics [11, 57, 72, 81, 98, 114, 134, 141, 142, 151]. Meanwhile, there is a growing interest in consensus algorithms where all agents are governed by second-order dynamics [50, 51, 82, 93, 95, 97, 146]. More precisely, second-order consensus refers to the problem of reaching an agreement among a group of autonomous agents governed by second-order dynamics. A detailed analysis of second-order consensus algorithms is a key step to bring more realistic dynamics into the model of each individual agent based on the general framework of multi-agent systems, thus it can help control engineers to implement distributed cooperative control strategies for networked multi-agent systems. It has been shown that, in sharp contrast to the first-order consensus problem, consensus may fail to be achieved for agents with second-order dynamics even if the network topology has a directed spanning tree [97].

On the other hand, time delay is ubiquitous in biological, physical, chemical, and electrical systems [11, 114]. In biological and communication networks, time delays are usually inevitable due to the possibly slow process of interactions among agents. It has been observed from numerical experiments that consensus algorithms without considering time delays may lead to unexpected instability. In [11, 114], some sufficient conditions were derived for first-order consensus in delayed multi-agent systems.

Very recently, some higher-order consensus algorithms in cooperative control of multi-agent systems were studied, such as in [100] based on the results derived in [97]. However, only third-order consensus was discussed in detail therein. In this book, a general higher-order consensus protocol is designed and analyzed based on the transverse stability to the consensus manifold, which originates from the study of synchronization in complex networks [117]. A detailed analysis of the higher-order consensus algorithms is a prerequisite to introducing more realistic dynamics into the model of each individual agent.

As validated by biological field studies and engineering robotic experiments, swarm cohesion can be achieved in a distributed fashion despite the fact that each agent may only have local information about its nearest neighbors. An in-depth understanding of the principles behind swarming behaviors will help engineers to develop distributed cooperative control strategies and algorithms for networked dynamical systems, such as formations of UAVs, autonomous robotic teams, and mobile sensors networks. Synchronous distributed coordination rules for swarming groups in one- or two-dimensional spaces were studied in [58], where convergence and stability analysis were given. In [44, 45], stability properties of a continuous-time model for swarm aggregation in the n-dimensional space was discussed, and an asymptotic bound for the spatial size of the swarm was computed using the parameters of the swarm model.

In [101], three heuristic rules were suggested by Reynolds to animate flocking behavior: (1) velocity consensus, (2) center cohesion, and (3) collision avoidance. In order to embody the three Reynolds' rules, Tanner et al. designed flocking algorithms in [110, 111], where a collective potential function and a velocity consensus term were introduced. Later, in [82], Olfati-Saber proposed a general framework to investigate distributed flocking algorithms where, in particular, three algorithms were developed for free and constrained flocking. In [110, 111], it was pointed out that due to the time-varying network topology, the set of differential equations describing the flocking behavior of a multi-agent dynamical system is in general nonsmooth; therefore, several techniques from nonsmooth analysis, such as Filippov solutions [38], generalized gradient [25], and differential inclusion [87], were applied for analysis.

1.1.3 Network Control in Multi-agent Systems

In the case where the whole network cannot synchronize by itself, some controllers may be designed and applied to force the network to synchronize. However, it is literally impossible to add controllers to all nodes. To reduce the number of controlled nodes, some local feedback injections may be applied only to a fraction of network nodes, which is known as pinning control [23, 119, 151]. In [47], pinning control of spatiotemporal chaos, and later in [89], global and local control of spatiotemporal chaos in coupled map lattices, were discussed. Recently, in [119], both specific and random pinning schemes were studied, where specific pinning to the nodes with large degrees is shown to require a smaller number of controlled nodes than the random pinning scheme, but the former requires more information about the network than the latter.

Recently, hybrid systems, namely complex systems with both continuous-time and discrete-time event dynamics, have been extensively investigated in the literature, for example continuous-time systems with impulsive responses, sampled data, quantization, to name just a few. Some real-world applications can be modeled by continuous-time systems together with some discrete-time events, such as an A/C unit containing some discrete modes with on or off states, changing the temperature continuously over time. In practice, it is quite difficult to measure the continuous information transmission due to the unreliability of information channels, the capability of transmission bandwidths of networks, etc. Thus, it is more practical to apply sampled-data control, which has been widely studied recently and applied in many real-world systems such as radar tracking systems, power grids, and temperature control. It has been found that sampled-data control has many good properties such as robustness and low cost. Recently, many results have been established from investigating the second-order consensus in multi-agent systems with sampled data. For example, some conditions were derived for multi-agent systems with sampled control by using zero-order holds or direct discretization [18, 41, 66, 163]. On the other hand, consensus of continuous-time multi-agent systems with time-varying topologies and sampled-data control was discussed in [42], and communication delays were considered in multi-agent systems with sampled-data control in [43, 162].

It should be noted that most of the results on consensus problems in multi-agent systems are derived based on a common assumption that the information is transmitted continuously among the agents, that is, each agent shares information with its neighbors without any communication constraints. However, this may not be the case in reality. In some cases, the mobile agents can only communicate with their neighbors at some disconnected time intervals. In order to describe such multi-agent systems more appropriately, a second-order consensus protocol based on globally synchronous intermittent local information feedback was proposed in [124, 129, 126] to guarantee the states of agents converging to consensus asymptotically.

In the literature, many derived conditions for ensuring network synchronization were only sufficient but not necessary, thus are somewhat conservative. Lately, a lot of work has been devoted to using adaptive strategies to adjust network parameters so as to derive better conditions for reaching network synchronization, which employed some existing results from adaptive synchronization in nonlinear systems [139, 155]. For example, in [23, 151, 165, 166], adaptive laws were applied to the control gains from the leader to the followers, and in [23, 151] centralized adaptive schemes were designed on the network coupling strengths. However, there is not much work on how to update the coupling weights of the network for reaching synchronization. In addition, in order to reach consensus or synchronization, some additional global conditions in terms of the spectrum of the Laplacian matrix or its eigenvalues must be satisfied even if the multi-agent system is linear, which actually did not take full advantage of the powerful distributed protocol technology. For example, in [50, 95, 146, 149], in order to reach second-order consensus, the Laplacian matrix or its eigenvalues must be known a priori. To overcome the disadvantage for checking the global information under the local distributed protocol, fully distributed adaptive control in multi-agent systems was investigated recently, which will be introduced in this book.

1.1.4 Distributed Consensus Filtering in Sensor Networks

Sensor networks have attracted increasing attention due to their wide applications in robotics, surveillance and environment monitoring, information collection, wireless communication networks, and so on. A sensor network consists of a large number of sensor nodes distributing over a spatial region. Each sensor has some levels of communication, intelligence for signal processing, and data fusion, which build up a sensing network. Due to the limited energy, computational ability, and communication capability, typically a large number of sensor nodes are used in a wider region so as to achieve higher accuracy of estimating the quantities of interest. Each sensor node is equipped with a microelectronic device having limited power source, which might not be able to transmit messages over a large sensor network. In order to save energy, a natural way is to carry out data fusion to reduce the communication overhead. Therefore, distributed estimation and tracking is one of the most important problems in large-scale sensor networks today. From a network-theoretic perspective, a large-scale sensor network can be viewed as a complex network or a multi-agent system with each node representing a sensor and each edge carrying the information exchange between two sensors. It would be interesting to see how synchronization of complex networks [70, 117, 118, 142] and consensus of multi-agent systems [57, 81, 98] can be used in distributed consensus filtering design. In a complex network, each node communicates with its neighboring nodes to exchange information, so that all the states could finally reach the synchronization or consensus manifold. Therefore, it is quite natural to use the synchronization fundamentals of complex networks and consensus in multi-agent systems as the theoretic basis for distributed consensus filtering design. Practically, it is very difficult to observe all the states of the target, so pinning observers may be designed in the case where the informed sensors can only measure partial states of the target. This notion will also be introduced in the present book.

1.2 Organization

This book focuses on distributed cooperative control in multi-agent systems, which includes complex dynamics, hybrid control, adaptive control, distributed filtering, etc. The contents of the book are summarized as follows.

In Chapter 2, first-order consensus for cooperative agents with nonlinear dynamics in a directed network is discussed. Both local and global consensus are defined and investigated. Techniques for studying synchronization in such complex networks are exploited to establish various sufficient conditions for reaching consensus. The local consensus problem is first studied by combining tools of complex analysis, local consensus manifold approach, and Lyapunov methods. A generalized algebraic connectivity is then derived for studying the global consensus problem in strongly connected networks and also in a broad class of networks containing spanning trees, for which ideas from algebraic graph theory, matrix theory, and Lyapunov methods are utilized. The concept of a virtual network, which has the same spectrum as the original one, is formulated to simplify the analysis.

In Chapter 3, some necessary and sufficient conditions for second-order consensus in multi-agent dynamical systems are presented. Here, theoretical analysis is carried out for the case where each agent with second-order dynamics is governed by the position and velocity terms and the asymptotic velocity is constant. A necessary and sufficient condition is given to ensure second-order consensus and it is found that both the real and imaginary parts of the eigenvalues of the Laplacian matrix of the corresponding network play key roles in reaching consensus. Based on this result, a second-order consensus algorithm is derived for the multi-agent system with communication delays. A necessary and sufficient condition is established, which shows that consensus can be achieved in a multi-agent system whose network topology contains a directed spanning tree if and only if the time delay is less than a critical value. Then, the second-order consensus problem is extended to multi-agent systems with nonlinear dynamics and directed topologies where the final asymptotic velocity is time-varying. Some sufficient conditions are derived for reaching second-order consensus in multi-agent systems with nonlinear dynamics based on algebraic graph theory, matrix theory, and the Lyapunov control approach.

Next, some general higher-order distributed consensus protocols in multi-agent dynamical systems are designed in Chapter 4. The notion of network synchronization is first introduced, with some necessary and sufficient conditions derived for higher-order consensus. It is found that consensus can be reached if and only if all subsystems are asymptotically stable. Based on this result, consensus regions are characterized. It is proved that for the mth-order consensus, there are at most $\left\lfloor \frac{m+1}{2} \right\rfloor$ disconnected stable and unstable consensus regions. It is shown that consensus can be achieved if and only if all the nonzero eigenvalues of the Laplacian matrix lie in the stable consensus regions. Moreover, since the ratio of the largest to the smallest nonzero eigenvalues of the Laplacian matrix plays a key role in reaching consensus, a scheme for choosing the coupling strength is derived, which determines the eigen-ratio. Furthermore, a leader-follower control problem with full-state or partial-state observations in multi-agent dynamical systems is considered respectively, which reveals that the agents with very small degrees must be informed.

In Chapter 5, the stability of a continuous-time swarm model with nonlinear profiles is investigated. It is shown that, under some mild conditions, all agents in a swarm can reach cohesion within a finite time, where some upper bounds for the cohesion are derived in terms of the parameters of the swarm model. The results are then generalized by allowing stochastic noise and switching between nonlinear profiles. Furthermore, swarm models with changing communication topologies and unbounded repulsive interactions among agents are studied by nonsmooth analysis, where the sensing range of each agent is limited and the possibility of collision among nearby agents can be high.

In Chapter 6, using tools from algebraic graph theory and nonsmooth analysis in combination with the ideas of collective potential functions, velocity consensus and navigation feedback, a distributed leader-follower flocking algorithm for multi-agent dynamical systems with time-varying velocities is developed, where each agent is governed by second-order dynamics. The distributed leader-follower flocking algorithm deals with the case where the group has one virtual leader with time-varying velocity. For each agent, this algorithm consists of four terms: the first term is the self nonlinear dynamics, which determines the final time-varying velocity; the second term is determined by the gradient of the collective potential between this agent and all of its neighbors; the third term is the velocity consensus term; the fourth term is the navigation feedback term of the leader. To avoid an unpractical assumption that the informed agents sense all the states of the leader, the new distributed algorithm is developed by making use of observer-based pinning navigation feedback. In this case, each informed agent only requires partial information about the leader, yet the velocity of the whole group can still converge to that of the leader; furthermore, the centroid of those informed agents, having the leader's position information, follows the trajectory of the leader asymptotically.

In Chapter 7, based on full sampled-data information, a distributed linear consensus protocol with second-order dynamics is first designed, where both sampled position and velocity data are utilized. A necessary and sufficient condition based on the sampling period, coupling gains, and spectrum of the Laplacian matrix, is established for reaching consensus of the system. It is found that second-order consensus in such a multi-agent system can be achieved by appropriately choosing the sampling period determined by a third-order polynomial function. Interestingly, second-order consensus cannot be reached for a sufficiently large sampling period while it can be reached for a sufficiently small one under some conditions. Then, the coupling gains are designed under the given network structure and the sampling period. Furthermore, the consensus regions are characterized based on the spectrum of the Laplacian matrix. On the other hand, second-order consensus in delayed undirected networks with sampled position and velocity data is discussed. A necessary and sufficient condition is also given, by which appropriate sampling periods can be chosen to achieve consensus in multi-agent systems. Then, a distributed linear consensus protocol for second-order dynamics is designed, where both the current and some sampled past position data are utilized. In particular, consensus regions are characterized. In such cases, it may even be possible to find some disconnected stable consensus regions determined by choosing appropriate sampling periods. Furthermore, the problem of consensus in directed networks of multiple agents with intrinsic nonlinear dynamics and sampled-data information is discussed. A new protocol is deduced from a class of continuous-time linear consensus protocols by implementing data-sampling technique and a zero-order hold circuit. On the basis of a delayed-input approach, the sampled-data multi-agent system is converted to an equivalent nonlinear system with a time-varying delay. Theoretical analysis on this time-delayed system shows that consensus with asymptotic time-varying velocities in a strongly connected network can be achieved over some suitable sampled-data intervals. A multi-step procedure is further presented to estimate the upper bound of the maximal allowable sampling intervals. The results are then extended to a network topology with a directed spanning tree. For the case of the topology without a directed spanning tree, it is shown that the new protocol can still guarantee the system to achieve consensus by appropriately informing a fraction of agents.

In Chapter 8, the problem of second-order consensus is investigated for a class of multi-agent systems with a fixed directed topology and some communication constraints, where each agent is assumed to share information only with its neighbors on some disconnected time intervals. A consensus protocol is designed based on synchronous intermittent local information, to coordinate the states of agents to converge to second-order consensus under a fixed strongly connected topology, which is then extended to the case where the communication topology contains a directed spanning tree. By using tools from algebraic graph theory and Lyapunov control methods, it is proved that second-order consensus can be reached if the general algebraic connectivity of the communication topology is larger than a threshold value, and the mobile agents communicate with their neighbors frequently enough as the network evolves. Furthermore, consensus of second-order multi-agent systems with nonlinear dynamics and intermittent communication is investigated.

In Chapter 9, distributed adaptive control in multi-agent systems is studied. First, distributed adaptive control of synchronization in complex networks is discussed. An effective distributed adaptive strategy to tune the coupling weights of a network is designed, based on local information of the node dynamics. The analysis is then extended to the case where only a small fraction of coupling weights are adjusted. A general criterion is derived and it is found that synchronization can be reached if the subgraph consisting of the edges and nodes corresponding to the updated coupling weights contains a spanning tree. Then, pinning control in complex networks is investigated. The design of distributed control gains for consensus in multi-agent systems with second-order nonlinear dynamics is discussed. First, an effective distributed adaptive gain-design strategy is proposed, based only on local information of the network structure. Then, a leader-follower consensus problem in multi-agent systems with updated control gains is studied. A distributed adaptive law is derived for each follower, based on local information of its neighboring agents and the leader if this follower is an informed agent. Finally, a distributed leader-follower consensus problem in multi-agent systems with unknown nonlinear dynamics is investigated by combining the variable structure approach and the adaptive method.

In Chapter 10, some applications of collective behaviors in multi-agent systems are studied. A new filtering problem for sensor networks is investigated. A new type of distributed consensus filters is designed, where each sensor can communicate with the neighboring sensors, and filtering can be performed in a distributed way. In the pinning control approach, only a small fraction of sensors need to measure the target information, with which the whole network can be controlled. Furthermore, pinning observers are designed in the case that the sensors can only observe partial target information.

Finally, in Chapter 11, delay-induced consensus and quasi-consensus protocols in multi-agent dynamical systems are designed, where both the current and delayed position information are utilized. Time delays, in a common perspective, can induce periodic oscillations or even chaos in dynamical systems. However, it is found that consensus and quasi-consensus in a multi-agent system cannot be reached without the delayed position information under the given protocol while they can be achieved with a relatively small time delay by appropriately choosing the coupling strengths. A necessary and sufficient condition for reaching consensus in time-delayed multi-agent dynamical systems is established. It is shown that consensus and quasi-consensus can be achieved if and only if the time delay is bounded by some critical values which depend on the coupling strengths and the largest eigenvalue of the Laplacian matrix of the network. The motivation for studying quasi-consensus is revealed, and a potential relationship between the second-order multi-agent system with delayed positive feedback and the first-order system with distributed-delay control input is discussed.

Finally in Chapter 12, conclusions are drown and future research outlook is presented with some brief discussions.

2

Consensus in Multi-agent Systems

In this chapter, consensus in multi-agent systems with both linear and nonlinear dynamics are discussed [147]. First, some classical works on first-order consensus in multi-agent systems with linear dynamics are briefly introduced. Then, first-order consensus for cooperative agents with nonlinear dynamics in a directed network is discussed. Both local and global consensus are defined and investigated. Techniques for studying the synchronization in such complex networks are exploited to establish various sufficient conditions for reaching consensus. The local consensus problem is first studied by combining tools from complex analysis, local consensus manifold approach, and Lyapunov methods. A generalized algebraic connectivity is then proposed to study the global consensus problem in strongly connected networks and also in a broad class of networks containing spanning trees, for which ideas from algebraic graph theory, matrix theory, and Lyapunov methods are utilized. The concept of a virtual network, which has the same spectrum as the original one, is formulated to simplify the analysis.

2.1 Consensus in Linear Multi-agent Systems

2.1.1 Preliminaries

In order to study cooperative control in multi-agent systems, some preliminary results for graphs are first introduced, which will be useful throughout the book.

Let $\mathscr{G} = (\mathscr{V}, \mathscr{E}, G)$ be a weighted directed graph of order N, with the set of nodes $\mathscr{V} = \{v_1, v_2, \cdots, v_N\}$, the set of directed edges $\mathscr{E} \subseteq \mathscr{V} \times \mathscr{V}$, and a weighted adjacency matrix $G = (G_{ij})_{N\times N}$. A directed edge $\mathscr{E}_{ij}$ in network $\mathscr{G}$ is denoted by the ordered pair of nodes (v_i, v_j), where v_i and v_j are called the child and parent nodes, respectively, which means that node v_i can receive information from node v_j, but not another way around. In this chapter, only positively weighted directed graphs are considered, i.e., $G_{ij} > 0$ if and only if there is a directed edge (v_i, v_j) in $\mathscr{G}$. As usual, assume that there is no self-loop in G.

Distributed Cooperative Control of Multi-agent Systems, First Edition.
Wenwu Yu, Guanghui Wen, Guanrong Chen, and Jinde Cao.

The Laplacian matrix $L = (L_{ij})_{N\times N}$ of graph $\mathcal{G}$ is defined by $L_{ij} = -G_{ij}$ for $i \neq j$, $i,j \in \{1,\dots,N\}$ and $L_{ii} = k_i^{in}$ for $i \in \{1,\dots,N\}$, where $k_i^{in} = \sum_{j=1,j\neq i}^{N} G_{ij}$ is the sum of the weights of the edges ending at node v_i. It is easy to check that $\sum_{j=1}^{N} L_{ij} = 0$ for all $i = 1,2,\dots,N$.

Definition 2.1 [52] A network $\mathcal{G}$ is called *undirected* if there is a connection between two nodes v_i and v_j in $\mathcal{G}$, then $G_{ij} = G_{ji} > 0$; otherwise, $G_{ij} = G_{ji} = 0$ $(i \neq j, i,j = 1,2,\dots,N)$. A network $\mathcal{G}$ is *directed* if there is a connection from node v_j to v_i in $\mathcal{G}$, then $G_{ij} > 0$; otherwise $G_{ij} = 0$ $(i \neq j, i,j = 1,2,\dots,N)$.

Note that undirected networks are special cases of directed networks with $G_{ij} = G_{ji}$ for all $i,j = 1,2,\dots,N$.

Definition 2.2 [52] A directed (undirected) path from node v_j to v_i is a sequence of edges $(v_i, v_{i_1}), (v_{i_1}, v_{i_2}), \dots, (v_{i_l}, v_j)$ in the directed (undirected) network with distinct nodes v_{i_k}, $k = 1,2,\dots,l$. A directed (undirected) network $\mathcal{G}$ is *strongly connected* (connected) if between any pair of distinct nodes v_i and v_j in $\mathcal{G}$, there exists a directed (undirected) path from v_i to v_j, $i,j = 1,2,\dots,N$.

Definition 2.3 [52] A matrix G in a directed (undirected) network $\mathcal{G}$ is *reducible* if there is a permutation matrix $P \in R^{N\times N}$ and an integer m, $1 \leq m \leq N-1$, such that

$$P^T G P = \begin{pmatrix} \widetilde{G}_{11} & 0 \\ \widetilde{G}_{21} & \widetilde{G}_{22} \end{pmatrix},$$

where $\widetilde{G}_{11} \in R^{m\times m}$, $\widetilde{G}_{21} \in R^{(n-m)\times m}$, and $\widetilde{G}_{22} \in R^{(n-m)\times(n-m)}$. Otherwise, G is called *irreducible*.

Definition 2.4 [13] A directed network is called a *directed tree* if the underlying network is a tree when the direction of the network is ignored. *A directed rooted tree* is a directed network with at least one root r having the property that for each node v different from r, there is a unique directed path from r to v. *A directed spanning tree* of a network $\mathcal{G}$ is a directed rooted tree, which contains all the nodes and some edges in $\mathcal{G}$.

Definition 2.5 The graph $\mathcal{G}$ is said to have a directed spanning tree if there is a node that can reach all the other nodes following the edge directions in graph $\mathcal{G}$.

The following lemmas are needed to present the main results throughout the book.

Lemma 2.6 *[52]*

1. *The Laplacian matrix L in an undirected network $\mathcal{G}$ has a simple eigenvalue 0 and all the other eigenvalues are positive if and only if the directed network is connected;*
2. *The second smallest eigenvalue $\lambda_2(L)$ of the Laplacian matrix L in an undirected network $\mathcal{G}$ satisfies*

$$\lambda_2(L) = \min_{x^T \mathbf{1}_N = 0, x \neq 0} \frac{x^T L x}{x^T x}.$$

3. *For any $\eta = (\eta_1, \ldots, \eta_N)^T \in \mathbb{R}^N$, $\eta^T L \eta = \frac{1}{2} \sum_{i=1}^{N} \sum_{j=1}^{N} G_{ij}(\eta_i - \eta_j)^2$.*

Lemma 2.7 *[98] Assume that there is a directed spanning tree in graph $\mathcal{G}$. Then the Laplacian matrix L of G has an eigenvalue 0 with algebraic multiplicity one, and the real parts of all the other eigenvalues are positive, i.e., the eigenvalues of L satisfy $0 = \lambda_1(G) < \mathcal{R}(\lambda_2(G)) \leq \ldots \leq \mathcal{R}(\lambda_N(G))$. In addition, rank$(L) = N - 1$.*

The Laplacian matrix L in an undirected network is symmetric and positive semi-definite. Moreover, L has a simple eigenvalue 0 and all the other eigenvalues are positive if and only if the undirected network is connected.

Lemma 2.8 *[53] The Kronecker product has the following properties: For matrices A, B, C and D of appropriate dimensions,*

1. $(\gamma A) \otimes B = A \otimes (\gamma B)$, *where γ is a constant;*
2. $(A + B) \otimes C = A \otimes C + B \otimes C$;
3. $(A \otimes B)(C \otimes D) = (AC) \otimes (BD)$;
4. $(A \otimes B)^T = A^T \otimes B^T$.

2.1.2 Model Formulation and Results

The classical consensus protocol in a multi-agent system is described by [81]

$$\begin{aligned} \dot{x}_i(t) &= u_i(t), \\ u_i(t) &= \sum_{j=1, j \neq i}^{N} G_{ij}(x_j(t) - x_i(t)) = -\sum_{j=1}^{N} L_{ij} x_j(t), i = 1, 2, \cdots, N, \end{aligned} \tag{2.1}$$

where $u_i(t)$ is the control input, $x_i(t) = (x_{i1}(t), x_{i2}(t), \cdots, x_{in}(t))^T \in R^n$ is the state vector of the ith node which can represent any physical quantity including voltage, output power, or incremental cost, and $G = (G_{ij})_{N\times N}$ is the coupling configuration matrix representing the topological structure of the network. Note that the consensus protocol in (2.1) is distributed where each agent utilizes information of its neighboring agents locally. If the network topology changes with time [15, 57, 77, 98], then in (2.1) $L = L(t)$ is a time-varying matrix.

Now, the definition for consensus in multi-agent systems is given.

Definition 2.9 Consensus in multi-agent system (2.1) is said to be achieved if for any initial conditions,

$$\lim_{t\to\infty} \| x_i(t) - x_j(t) \| = 0, \forall i, j = 1, 2, \cdots, N.$$

Throughout, $\| \cdot \|$ is the Eucleadian norm. A well-known result for consensus in multi-agent system (2.1) is established in [81, 99] as follows.

Lemma 2.10 *Consensus in multi-agent system (2.1) can be reached if the undirected network is connected, and in addition,* $\lim_{t\to\infty} \| x_i(t) - x^*(t) \| = 0, \forall i, j = 1, 2, \cdots, N,$ *where* $x^*(t) = \frac{1}{N}\sum_{k=1}^{N} x_k(t)$ *is the average center of the states of all agents.*

Proof. A simple proof is given here. Let the error state of node i be $e_i(t) = x_i(t) - x^*(t)$ and consider the following Lyapunov candidate:

$$V(t) = \sum_{i=1}^{N} e_i^T(t)e_i(t).$$

Since the Laplacian matrix is a zero-row-sum matrix and one has

$$\dot{x}^*(t) = -\sum_{i=1}^{N}\sum_{j=1}^{N} L_{ij}x_j(t) = 0.$$

Thus, $\dot{e}_i(t) = \dot{x}_i(t) = \sum_{j=1}^{N} L_{ij}x_j(t) = \sum_{j=1}^{N} L_{ij}e_j(t)$. Then, by Lemma 2.6, the derivative along the trajectory of $V(t)$ is

$$\begin{aligned}\dot{V}(t) &= 2\sum_{i=1}^{N} e_i^T(t)\dot{e}_i(t)\\ &= -2\sum_{i=1}^{N}\sum_{j=1}^{N} L_{ij}e_i^T(t)e_j(t)\\ &= -e^T(t)(L\otimes I_n)e(t) \le -\lambda_2 V(t),\end{aligned}$$

where $e(t) = (e_1^T(t), \ldots, e_1^T(t))^T$ and I_n is an n-dimensional identity matrix. This completes the proof. □

2.2 Consensus in Nonlinear Multi-agent Systems

2.2.1 Preliminaries and Model Formulation

Let $\mathcal{S} = \{(x_1, x_2, \cdots, x_N) : x_1 = x_2 = \cdots = x_N\}$ be the *consensus manifold*. It is clear that since $\sum_{j=1}^{N} L_{ij} = 0$ for all $i = 1, \ldots, N$, x_i must be time-invariant on the consensus manifold in (2.1). In other words, the values of x_i will not change with time once the consensus $x_1(t) = x_2(t) = \ldots = x_N(t)$ is achieved. However, as has been repeatedly demonstrated in complex physical networks, the state of each agent is, more often than not, a dynamical variable because of the intrinsic nonlinear dynamics of each agent and the possibly complicated ways in which the network is evolving. To study the synchronization of complex networks for more general cases, particularly when the synchronized state is a time-varying function rather than a constant equilibrium, consider the following general consensus protocol:

$$\dot{x}_i(t) = f(x_i(t)) - c\sum_{j=1}^{N} L_{ij}\Gamma x_j(t), \tag{2.2}$$

where $x_i \in R^n$ is the state of agent i, $f(x_i) = (f_1(x_i), f_2(x_i), \ldots, f_n(x_i))^T$ is a nonlinear function, c is the coupling strength, and $\Gamma = \mathrm{diag}(\gamma_1, \ldots, \gamma_n) \in R^{n\times n}$ is a semi-positive definite diagonal matrix where $\gamma_j > 0$ means that the agents can communicate through their jth state [117, 118]. Here, the state of each agent is an n-dimensional vector as compared to one-dimensional variables considered in previous works, e.g. [57, 77, 98, 103]. Note also that it is straightforward to construct the discrete-time counterpart of system (2.2), but only the continuous-time case is investigated in this chapter.

Clearly, since $\sum_{j=1}^{N} L_{ij} = 0$, if consensus can be achieved, in many applications, the solution $s(t)$ of system (2.2) is expected to be a trajectory of an isolated node satisfying

$$\dot{s}(t) = f(s(t)). \tag{2.3}$$

Here, $s(t)$ may be an isolated equilibrium point [57, 77, 98, 103], a periodic orbit, or even a chaotic orbit [142].

Let $A \otimes B$ denote the Kronecker product [53] of matrices A and B, $x(t) = (x_1^T(t), x_2^T(t), \ldots, x_N^T(t))^T$, $f(x(t)) = (f^T(x_1(t)), f^T(x_2(t)), \ldots, f^T(x_N(t)))^T$, I_n be the n-dimensional identity matrix, $\mathbf{1}_N$ be the N-dimensional column vector with all entries being 1, and $A^s = \frac{1}{2}(A + A^T)$. Then, system (2.2) can be written as

$$\dot{x}(t) = f(x(t)) - c(L \otimes \Gamma)x(t). \tag{2.4}$$

A few definitions and some results are given here, which will be useful in the development of the next few sections.

Definition 2.11 The consensus in system (2.4) is said to be local if, for any $\epsilon > 0$, there exist a $\delta(\epsilon)$ and a $T > 0$, such that $\|x_i(0) - x_j(0)\| \leq \delta(\epsilon)$ implies $\|x_i(t) - x_j(t)\| \leq \epsilon$ for all $t > T$ and $i, j = 1, 2, \ldots, N$.

Definition 2.12 The consensus in system (2.4) is said to be global if, for any $\epsilon > 0$, there exists a $T > 0$ such that $\|x_i(t) - x_j(t)\| \leq \epsilon$ for any initial conditions and all $t > T$, $i,j = 1, 2, \ldots, N$.

2.2.2 *Local Consensus of Multi-agent Systems*

In this section, local consensus of multi-agent systems is investigated. Subtracting (2.3) from (2.2) yields the following error dynamical system:

$$\dot{y}_i(t) = f(x_i(t)) - f(s(t)) - c \sum_{j=1}^{N} L_{ij}\Gamma y_j(t), \tag{2.5}$$

where $y_i = x_i - s$, $i = 1, 2, \ldots, N$. Linearizing (2.5) around $s(t)$ leads to

$$\dot{y}(t) = (I_N \otimes Df(s(t)))y(t) - c(L \otimes \Gamma)y(t), \tag{2.6}$$

where $y(t) = (y_1^T(t), y_2^T(t), \ldots, y_N^T(t))$ and $Df(s) \in R^{n\times n}$ is the Jacobian matrix of f at $s(t)$. Let P be the Jordan form associated with the Laplacian matrix L, i.e., $L = PJP^{-1}$ where J is the Jordan form of L. Then one has

$$\dot{z}(t) = (I_N \otimes Df(s(t)))z(t) - c(J \otimes \Gamma)z(t), \tag{2.7}$$

where $z(t) = (P^{-1} \otimes I_n)y(t)$. If L is symmetric, i.e., graph $\mathcal{G}$ is undirected, then J is a diagonal matrix with real eigenvalues. However, when G is directed, some eigenvalues of L may be complex, and $J = \mathrm{diag}(J_1, J_2, \ldots, J_r)$, where

$$J_l = \begin{pmatrix} \lambda_l & 0 & 0 & 0 \\ 1 & \ddots & 0 & 0 \\ 0 & \ddots & \ddots & 0 \\ 0 & 0 & 1 & \lambda_l \end{pmatrix}_{N_l \times N_l}. \tag{2.8}$$

Here, it is assumed that the Laplacian matrix L has eigenvalues λ_l with multiplicity N_l, $l = 1, 2, \ldots, r$, $N_1 + N_2 + \ldots + N_r = N$. Let $\mathcal{R}(u)$ and $\mathcal{I}(u)$ be the real and imaginary parts of a complex number u, and $\mathcal{R}(A)$ and $\mathcal{I}(A)$ be the real and imaginary parts of matrix $A = (A_{ij})$, where $\mathcal{R}(A)_{ij} = \mathcal{R}(A_{ij})$ and $\mathcal{I}(A)_{ij} = \mathcal{I}(A_{ij})$, respectively. Let $\widetilde{N}_i = N_1 + \ldots + N_i$, $i = 1, 2, \ldots, r$. Then, separating the real and imaginary parts of (2.7), one obtains

$$\begin{aligned}
\mathcal{R}(\dot{\widetilde{z}}_i(t)) = {} & (I_{N_i} \otimes Df(s(t)))\mathcal{R}(\widetilde{z}_i(t)) - c(\mathcal{R}(J_i) \otimes \Gamma)\mathcal{R}(\widetilde{z}_i(t)) \\
& + c(\mathcal{I}(J_i) \otimes \Gamma)\mathcal{I}(\widetilde{z}_i(t)), \\
\mathcal{I}(\dot{\widetilde{z}}_i(t)) = {} & (I_{N_i} \otimes Df(s(t)))\mathcal{I}(\widetilde{z}_i(t)) - c(\mathcal{R}(J_i) \otimes \Gamma)\mathcal{I}(\widetilde{z}_i(t)) \\
& - c(\mathcal{I}(J_i) \otimes \Gamma)\mathcal{R}(\widetilde{z}_i(t)),
\end{aligned} \tag{2.9}$$

where $\widetilde{z}_i \in R^{N_i}$, $i = 1, \ldots, r$.

Lemma 2.13 *Suppose that graph $\mathscr{G}$ has a spanning tree. If system (2.9) is asymptotically stable for $i = 2, \ldots, r$, then local consensus can be reached in system (2.4).*

Proof. According to Lemma 2.4, zero is a simple eigenvalue of the Laplacian matrix L. From (2.7), one has $y(t) = (P \otimes I_n)z(t)$, where $LP = PJ$. Let $P = (p_1, p_2, \ldots, p_N)$. Then p_1 is the right eigenvector of L associated with eigenvalue 0, i.e., $Lp_1 = 0$. Since $\sum_{j=1}^{N} L_{ij} = 0$ and $\text{rank}(L) = N - 1$, one has $p_1 = \eta(1, 1, \ldots, 1)^T$, where η is a constant. If system (2.9) is asymptotically stable for $i = 2, \ldots, r$, then $\|\mathscr{R}(z_i(t))\| \to 0$ and $\|\mathscr{I}(z_i(t))\| \to 0$ as $t \to \infty$, $i = 2, \ldots, N$. Therefore,

$$y(t) \to \eta(z_1(t), z_1(t), \ldots, z_1(t))^T,$$

where $\dot{z}_1(t) = Df(s(t))z_1$. □

Lemma 2.14 *Let*

$$\widetilde{L}^* = \begin{pmatrix} \widetilde{L}^*_{11} & O & \cdots & O \\ \widetilde{L}^*_{21} & \widetilde{L}^*_{22} & \cdots & O \\ \vdots & \vdots & \ddots & O \\ \widetilde{L}^*_{p1} & \widetilde{L}^*_{p2} & \cdots & \widetilde{L}^*_{pp} \end{pmatrix}. \tag{2.10}$$

where O is a zero matrix with appropriate dimension, $\widetilde{L}_{kk} \in R^{m_k m_k}$, and m_k are positive integers for all $k = 1, 2, \ldots, p$. If there exist positive definite diagonal matrices $Q^*_k \in R^{m_k m_k}$, such that

$$Q^*_k \widetilde{L}^*_{kk} + \widetilde{L}^{*T}_{kk} Q^*_k < 0, \tag{2.11}$$

then, there exists a positive definite diagonal matrix $\Delta = \text{diag}(\Delta_1 I_{m_1}, \ldots, \Delta_p I_{m_p})$, such that

$$\Delta \widetilde{Q}^* \widetilde{L}^* + \widetilde{L}^{*T} \widetilde{Q}^* \Delta < 0, \tag{2.12}$$

where $\widetilde{Q}^* = diag(\widetilde{Q}^*_1, \ldots, \widetilde{Q}^*_p)$.

Proof. Let

$$\Phi_i = \begin{pmatrix} \Delta_1(Q^*_1 \widetilde{L}^*_{11} + \widetilde{L}^{*T}_{11} Q^*_1) & \Delta_2 \widetilde{L}^{*T}_{21} Q^*_2 & \cdots & \Delta_i \widetilde{L}^{*T}_{i1} Q^*_i \\ \Delta_2 Q^*_2 \widetilde{L}^*_{21} & \Delta_2(Q^*_2 \widetilde{L}^*_{22} + \widetilde{L}^{*T}_{22} Q^*_2) & \cdots & \Delta_i \widetilde{L}^{*T}_{i2} Q^*_i \\ \vdots & \vdots & \ddots & \vdots \\ \Delta_i Q^*_i \widetilde{L}^*_{i1} & \Delta_i Q^*_i \widetilde{L}^*_{i2} & \cdots & \Delta_i(Q^*_i \widetilde{L}^*_{ii} + \widetilde{L}^{*T}_{ii} Q^*_i) \end{pmatrix}.$$

Then $\Phi_p = \Delta \widetilde{Q}^* \widetilde{L}^* + \widetilde{L}^{*T} \widetilde{Q}^* \Delta$. From (2.10), it is easy to see that $\Phi_1 < 0$. Now, the lemma is proved by induction. Suppose that $\Phi_i < 0$. It suffices to show that $\Phi_{i+1} < 0$.

By using Schur complement [12], one has $\Phi_{i+1} < 0$ is equivalent to $Q^*_{i+1}\widetilde{L}^*_{i+1,i+1} + \widetilde{L}^{*T}_{i+1,i+1}Q^*_{i+1} < 0$ and

$$\Phi_i - \Delta_{i+1}\Pi_{i+1}(Q^*_{i+1}\widetilde{L}^*_{i+1,i+1} + \widetilde{L}^{*T}_{i+1,i+1}Q^*_{i+1})^{-1}\Pi^T_{i+1} < 0, \tag{2.13}$$

where $\Pi^T_{i+1} = Q^*_{i+1}(\widetilde{L}^*_{(i+1)1}, \widetilde{L}^*_{(i+1)2}, \ldots, \widetilde{L}^*_{(i+1)i})$. If Δ_{i+1} is sufficiently smaller than Δ_j for $j < i+1$, then (2.13) is satisfied. Therefore, by choosing Δ_{j+1} sufficiently smaller than Δ_j for $j < i+1$, the condition in (2.12) can be satisfied and the proof is completed. □

Theorem 2.15 *Suppose that graph $\mathcal{G}$ has a spanning tree. Then the local consensus of system (2.4) can be reached if*

$$(Df(s(t)))^s - c\mathcal{R}(\lambda_2)\Gamma < 0, \forall t > 0. \tag{2.14}$$

Proof. In view of Lemma 2.13, one only needs to prove that under condition (2.14), system (2.9) is asymptotically stable for $i = 2, \ldots, r$.

Consider the Lyapunov function candidate

$$V(t) = \frac{1}{2}\sum_{i=2}^{r} \Delta_i\{\mathcal{R}^T(\widetilde{z}_i(t))\mathcal{R}(\widetilde{z}_i(t)) + \mathcal{I}^T(\widetilde{z}_i(t))\mathcal{I}(\widetilde{z}_i(t))\},$$

where Δ_i is positive for $i = 2, \ldots, r$.

Taking the derivative of $V(t)$ along the trajectories of (2.9) gives

$$\begin{aligned}\dot{V} = &\sum_{i=2}^{r} \Delta_i\mathcal{R}^T(\widetilde{z}_i(t))\{(I_{N_i} \otimes Df(s(t))) - c(\mathcal{R}(J_i) \otimes \Gamma)\}\mathcal{R}(\widetilde{z}_i(t)) \\ &+ \sum_{i=2}^{r} \Delta_i\mathcal{I}^T(\widetilde{z}_i(t))\{(I_{N_i} \otimes Df(s(t))) - c(\mathcal{R}(J_i) \otimes \Gamma)\}\mathcal{I}(\widetilde{z}_i(t)).\end{aligned} \tag{2.15}$$

From Lemma 2.14, it is easy to see that if $(Df(s(t)))^s - c\mathcal{R}(\lambda_i)\Gamma < 0$, then by choosing appropriate positive constants Δ_i, one can obtain that $[(I_{N_i} \otimes Df(s(t))) - c(\mathcal{R}(J_i) \otimes \Gamma)]^s < 0$. Under condition (2.14), system (2.9) is asymptotically stable for $i = 2, \ldots, r$. Therefore, by Lemmas 2.13 and 2.14, the local consensus of system (2.4) can be reached. The proof is completed. □

Note that system (2.4) is linearized around the state of a single node $s(t)$ to obtain system (2.6). Thus, only local consensus is ensured. If $s(t)$ does not contain any asymptotical attractor or the state x_i of each agent system is not in the neighborhood of $s(t)$, then local consensus may not be reached. The limitation of the result in Theorem 1 motivates the following study of the global properties of system (2.4).

2.2.3 Global Consensus of Multi-agent Systems in General Networks

The following result is widely used to compute the algebraic connectivity of an undirected graph.

Lemma 2.16 *[37, 46] For an undirected graph with Laplacian matrix L, the algebraic connectivity of the network is given by*

$$\lambda_2(L) = \min_{x^T \mathbf{1}_N = 0, x \neq 0} \frac{x^T L x}{x^T x}. \tag{2.16}$$

Let the generalized in-degree and out-degree of a node in a network be the sum of the weights of the edges pointing to or leaving from the node, respectively.

Definition 2.17 (Balanced graphs [81]) The node in a directed graph $\mathcal{G}$ is said to be balanced if its in-degree is equal to its out-degree. A graph $\mathcal{G}$ is called balanced if and only if all its nodes are balanced, i.e., $\sum_{j=1}^{N} L_{ji} = \sum_{j=1}^{N} L_{ij} = 0$, $i = 1, 2, \ldots, N$.

In [81], the consensus problem of strongly connected balanced graphs was investigated. Let $\hat{L} = (L + L^T)/2$. Then $\hat{L}$ is symmetric, and the sums of the entries in each row and each column are 0. Thus, $\mathbf{1}_N$ is the eigenvector associated with the simple eigenvalue 0. From Lemma 2.16, one has $x^T L x = x^T \hat{L} x \geq \lambda_2(\hat{L}) x^T x$, where $x^T \mathbf{1}_N = 0$. Now the notion of algebraic connectivity is generalized to directed graphs.

Definition 2.18 For a strongly connected network G with Laplacian matrix L, the general algebraic connectivity is defined to be the real number

$$a_\xi(L) = \min_{x^T \xi = 0, x \neq 0} \frac{x^T \hat{L} x}{x^T \Xi x}, \tag{2.17}$$

where $\hat{L} = \frac{\Xi L + L^T \Xi}{2}$, $\Xi = \mathrm{diag}(\xi_1, \ldots, \xi_N)$, $\xi = (\xi_1, \xi_2, \ldots, \xi_N)^T$ with $\xi_i > 0$ for $i = 1, 2, \ldots, N$ and $\sum_{i=1}^{N} \xi_i = 1$.

Note that if $\Xi = \eta I_N$, $a_\xi(L) = \lambda_2(\hat{L})$.

Definition 2.19 [134, 142] Let $T(\epsilon)$ be the set of matrices with real entries such that the sum of the entries in each row is equal to the real number ϵ. The set $M \in M^N(1)$ if and only if $M = I_N - \mathbf{1}_N \xi^T$ and $M \in T(0)$. The set $M^N(n) = \{\mathbf{M} = M \otimes I_n : M \in M^N(1), I_n$ is the n-dimensional identity matrix$\}$.

Lemma 2.20 *[142, 141] Let $x = (x_1, x_2, \cdots, x_N)^T$, where $x_i \in R^n, i = 1, 2, \cdots, N$. Then the global consensus in system (2.4) can be reached if there exists an $\mathbf{M} \in M^N(n)$ satisfying $\| \mathbf{M}x \| \to 0$ as $t \to \infty$.*

Now, define a nonnegative distance function by

$$d(x) = \| \mathbf{M}x\|^2 = x^T\mathbf{M}^T\mathbf{M}x, \mathbf{M} \in M^N(n). \tag{2.18}$$

From the assumptions on $\mathbf{M}$, one has $d(x) \to 0$ if and only if $\| x_i(t) - \overline{x} \| \to 0$ for all $i = 1, 2, \cdots, N$, where $\overline{x} = \sum_{j=1}^{N} \xi_j x_j(t)$ is the objective consensus state.

Assumption 2.21 *There exist constants θ and $\epsilon > 0$ such that*

$$\begin{aligned}&(x-y)^T(f(x)-f(y)) - \theta(x-y)^T\Gamma(x-y)\\ \leq& -\epsilon(x-y)^T(x-y), \forall x, y \in R^n.\end{aligned} \tag{2.19}$$

Note that the condition (2.19) is very mild: If $\partial f_j/\partial x_{ij}, i = 1, 2, \ldots, N, j = 1, 2, \ldots, n$, are bounded, then this condition is automatically satisfied. So systems satisfying (2.19) include many well-known systems, such as the Lorenz system, Chen system, Lü system, various neural networks, Chua's circuit, to name just a few.

Theorem 2.22 *Under Assumption 2.21, the global consensus of system (2.4) can be reached if*

$$\theta - ca_{\xi}\ (L) < 0. \tag{2.20}$$

Proof. Consider the following Lyapunov function candidate defined by the distance function in Lemma 2.20:

$$V(t) = \frac{1}{2}x^T\mathbf{M}^T\mathbf{\Xi}\mathbf{M}x,$$

where $\mathbf{M} = (I_N - \mathbf{1}_N\xi^T) \otimes I_n$, $\Xi = \text{diag}(\xi_1, \ldots, \xi_N)$, $\xi_i > 0$, $i = 1, 2, \ldots, N$, with $\sum_{i=1}^{N} \xi_i = 1$, and $\mathbf{\Xi} = \Xi \otimes I_n$. Let $\overline{x} = \sum_{j=1}^{N} \xi_j x_j(t)$ and $\mathbf{M}x = x - \mathbf{1}_N \otimes \overline{x}$.

Taking the derivative of $V(t)$ along the trajectories of (2.4) gives

$$\begin{aligned}\dot{V} =& x^T\mathbf{M}^T\mathbf{\Xi}\mathbf{M}[f(x(t)) - c(L \otimes \Gamma)x(t)]\\ =& x^T\mathbf{M}^T\mathbf{\Xi}[f(x(t)) - ((\mathbf{1}_N\xi^T) \otimes I_n)f(x(t))]\\ &-cx^T\mathbf{M}^T\mathbf{\Xi}[(I_N - \mathbf{1}_N\xi^T) \otimes I_n](L \otimes \Gamma)x(t)\\ =& x^T\mathbf{M}^T\mathbf{\Xi}[f(x(t)) - \mathbf{1}_N \otimes f(\overline{x})]\\ &-cx^T\mathbf{M}^T\mathbf{\Xi}(L \otimes \Gamma)[(I_N - \mathbf{1}_N\xi^T) \otimes I_n]x(t)\\ &+x^T\mathbf{M}^T\mathbf{\Xi}[\mathbf{1}_N \otimes f(\overline{x}) - ((\mathbf{1}_N\xi^T) \otimes I_n)f(x(t))]\\ &+cx^T\mathbf{M}^T\mathbf{\Xi}[(\mathbf{1}_N\xi^T) \otimes I_n](L \otimes \Gamma)x(t).\end{aligned} \tag{2.21}$$

The third equality is satisfied due to the fact that $(L \otimes \Gamma)[(\mathbf{1}_N\xi^T) \otimes I_n] = 0$, since $\mathbf{1}_N$ is the right eigenvector of L associated with eigenvalue 0. From $\xi^T\mathbf{1}_N = 1$, one has

$$
\begin{aligned}
x^T\mathbf{M}^T\Xi[\mathbf{1}_N \otimes f(\bar{x})] &= [\mathbf{1}_N^T \otimes f^T(\bar{x})](\Xi \otimes I_n)[(I_N - \mathbf{1}_N\xi^T)\otimes I_n]x(t) \\
&= \{[\xi^T(I_N - \mathbf{1}_N\xi^T)] \otimes f^T(\bar{x})\}x(t) = 0, \qquad (2.22)
\end{aligned}
$$

$$
x^T\mathbf{M}^T\Xi((\mathbf{1}_N\xi^T)\otimes I_n)f(x(t)) - \{[\xi\xi^T(I_N - \mathbf{1}_N\xi^T)] \otimes f^T(\bar{x})\}x(t) = 0, \qquad (2.23)
$$

and

$$
x^T\mathbf{M}^T\Xi((\mathbf{1}_N\xi^T)\otimes I_n)(L\otimes\Gamma)x(t) = 0. \qquad (2.24)
$$

Combining (2.21)–(2.24), one obtains

$$
\begin{aligned}
\dot{V} &= x^T\mathbf{M}^T\Xi[f(x(t)) - \mathbf{1}_N \otimes f(\bar{x})] - cx^T\mathbf{M}^T\Xi(L\otimes\Gamma)\mathbf{M}x(t) \\
&\le -\epsilon x^T\mathbf{M}^T\Xi\mathbf{M}x(t) + \theta x^T\mathbf{M}^T\Xi\Gamma\mathbf{M}x(t) - cx^T\mathbf{M}^T\Xi(L\otimes\Gamma)\mathbf{M}x(t) \\
&= -\epsilon x^T\mathbf{M}^T\Xi\mathbf{M}x(t) + x^T\mathbf{M}^T[(\theta\Xi - c\Xi L)\otimes\Gamma]\mathbf{M}x(t) \\
&\le -\epsilon x^T\mathbf{M}^T\Xi\mathbf{M}x(t) + (\theta - ca_\xi(L))x^T\mathbf{M}^T\Xi\Gamma\mathbf{M}x(t). \qquad (2.25)
\end{aligned}
$$

Under condition (2.20), global consensus of system (2.4) is reached. This completes the proof. □

It is still not straightforward to verify whether the condition in (2.20) is satisfied by a properly chosen positive vector ξ. If $\theta < 0$, then it is possible that $a_\xi(L) = 0$; if $\theta = 0$, the condition depends only on $a_\xi(L)$; and if $\theta > 0$, then $a_\xi(L) > 0$ must be satisfied. From condition (2.20), global consensus can be reached even if the network is disconnected with $\theta < 0$. For periodic and chaotic nodes with $\theta > 0$, one may be interested in the condition under which $a_\xi(L) > 0$. However, when is it possible to have $a_\xi(L) > 0$? In what follows, an answer is given to this question.

Lemma 2.23 *(Theorem 8.4.4 in [52]) Suppose that A is irreducible and nonnegative. Then there is a positive vector x such that $Ax = \rho(A)x$, where $\rho(A)$ is the spectral radius of matrix A.*

Lemma 2.24 *Suppose that the Laplacian matrix L is irreducible. Then there is a positive vector x such that $L^Tx = 0$.*

Proof. Choose a positive integer l such that $l - \lambda_N(L) > 0$ and $l - L_{ii} > 0$ for all $i = 1, 2, \ldots, N$. Then matrix $lI_N - L$ is positive definite and, from Lemma 2.7, $\rho(lI_N - L) = l$. The matrix $(lI_N - L)^T = lI_N - L^T$ is also positive definite and $\rho(lI_N - L^T) = l$. By Lemma 2.23, there is a positive vector x such that $(lI_N - L^T)x = lx$, and one obtains $L^Tx = 0$. The proof is thus completed. □

Lemma 2.25 *Suppose that the Laplacian matrix L is irreducible. Then there exists a positive-definite diagonal matrix $\Xi = \mathrm{diag}(\xi_1, \xi_2, \ldots, \xi_N)$, such that $\widehat{L} = \frac{1}{2}(\Xi L + L^T\Xi)$ is symmetric and $\sum_{j=1}^{N}\widehat{L}_{ij} = 0$, $i = 1, 2, \ldots, N$.*

Proof. It is easy to see that $\widehat{L}$ is symmetric. From Lemma 2.24, there is a positive vector $\xi = (\xi_1, \xi_2, \dots, \xi_N)^T$ such that $\xi^T L = 0$, i.e., ξ is the left eigenvector of the Laplacian matrix associated with eigenvalue 0. Then one has $\xi = \Xi \mathbf{1}_N$ and thus $L^T \Xi \mathbf{1}_N = 0$. Therefore, $L^T \Xi$ is a matrix in which the sum of the entries in each row is zero. Since $\sum_{j=1}^{N} L_{ij} = 0$, one has $\Xi L \mathbf{1}_N = 0$, and hence the sum of the entries in each row in ΞL is zero. Also, the sum of the entries in each row in matrix $\widehat{L}$ is zero. In addition, since $\widehat{L}$ is symmetric, the sum of the entries in each column in matrix $\widehat{L}$ is also zero. The proof is thus completed. □

Lemma 2.26 *Suppose that the matrix $\widehat{L}$ is symmetric and irreducible, and satisfies $\sum_{j=1}^{N} \widehat{L}_{ij} = 0$ with $\widehat{L}_{ij} \leq 0, i \neq j,\ i,j = 1, 2, \dots, N$. Let*

$$\widehat{a}_\xi(\widehat{L}) = \min_{x^T\xi=0, x\neq 0} \frac{x^T \widehat{L} x}{x^T x}. \tag{2.26}$$

Then $\lambda_2(\widehat{L}) \geq \widehat{a}_\xi(\widehat{L}) \geq 0$. In addition, $\widehat{a}_\xi(\widehat{L}) = 0$ if and only if ξ is orthogonal to the left eigenvector of $\widehat{L}$ associated with eigenvalue 0; $\widehat{a}_\xi(\widehat{L}) = \lambda_2(\widehat{L})$ if ξ is the left eigenvector of $\widehat{L}$ associated with eigenvalue 0.

Proof. From the Courant–Fischer min-max theorem [52], one has $\lambda_2(\widehat{L}) \geq \widehat{a}_\xi(\widehat{L}) \geq 0$. Let Λ be the diagonal matrix associated with $\widehat{L}$, i.e., there exists a $P = (p_1, p_2, \dots, p_N)$, such that $\widehat{L} = P\Lambda P^T$ and $y = P^T x$. Then,

$$\begin{aligned}
\widehat{a}_\xi(\widehat{L}) &= \min_{x^T\xi=0, x^T x=1} x^T \widehat{L} x = \min_{x^T\xi=0, x^T x=1} x^T P\Lambda P^T x = \min_{x^T\xi=0, x^T x=1} \sum_{i=1}^{N} \lambda_i(\widehat{L}) y_i^2 \\
&= \min_{y^T P^T \xi=0, y^T y=1} \sum_{i=1}^{N} \lambda_i(\widehat{L}) y_i^2 \leq \min_{y^T P^T \xi=0, y^T y=1, y_3=\cdots=y_N=0} \sum_{i=1}^{2} \lambda_i(\widehat{L}) y_i^2 \\
&\leq \lambda_2(\widehat{L}).
\end{aligned} \tag{2.27}$$

The inequalities hold under the conditions $y^T P^T \xi = 0$, $y_1 = 0$, $p_2^T \xi = 0$, and $p_1^T \xi \neq 0$, $\forall y \in R^N$. If ξ is the left eigenvector of $\widehat{L}$ associated with eigenvalue 0, one has $p_1^T \xi \neq 0$ and $\xi \perp p_i$, $i = 2, \dots, N$, for any y with $y_1 = 0$. Therefore, $\widehat{a}_\xi(\widehat{L}) = \lambda_2(\widehat{L})$ if ξ is the left eigenvector of eigenvalue 0. Similarly,

$$\widehat{a}_\xi(\widehat{L}) = \min_{y^T P^T \xi=0, y^T y=1} \sum_{i=1}^{N} \lambda_i(\widehat{L}) y_i^2 \geq \lambda_1(\widehat{L}) = 0. \tag{2.28}$$

The above inequality holds if and only if $y^T P^T \xi = 0$ i.e., $y^T P^T \xi = \sum_{i=1}^{N} y_i p_i^T \xi$, and $y_2 = \dots, y_N = 0$. It then follows that $p_1^T \xi = 0$. This completes the proof. □

Corollary 2.27 *If the Laplacian matrix L is irreducible, then $a_\xi(L) > 0$, where the chosen positive vector ξ satisfies $\xi^T L = 0$.*

Proof. From Lemma 2.18, there exist a positive vector $\xi = (\xi_1, \xi_2, \dots, \xi_N)$ and a definite diagonal matrix $\Xi = \text{diag}(\xi_1, \xi_2, \dots, \xi_N)$, such that $\hat{L} = \frac{1}{2}(\Xi L + L^T \Xi)$ is symmetric and $\sum_{j=1}^{N} \hat{L}_{ij} = 0$, $i = 1, 2, \dots, N$. It is easy to see that $\mathbf{1}$ is the left eigenvector of eigenvalue 0, and $\mathbf{1}^T \xi = 1 \neq 0$. From Lemma 2.26, $a_\xi(L) \geq \hat{a}_\xi(\hat{L})/\max_i \xi_i > 0$. The proof is thus completed. □

Lemma 2.28 *The general algebraic connectivity of a strongly connected network can be computed by the following LMI:*

$$\max \delta \qquad \text{Subject to } \ Q^T(\hat{L} - \delta \Xi) Q \geq 0,$$

where $Q = \begin{pmatrix} I_{N-1} \\ -\frac{\hat{\xi}^T}{\xi_N} \end{pmatrix} \in R^{N\times(N-1)}$ *and* $\hat{\xi} = (\xi_1, \dots, \xi_{N-1})^T$.

Proof. It is easy to see that the columns of Q form a basis for the orthogonal subspace of the vector ξ. Thus, by letting $x = Qz$, one has

$$a_\xi(L) = \min_{z \neq 0} \frac{z^T Q^T \hat{L} Q z}{z^T Q^T \Xi Q z}.$$

The proof is completed. □

Remark 2.29 The definition of the algebraic connectivity presented here is motivated by a similar definition in [133], where it is assumed that $\xi^T L = 0$. In this chapter, a more general case is considered, where ξ is a positive vector. Based on this general definition of the algebraic connectivity, the above theoretical analysis for reaching global consensus can be carried out.

Next, analysis on the global consensus is presented assuming that graph $\mathscr{G}$ has a spanning tree. Let the Laplacian matrix L of graph $\mathscr{G}$ be in its Frobenius normal form [13]:

$$L = \begin{pmatrix} \widetilde{L}_{11} & O & \dots & O \\ \widetilde{L}_{21} & \widetilde{L}_{22} & \dots & O \\ \vdots & \vdots & \ddots & \vdots \\ \widetilde{L}_{p1} & \widetilde{L}_{p2} & \dots & \widetilde{L}_{pp} \end{pmatrix}, \tag{2.29}$$

where $\widetilde{L}_{kk} \in R^{m_k m_k}$ is irreducible for all $k = 1, 2, \dots, p$. Matrix (2.29) can be interpreted as follows: the nodes and their adjacent edges in $\widetilde{L}_{kk}$ constitute an irreducible subgraph of $\mathscr{G}$, and $\widetilde{L}_{kj}$ $(j < k)$ represents the influence from subgraph $\widetilde{L}_{jj}$ to subgraph $\widetilde{L}_{kk}$.

Definition 2.30 [13] Let $\mathscr{G}$ be a directed network and let $\mathscr{G}_1, \mathscr{G}_2, \dots, \mathscr{G}_p$ be the strongly connected components of $\mathscr{G}$ with connection matrices $\widetilde{L}_{11}, \widetilde{L}_{22}, \dots, \widetilde{L}_{pp}$. $\mathscr{G}^*$ is a *condensation network* of $\mathscr{G}$ if there is a connection from a node in $\mathscr{V}(\mathscr{G}_j)$ to a node in $\mathscr{V}(\mathscr{G}_i)$ $(i \neq j)$, then the weight $G^*_{ij} > 0$; otherwise, $G^*_{ij} = 0$ for $i, j = 1, 2, \dots, p$; $G^*_{ii} = 0$ for $i = 1, 2, \dots, p$.

Note that the condensation network $\mathscr{G}^*$ of a directed network $\mathscr{G}$ has no closed directed walks [13].

Lemma 2.31 *For every $i = 2, 3, \dots, p$, there is an integer $j < i$ such that $G^*_{ij} > 0$ if and only if the directed network $\mathscr{G}$ contains a directed spanning tree.*

Proof. If the directed network $\mathscr{G}$ contains a directed spanning tree, then there is a directed path from root (a node in $\mathscr{G}_1$) to every other node. Suppose that for an integer $2 \leq i \leq m$, $G^*_{ij} = 0$ for all $j < i$. Because of the network structure in (2.29), $G^*_{ij} = 0$ for all $j \neq i$, which means that there are no paths from the root to the ith strongly connected components. This contradicts with the fact that $\mathscr{G}$ contains a directed spanning tree.

Suppose that for every $i = 2, 3, \dots, m$, there is an integer $j < i$ such that $G^*_{ij} > 0$, which implies that there exists a directed edge from node j to node i in $\mathscr{G}^*$. When $i = 2$, one knows that there is a directed path from node 1 to node 2. Suppose there is a directed path from node 1 to all the nodes $2, 3, \dots, k$. Then, it suffices to prove that there is a directed path from node 1 to node $k + 1$. By assumption, there is an integer $j < k + 1$ and a directed path from j to $k + 1$. Hence, a directed path from node 1 to node $k + 1$ exists from node 1 to node j and then to node $k + 1$. Therefore, G^* contains a directed spanning tree with root 1 which indicates that G has a directed spanning tree with every node in $\mathscr{G}_1$ as a root. □

Let $\widetilde{L}_{kk} = \widetilde{A}^k + \widetilde{D}^k$, where $\sum_{j=1}^{m_k} \widetilde{A}^k_{ij} = 0$ and $\widetilde{D}^k$ is a diagonal matrix for all $i = 1, 2, \dots, m_k$. From Lemma 2.31, it is easy to see that $\widetilde{D}^k \geq 0$ and $\widetilde{D}^k \neq 0$ for all $k = 2, \dots, p$.

Lemma 2.32 *[23] If L is irreducible, $L_{ij} = L_{ji} \geq 0$ for $i \neq j$, and $\sum_{j=1}^{N} L_{ij} = 0$, for all $i = 1, 2, \dots, N$, then all eigenvalues of the matrix*

$$\begin{pmatrix} L_{11} + \epsilon & L_{12} & \dots & L_{1N} \\ L_{21} & L_{22} & \dots & L_{2N} \\ \vdots & \vdots & \ddots & \vdots \\ L_{N1} & L_{N2} & \dots & L_{NN} \end{pmatrix}$$

are positive for any positive constant ϵ.

Definition 2.33 For a network with a directed spanning tree and the Laplacian matrix in the form of (2.29), the general algebraic connectivity of the ith strongly connected component ($2 \leq i \leq p$) is defined to be the real number

$$b_{\widetilde{\xi}_i}(\widetilde{L}_{ii}) = \min_{x \neq 0} \frac{x^T \widehat{\widetilde{L}}_{ii} x}{x^T \widetilde{\Xi}_i x} = \min_{x \neq 0} \frac{\left(\sqrt{\widetilde{\Xi}_i} x\right)^T \sqrt{\widetilde{\Xi}_i}^{-1} \widehat{\widetilde{L}}_{ii} \sqrt{\widetilde{\Xi}_i}^{-1} \left(\sqrt{\widetilde{\Xi}_i} x\right)}{\left(\sqrt{\widetilde{\Xi}} x\right)^T \left(\sqrt{\widetilde{\Xi}_i} x\right)}$$

$$= \lambda_{\min} \sqrt{\widetilde{\Xi}_i}^{-1} \widehat{\widetilde{L}}_{ii} \sqrt{\widetilde{\Xi}_i}^{-1}, \tag{2.30}$$

where $\widehat{\widetilde{L}}_{ii} = \frac{\widetilde{\Xi}_i \widetilde{L}_i + \widetilde{L}_i^T \widetilde{\Xi}_i}{2}$, $\widetilde{\Xi}_i = \text{diag}(\widetilde{\xi}_{i1}, \ldots, \widetilde{\xi}_{im_i})$, $\sqrt{\widetilde{\Xi}_i} = \text{diag}\left(\sqrt{\widetilde{\xi}_{i1}}, \ldots, \sqrt{\widetilde{\xi}_{im_i}}\right)$, $\widetilde{\xi}_i = (\widetilde{\xi}_{i1}, \ldots, \widetilde{\xi}_{im_i})^T > 0$, and $\widetilde{\xi}_i^T \widetilde{A}^i = 0$, $\sum_{j=1}^{m_i} \widetilde{\xi}_{ij} = 1$.

Lemma 2.34 *If the Laplacian matrix L has a directed spanning tree, then* $\min_{2 \leq j \leq p} \{a_\xi(\widetilde{L}_{11}), b_{\widetilde{\xi}_i}(\widetilde{L}_{ii})\} > 0$, *where the chosen positive vector ξ in $a_\xi(\widetilde{L}_{11})$ satisfies $\xi^T \widetilde{L}_{11} = 0$ and the positive vectors $\widetilde{\xi}_i$ in $b_{\widetilde{\xi}_i}(\widetilde{L}_{ii})$ satisfy $\widetilde{\xi}_i^T \widetilde{A}^i = 0$ for $i = 2, \ldots, p$.*

Proof. From Corollary 2.27, one knows that $a_\xi(\widetilde{L}_{11}) > 0$. It suffices to prove that $\min_{2 \leq j \leq p} b_{\widetilde{\xi}_j}(\widetilde{L}_{jj}) > 0$. Note that

$$b_{\widetilde{\xi}_i}(\widetilde{L}_{ii}) = \lambda_{\min} \sqrt{\widetilde{\Xi}_i}^{-1} \widehat{\widetilde{L}}_{ii} \sqrt{\widetilde{\Xi}_i}^{-1} = \lambda_{\min} \sqrt{\widetilde{\Xi}_i}^{-1} \left(\frac{\widetilde{\Xi}_i \widetilde{A}^i + \widetilde{A}^{iT} \widetilde{\Xi}_i}{2} + \widetilde{D}^i \widetilde{\Xi}_i \right) \sqrt{\widetilde{\Xi}_i}^{-1},$$

where $\frac{1}{2}(\widetilde{\Xi}_i \widetilde{A}^i + \widetilde{A}^{iT} \widetilde{\Xi}_i)$ is a zero sums symmetric matrix and $\widetilde{D}^i \geq 0$. By Lemma 2.31, there is at least one positive diagonal entry in $\widetilde{D}^i$. According to Lemma 2.32, $b_{\widetilde{\xi}_i}(\widetilde{L}_{ii}) > 0$ for $2 \leq i \leq p$. □

Theorem 2.35 *Suppose that Assumption 2.21 holds and graph $\mathcal{G}$ has a spanning tree. Then global consensus of system (2.4) can be reached if*

$$\theta - c \min_{2 \leq j \leq p} \left\{ a_{\widetilde{\xi}}(\widetilde{L}_{11}), b_{\widetilde{\xi}_j}(\widetilde{L}_{jj}) \right\} < 0. \tag{2.31}$$

Proof. From (2.29), one knows that $\widetilde{L}_{11}$ is irreducible, and in view of Theorem 2.22, the consensus of agents in the subgraph $\widetilde{L}_{11}$ can be reached. Suppose that agents $1, \ldots, m_1$ are synchronized to the state of the following system:

$$\dot{s}(t) = f(s(t)) + \mathcal{O}(e^{-\epsilon t}), \tag{2.32}$$

where ϵ is a positive constant. First, one has

$$\dot{x}_i(t) = f(x_i(t)) - c\sum_{j=1}^{N} L_{ij}x_j(t) = f(x_i(t)) - c\sum_{j=1}^{q_{k-1}} L_{ij}\Gamma x_j(t)$$
$$-c\sum_{j=q_{k-1}+1}^{q_{k-1}+m_k} L_{ij}\Gamma x_j(t), \tag{2.33}$$

where $q_k = m_1 + \ldots + m_k$. Subtracting (2.32) from (2.33) yields the following error dynamical system:

$$\dot{e}_i(t) = f(x_i(t)) - f(s(t)) - c\sum_{j=1}^{q_k} L_{ij}\Gamma[x_j(t) - s(t)] + \mathcal{O}(e^{-\epsilon t}), \tag{2.34}$$

where $i = q_{k-1} + 1, \ldots, q_{k-1} + m_k$ and $e_i(t) = x_i(t) - s(t)$. Choose the following Lyapunov function candidate:

$$V(t) = \frac{1}{2}\sum_{k=2}^{p}\sum_{i=q_{k-1}+1}^{q_{k-1}+m_k} \Delta_k\widetilde{\xi}_{ki}e_i^T(t)e_i(t), \tag{2.35}$$

where Δ_k are positive constants to be determined and $\widetilde{\xi}_{ki}$ are defined in Definition 2.33, $k = 2, \ldots, p$ and $i = q_{k-1} + 1, \ldots, q_{k-1} + m_k$.

The derivative of $V(t)$ along the trajectories (2.34) gives

$$\dot{V}(t) = \sum_{k=2}^{p}\sum_{i=q_{k-1}+1}^{q_{k-1}+m_k} \Delta_k\widetilde{\xi}_{ki}e_i^T(t)\left[f(x_i(t)) - f(s(t)) - c\sum_{j=1}^{q_k} L_{ij}\Gamma e_j(t) + \mathcal{O}(e^{-\epsilon t})\right]$$
$$\leq \sum_{k=2}^{p}\{(\theta - cb_{\widetilde{\xi}_k}(\widetilde{L}_{kk}))\widetilde{e}_k^T(t)(\widetilde{\Xi}_k \otimes \Gamma)\widetilde{e}_k(t) - c\Delta_k\sum_{l=2}^{k-1}\widetilde{e}_i^T(t)(\widetilde{\Xi}_k\widetilde{L}_{kl} \otimes \Gamma)\widetilde{e}_j(t)\}$$
$$+\mathcal{O}(e^{-\epsilon t}) - \epsilon\sum_{k=2}^{p}\widetilde{e}_k^T(t)(\widetilde{\Xi}_k \otimes I_n)\widetilde{e}_k(t), \tag{2.36}$$

where $\widetilde{e}_k(t) = (e^T_{q_{k-1}+1}, \ldots, e^T_{q_{k-1}+m_k})^T$ and $\widetilde{\Xi}_k = \mathrm{diag}(\widetilde{\xi}_{k1}, \ldots, \widetilde{\xi}_{km_k})$.

Under condition (2.31) and by Lemma 2.14, one has $\dot{V}(t) < 0$. Since graph $\mathcal{G}$ has a spanning tree, $\widetilde{L}_{kk} = \widetilde{A}^k + \widetilde{D}^k$, where the sum of the entries in $\widetilde{A}^k$ is zero, and $\widetilde{D}^k \neq 0$. Now, Lemmas 2.32 and 2.34 together lead to the conditions given in (2.31) where $\theta > 0$. The proof is thus completed. □

2.2.4 *Global Consensus of Multi-agent Systems in Virtual Networks*

In the previous section, the general average consensus state $\bar{x}(t) = \sum_{j=1}^{N}\xi_j x_j = (\xi^T \otimes I_n)x(t)$ is used as a reference, where $\xi \geq 0$ and $\sum_{j=1}^{N}\xi_j = 1$. In this section, a different consensus state is considered as the reference. Without loss of generality, let $\bar{x} = x_1$ and define $\widetilde{e}_i(t) = x_{i+1} - x_1$, for $i = 1, 2, \ldots, N-1$. Then one has

$$
\begin{aligned}
\dot{\widetilde{e}}_i(t) &= f(x_{i+1}(t)) - f(x_1(t)) - c\sum_{j=1}^{N}(L_{i+1,j} - L_{1j})x_j(t) \\
&= f(x_{i+1}(t)) - f(x_1(t)) - c\sum_{j=1}^{N-1}\overline{L}_{ij}\widetilde{e}_j(t),
\end{aligned} \tag{2.37}
$$

where $\overline{L}_{ij} = L_{i+1,j+1} - L_{1,j+1}$ for $i,j = 1,\ldots,N-1$. It is easy to see that

$$
\overline{L} = \begin{pmatrix}
L_{22} - L_{12} & L_{23} - L_{13} & \cdots & L_{2N} - L_{1N} \\
L_{32} - L_{12} & \ddots & \cdots & L_{3N} - L_{1N} \\
\vdots & \vdots & \ddots & \vdots \\
L_{N2} - L_{12} & L_{N3} - L_{13} & \cdots & L_{NN} - L_{1N}
\end{pmatrix}, \tag{2.38}
$$

and $\sum_{j=1}^{N-1}\overline{L}_{ij} = L_{11} - L_{i+1,1}$ for $i = 1,2,\ldots,N-1$. Consider $\widetilde{e}_i(t)$ as the state of virtual node i, and the virtual network of interactions among these nodes are represented by $\overline{L}$. Assume that $\overline{L}$ has $N-1$ eigenvalues: $\lambda_1(\overline{L}) \geq \lambda_2(\overline{L}) \geq \ldots \geq \lambda_{N-1}(\overline{L})$. If the virtual network (2.37) is stable, then the global consensus in the multi-agent system (2.4) can be reached.

One interesting question arises: Is there any relation between the spectrum of the virtual network topology $\overline{L}$ and the spectrum of the original Laplacian matrix L?

Lemma 2.36 *Suppose that $\mathcal{G}$ has a spanning tree. Then, all the eigenvalues of $\overline{L}$ are the same as that of L except the simple eigenvalue 0, i.e., $\lambda_i(\overline{L}) = \lambda_{i+1}(L)$ for $i = 1,2,\ldots,N-1$.*

Proof. Assume that λ is an eigenvalue of L. Then, one has

$$
\begin{aligned}
|\lambda I_N - L| &= \begin{vmatrix}
\lambda - L_{11} & -L_{12} & \cdots & -L_{1N} \\
-L_{21} + L_{11} - \lambda & \lambda - L_{22} + L_{12} & \cdots & -L_{2N} + L_{1N} \\
\vdots & \vdots & \ddots & \vdots \\
-L_{N1} + L_{11} - \lambda & -L_{N2} + L_{12} & \cdots & \lambda - L_{NN} + L_{1N}
\end{vmatrix} \\
&= \begin{vmatrix}
\lambda - \sum_{j=1}^{N} L_{1j} & -L_{12} & \cdots & -L_{1N} \\
0 & \lambda - L_{22} + L_{12} & \cdots & -L_{2N} + L_{1N} \\
\vdots & \vdots & \ddots & \vdots \\
0 & -L_{N2} + L_{12} & \cdots & \lambda - L_{NN} + L_{1N}
\end{vmatrix} \\
&= \lambda |\lambda I_{N-1} - \overline{L}|,
\end{aligned} \tag{2.39}
$$

where the first equality is obtained by subtracting the first row from other rows and the third equality is satisfied by adding all the other columns into the first column. Since graph $\mathcal{G}$ has a spanning tree, 0 is a simple eigenvalue of L. This completes the proof. □

The global consensus problem can be studied with respect to the virtual network. In addition, ideas used in the previous two sections can also be applied to the virtual network system (2.37) and thus are omitted here.

Suppose that graph $\mathcal{G}$ has a spanning tree and there exists a node v such that there are no connections from other nodes to this node v. In fact, node v can be considered as a virtual leader here: all the other nodes receive information from node v but do not send any information back to node v. Without loss of generality, assume $v = 1$. Then it follows that $L_{1j} = 0$ for $j = 1, 2, \ldots, N$. From (2.38), one has $\overline{L}_{ij} = L_{i+1,j+1} \leq 0$ $(i \neq j)$ and $\sum_{j=1}^{N-1} \overline{L}_{ij} = -L_{i+1,1}$. If there is an edge from node v to node $i+1$, then $\sum_{j=1}^{N-1} \overline{L}_{ij} > 0$; otherwise, $\sum_{j=1}^{N-1} \overline{L}_{ij} = 0$. Let $\overline{L} = L1 + L2$, where $\sum_{j=1}^{N-1} L1_{ij} = 0$ and $L2 = \text{diag}(-L_{21}, -L_{31}, \ldots, -L_{N1})$.

Corollary 2.37 *Suppose that Assumption 2.21 holds and the subgraph graph with node set $\mathcal{V} - \{v\}$ is irreducible. Then global consensus of system (2.37) can be reached if*

$$\theta\Xi - c\left(\Xi\overline{L}\right)^s < 0, \tag{2.40}$$

where $\Xi = \text{diag}(\xi_1, \xi_2, \ldots, \xi_{N-1})$ is a positive diagonal matrix and satisfies $\xi^T L1 = 0$.

Proof. Consider the Lyapunov function candidate:

$$V(t) = \frac{1}{2}\sum_{i=1}^{N-1} \xi_i \widetilde{e}_i^T(t)\widetilde{e}_i(t). \tag{2.41}$$

The derivative of $V(t)$ along the trajectories of (2.37) gives

$$\begin{aligned}
\dot{V} &= \sum_{i=1}^{N-1} \xi_i \widetilde{e}_i^T(t) \left[f(x_{i+1}(t)) - f(x_1(t)) - c\sum_{j=1}^{N-1} \overline{L}_{ij}\widetilde{e}_j(t) \right] \\
&\leq \sum_{i=1}^{N-1} \xi_i \widetilde{e}_i^T(t) \left[\theta\widetilde{e}_i(t) - c\sum_{j=1}^{N-1} \overline{L}_{ij}\widetilde{e}_j(t) \right] \\
&= \widetilde{e}^T(t) \left[\left(\theta\Xi - c\Xi\overline{L} \right) \otimes I_n \right] \widetilde{e}(t),
\end{aligned} \tag{2.42}$$

where $\widetilde{e}(t) = (\widetilde{e}_1^T(t), \ldots, \widetilde{e}_{N-1}^T(t))^T$.

Since $\overline{L} = L1 + L2$, one knows that $\sum_{j=1}^{N-1} L1_{ij} = 0$ and $L1$ is irreducible. In view of Lemma 2.24, there exists a positive vector $\xi = (\xi_1, \xi_2, \ldots, \xi_{N-1})^T$, such that $\xi^T L1 = 0$. Let $\Delta = \Xi = \text{diag}(\xi_1, \xi_2, \ldots, \xi_{N-1})$. Then, global consensus of system (2.37) will be reached if condition (2.40) is satisfied. Since $L2 \neq 0$, from Lemma 2.32, one has $(\Xi\overline{L})^s > 0$. Thus, when the coupling strength is large enough, global consensus of the Multi-agent system (2.4) is reached. This completes the proof. □

The method applied in Corollary 2.37 was used in [23, 151], where pinning control synchronization of complex networks was considered: a reference node v as an isolated node is first chosen, and some pinning controllers are then designed to ensure that the whole network is synchronized with the state of node v. In this section, a more general discussion on the consensus of a virtual network has been carried out. If the graph with node set $\mathscr{V} - \{v\}$ has a spanning tree, then similar analytical results can be obtained by using Theorem 2.35.

2.2.5 Simulation Examples

In this subsection, a simulation example is given to verify the theoretical analysis.

Consider the multi-agent system (2.2), where the network structure is shown in Fig. 2.1, the coupling strength $c = 11$, and the nonlinear function f is described by Chua's circuit [24]

$$f(x_i(t)) = \begin{pmatrix} \alpha(-x_{i1} + x_{i2} - l(x_{i1})), \\ x_{i1} - x_{i2} + x_{i3}, \\ -\beta x_{i2}, \end{pmatrix}, \tag{2.43}$$

where $l(x_{i1}) = bx_{i1} + 0.5(a-b)(|x_{i1}+1| - |x_{i1}-1|)$. The system (2.43) is chaotic when $\alpha = 10$, $\beta = 18$, $a = -4/3$, and $b = -3/4$, as shown in [142]. In view of Assumption 2.21, by computations, one obtains $\theta = 10.3246$. From Fig. 2.1, it is easy to see that the network contains a directed spanning tree where the nodes 1–4 and 5–7 belong to the first and second strongly connected components, respectively. By Lemma 2.28 and Definition 2.33, one has $a(\widetilde{L}_{11}) = 1.8118$ and $b(\widetilde{L}_{22}) = 1.0206$, where $\widetilde{\xi}_1 = (0.2727, 0.1818, 0.1364, 0.4091)^T$ and $\widetilde{\xi}_2 = (0.4615, 0.3077, 0.2308)^T$. By Theorem 2.35, one has that $c\min\{a(\widetilde{L}_{11}), b(\widetilde{L}_{22})\} = 11.2266 > \theta = 10.3246$. Therefore, consensus can be achieved in this example of the multi-agent system (2.2). The states of all the agents are shown in Fig. 2.2.

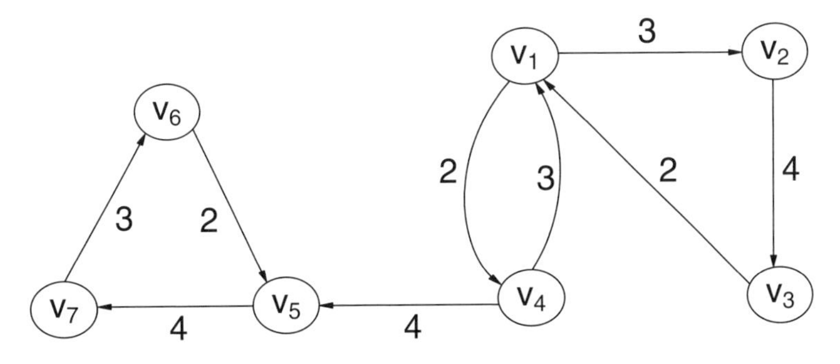

Figure 2.1 Network structure of a network with a directed spanning tree

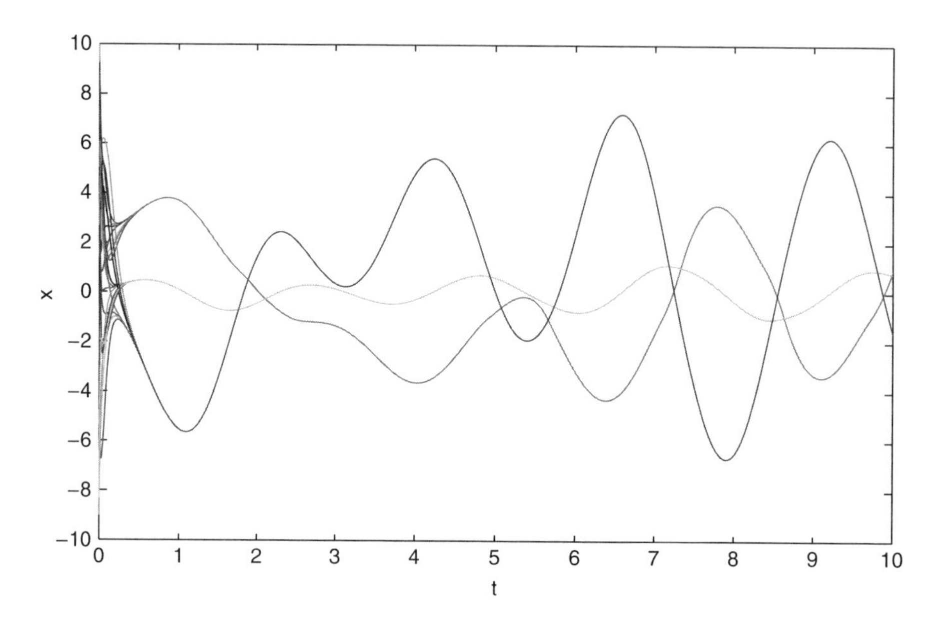

Figure 2.2 States of multiple agents in a network with time-varying velocities

2.3 Notes

In this chapter, consensus in linear multi-agent systems has been introduced. Then, both local and global consensus problems for multi-agent systems in weighted directed networks have been investigated. The main contents of this chapter include the following: (i) Local consensus in a directed network of agents with nonlinear dynamics is discussed, where the real part of the second smallest eigenvalue of the Laplacian matrix plays a key role in deriving the consensus conditions. (ii) A generalized algebraic connectivity, which can be used to describe the consensus ability of the network, is proposed to discuss global convergence properties of consensus in strongly connected networks and later in networks containing spanning trees. (iii) A new concept of virtual network, which has the same spectral structure as the original network, is introduced to simplify the analysis.

In the future, networks with nonlinear couplings and networks consisting of heterogeneous autonomous agents may be studied. A general stochastic framework may be helpful in bringing more ideas from the study of complex networks into the study of the consensus problem, which is also a promising approach deserving further investigation.

3

Second-Order Consensus in Multi-agent Systems

This chapter studies second-order consensus in multi-agent dynamical systems [146, 149].

In the first section, some necessary and sufficient conditions for second-order consensus in multi-agent systems with linear dynamics are derived. First, basic theoretical analysis is carried out for the case where for each agent the second-order dynamics are governed by the position and velocity terms and the asymptotic velocity is constant. A necessary and sufficient condition is given to ensure second-order consensus, and it is found that both the real and imaginary parts of the eigenvalues of the Laplacian matrix of the corresponding network play key roles in reaching consensus. Based on this result, a second-order consensus algorithm is derived for those multi-agent systems facing communication delays. A necessary and sufficient condition is provided, which shows that consensus can be achieved in a multi-agent system whose network topology contains a directed spanning tree if and only if the time delay is less than a critical value.

The second section considers a second-order consensus problem for multi-agent systems with nonlinear dynamics and directed topologies, where each agent is governed by both position and velocity consensus terms with a time-varying asymptotic velocity. To describe the system's ability for reaching consensus, the generalized algebraic connectivity is used for strongly connected networks and then extended to the strongly connected components of the directed network containing a spanning tree. Some sufficient conditions are derived for reaching second-order consensus in multi-agent systems with nonlinear dynamics based on algebraic graph theory, matrix theory, and Lyapunov control approach.

The following notations are used throughout Chapter 3. Let $\lambda_{\max}(F)$ be the largest eigenvalue of matrix F, I_N (O_N) be the N-dimensional identity (zero) matrix, $1_N \in R^N$ ($0_N \in R^N$) be a vector with each entry being 1 (0), $\mathscr{R}(u)$ and $\mathscr{I}(u)$ be the real and

Distributed Cooperative Control of Multi-agent Systems, First Edition.
Wenwu Yu, Guanghui Wen, Guanrong Chen, and Jinde Cao.

imaginary parts of a complex number u, and $\otimes$ be the Kronecker product [53]. For matrices $\widetilde{A}$ and $\widetilde{B}$ with the same order, $\widetilde{A} > \widetilde{B}$ means that $\widetilde{A} - \widetilde{B}$ is positive definite. A matrix $G \in R^{N\times N}$ is nonnegative if every entry $G_{ij} \geq 0$ $(1 \leq i,j \leq N)$ and a vector $x \in R^N$ is positive if every entry $x_i > 0$ $(1 \leq i \leq N)$. Finally, let $\rho(A)$ be the spectral radius of matrix A.

3.1 Second-Order Consensus in Linear Multi-agent Systems

3.1.1 Model Formulation

The first-order consensus protocol has been widely studied for networks consisting of N nodes with linearly diffusive coupling [57, 72, 81, 98, 134, 141, 142, 151, 165]:

$$\dot{x}_i(t) = \widetilde{c} \sum_{j=1,\ j\neq i}^{N} G_{ij}(x_j(t) - x_i(t)), i = 1, 2, \cdots, N, \tag{3.1}$$

where $x_i(t) = (x_{i1}(t), x_{i2}(t), \cdots, x_{in}(t))^T \in R^n$ is the state vector of the ith node and $\widetilde{c}$ is the coupling strength.

As to the second-order dynamics, the second-order consensus protocol is [50, 51, 93, 95, 97]:

$$\begin{aligned}
\dot{x}_i(t) &= v_i, \\
\dot{v}_i(t) &= \alpha \sum_{j=1,\ j\neq i}^{N} G_{ij}(x_j(t) - x_i(t)) + \beta \sum_{j=1,\ j\neq i}^{N} G_{ij}(v_j(t) - v_i(t)), \\
& \quad i = 1, 2, \cdots, N,
\end{aligned} \tag{3.2}$$

where $x_i \in R^n$ and $v_i \in R^n$ are the position and velocity states of the ith node, respectively, and $\alpha > 0$ and $\beta > 0$ are the coupling strengths.

Equivalently, system (3.2) can be rewritten as follows:

$$\begin{aligned}
\dot{x}_i(t) &= v_i, \\
\dot{v}_i(t) &= -\alpha \sum_{j=1}^{N} L_{ij} x_j(t) - \beta \sum_{j=1}^{N} L_{ij} v_j(t), i = 1, 2, \cdots, N.
\end{aligned} \tag{3.3}$$

Let $x = (x_1^T, x_2^T, \ldots, x_N^T)^T$, $v = (v_1^T, v_2^T, \ldots, v_N^T)^T$, and $y = (x^T, v^T)^T$. Then, network (3.3) can be rewritten in a compact matrix form as

$$\dot{y}(t) = (\widetilde{L} \otimes I_n) y, \tag{3.4}$$

where $\widetilde{L} = \begin{pmatrix} O_N & I_N \\ -\alpha L & -\beta L \end{pmatrix}$ and $\otimes$ is the Kronecker product [53].

Definition 3.1 Second-order consensus in the multi-agent system (3.4) is said to be achieved if for any initial conditions,

$$\lim_{t\to\infty} \| x_i(t) - x_j(t) \| = 0, \lim_{t\to\infty} \| v_i(t) - v_j(t) \| = 0, \forall i, j = 1, 2, \cdots, N.$$

3.1.2 Second-Order Consensus in Directed Networks

In this subsection, some second-order consensus algorithms for the multi-agent system (3.4) with directed topologies are developed.

For the linear model (3.4), eigenvalues of matrix $\widetilde{L}$ are very important in convergence analysis. Suppose that λ_{ij} $(i = 1, 2, \ldots, N,\ j = 1, 2)$ and μ_i $(i = 1, 2, \ldots, N)$ are eigenvalues of $\widetilde{L}$ and the Laplacian matrix L, respectively. First, some relationships between the eigenvalues of $\widetilde{L}$ and L are reviewed [95, 97].

Let λ be an eigenvalue of matrix $\widetilde{L}$. Then, one has $\det(\lambda I_{2N} - \widetilde{L}) = 0$. Note that

$$\det(\lambda I_{2N} - \widetilde{L}) = \det\begin{pmatrix} \lambda I_N & -I_N \\ \alpha L & \lambda I_N + \beta L \end{pmatrix} = \det(\lambda^2 I_N + (\alpha + \beta\lambda)L)$$

$$= \prod_{i=1}^{N} (\lambda^2 + (\alpha + \beta\lambda)\mu_i) = 0.$$

Hence,

$$\lambda_{i1} = \frac{-\beta\mu_i + \sqrt{\beta^2\mu_i^2 - 4\alpha\mu_i}}{2},$$

$$\lambda_{i2} = \frac{-\beta\mu_i - \sqrt{\beta^2\mu_i^2 - 4\alpha\mu_i}}{2}, i = 1, 2, \ldots, N. \tag{3.5}$$

From (3.5), it is easy to see that L has a zero eigenvalue of algebraic multiplicity m if and only if $\widetilde{L}$ has a zero eigenvalue of algebraic multiplicity $2m$. In the following, for the sake of simplicity, we simply write algebraic multiplicity as multiplicity.

Lemma 3.2 *Second-order consensus in multi-agent system (3.4) can be achieved if and only if matrix $\widetilde{L}$ has exactly a zero eigenvalue of multiplicity two and all the other eigenvalues have negative real parts. In addition, if second-order consensus is reached, $\| v_i(t) - \sum_{j=1}^{N} \xi_j v_j(0) \| \to 0$ and $\| x_i(t) - \sum_{j=1}^{N} \xi_j x_j(0) - \sum_{j=1}^{N} \xi_j v_j(0) t \| \to 0$ as $t \to \infty$, where ξ is the unique nonnegative left eigenvector of L associated with eigenvalue 0 satisfying $\xi^T 1_N = 1$.*

Proof. For $n = 1$ and $\alpha = 1$, a proof of this lemma was given in [97]. Now, this lemma is proved for any integer $n \geq 1$ and $\alpha > 0$.

(Sufficiency). Note that 0 is an eigenvalue of matrix $\widetilde{L}$ with multiplicity 2. From calculation of $\widetilde{L}\varphi = 0$, where φ is a unit right eigenvector of matrix $\widetilde{L}$ associated with eigenvalue 0, one can easily obtain that $\varphi = (1_N^T, 0_N^T)^T/\sqrt{N}$, which is unique. So, matrix $\widetilde{L}$ cannot be diagonal since there is only one unit eigenvector of matrix $\widetilde{L}$ associated with eigenvalue 0. Therefore, a Jordan form is used here. If $\widetilde{L}$ has exactly a zero eigenvalue of multiplicity two and all the other eigenvalues have negative real parts, then there exists a nonsingular matrix $P \in R^{2N\times 2N}$, such that $P^{-1}\widetilde{L}P = J$, where J is the Jordan canonical form associated with $\widetilde{L}$. Thus, one has

$$\widetilde{L} = PJP^{-1} = (\zeta_1, \cdots, \zeta_{2N}) \times \begin{pmatrix} 0 & 1 & 0_{1\times(2N-2)} \\ 0 & 0 & 0_{1\times(2N-2)} \\ 0_{(2N-2)\times 1} & 0_{(2N-2)\times 1} & \widetilde{J} \end{pmatrix} \begin{pmatrix} \eta_1^T \\ \vdots \\ \eta_{2N}^T \end{pmatrix}, \tag{3.6}$$

where ζ_j and η_j $(j = 1, 2, \ldots, 2N)$ are the right and left eigenvectors or generalized eigenvectors of $\widetilde{L}$, respectively, and $\widetilde{J}$ is the upper diagonal Jordan block matrix associated with the nonzero eigenvalues λ_{ij}, $i = 2, \ldots, N; j = 1, 2$. It follows that $e^{\widetilde{J}t} \to 0_{(2N-2)\times(2N-2)}$ as $t \to \infty$.

From $\widetilde{L}P = PJ$, one obtains

$$\widetilde{L}(\zeta_1, \zeta_2) = (\zeta_1, \zeta_2)\begin{pmatrix} 0 & 1 \\ 0 & 0 \end{pmatrix} = (0_{2N}, \zeta_1),$$

giving that

$$\widetilde{L}\zeta_1 = 0_{2N}, \widetilde{L}\zeta_2 = \zeta_1.$$

Therefore, ζ_2 is the generalized right eigenvector of matrix $\widetilde{L}$ associated with eigenvalue 0 and can be computed by

$$\widetilde{L}^2\zeta_2 = \widetilde{L}\zeta_1 = 0\zeta_2 = 0_{2N}.$$

One can easily obtain a right eigenvector $\zeta_1 = (1_N^T, 0_N^T)^T$ and a generalized right eigenvector $\zeta_2 = (0_N^T, 1_N^T)^T$ of matrix $\widetilde{L}$ associated with eigenvalue 0. Accordingly, $\eta_1 = (\xi^T, 0_N^T)^T$ and $\eta_2 = (0_N^T, \xi^T)^T$ are the generalized left eigenvector and the left eigenvector of matrix $\widetilde{L}$ associated with eigenvalue 0, respectively. Also, in view of the properties of the Kronecker product [53], one has

$$\begin{aligned} e^{(\widetilde{L}\otimes I_n)t} &= e^{(P\otimes I_n)(Jt\otimes I_n)(P^{-1}\otimes I_n)} \\ &= (P \otimes I_n)e^{(Jt\otimes I_n)}(P^{-1} \otimes I_n) \\ &= (P \otimes I_n)(e^{Jt} \otimes I_n)(P^{-1} \otimes I_n) \\ &= (Pe^{Jt}P^{-1}) \otimes I_n \end{aligned}$$

$$= \left\{ P \begin{pmatrix} 1 & t & 0_{1\times(2N-2)} \\ 0 & 1 & 0_{1\times(2N-2)} \\ 0_{(2N-2)\times 1} & 0_{(2N-2)\times 1} & e^{\widetilde{J}t} \end{pmatrix} P^{-1} \right\} \otimes I_n.$$

Therefore, one obtains

$$\lim_{t\to\infty} \left\| \begin{pmatrix} x(t) \\ v(t) \end{pmatrix} - 1_N \otimes \begin{pmatrix} \sum_{j=1}^{N} \xi_j(x_j(0) + v_j(0)t) \\ \sum_{j=1}^{N} \xi_j v_j(0) \end{pmatrix} \right\|$$

$$= \lim_{t\to\infty} \left\| e^{(\widetilde{L}\otimes I_n)t} \begin{pmatrix} x(0) \\ v(0) \end{pmatrix} - 1_N \otimes \begin{pmatrix} \sum_{j=1}^{N} \xi_j(x_j(0) + v_j(0)t) \\ \sum_{j=1}^{N} \xi_j v_j(0) \end{pmatrix} \right\|$$

$$= \lim_{t\to\infty} \left\| \left[\begin{pmatrix} 1_N\xi^T & t1_N\xi^T \\ O_N & 1_N\xi^T \end{pmatrix} \otimes I_n \right] \begin{pmatrix} x(0) \\ v(0) \end{pmatrix} \right.$$

$$\left. -1_N \otimes \begin{pmatrix} \sum_{j=1}^{N} \xi_j(x_j(0) + v_j(0)t) \\ \sum_{j=1}^{N} \xi_j v_j(0) \end{pmatrix} \right\| = 0, \tag{3.7}$$

which indicates that second-order consensus is achieved in system (3.4).

(Necessity). If the condition that matrix $\widetilde{L}$ has exactly one zero eigenvalue of multiplicity two and all the other eigenvalues have negative real parts is not satisfied, then $\lim_{t\to\infty} e^{\widetilde{L}t}$ has a rank greater than 2, which contradicts the assumption that second-order consensus is reached. (See [97] for a similar argument.) □

Although a necessary and sufficient condition is given in Lemma 3.2 to ensure the second-order consensus in multi-agent system (3.4), it does not show any relationship between the eigenvalues of matrix $\widetilde{L}$ and the Laplacian matrix L. A natural question is: on what kind of networks can second-order consensus be reached? In [97], an example is given where second-order consensus can be achieved in a network whose topology is a directed spanning tree but cannot be achieved after adding only one extra edge into the directed spanning tree. This is a bit surprising as it is inconsistent with the intuition that connections are helpful for reaching consensus. The following result addresses this issue.

Theorem 3.3 *Second-order consensus in multi-agent system (3.4) can be achieved if and only if the network contains a directed spanning tree and*

$$\frac{\beta^2}{\alpha} > \max_{2\le i\le N} \frac{\mathcal{I}^2(\mu_i)}{\mathcal{R}(\mu_i)[\mathcal{R}^2(\mu_i) + \mathcal{I}^2(\mu_i)]}, \tag{3.8}$$

where μ_i are the nonzero eigenvalues of the Laplacian matrix L, $i = 2, 3, \ldots, N$. In addition, if second-order consensus is reached, $\| v_i(t) - \sum_{j=1}^{N} \xi_j v_j(0) \| \to 0$

and $\|x_i(t) - \sum_{j=1}^{N} \xi_j x_j(0) - \sum_{j=1}^{N} \xi_j v_j(0)t\| \to 0$ *as* $t \to \infty$*, where* ξ *is the unique nonnegative left eigenvector of* L *associated with eigenvalue 0 satisfying* $\xi^T 1_N = 1$.

Proof. From Lemma 2.7, one knows that the Laplacian matrix L has a simple eigenvalue 0 and all the other eigenvalues have positive real parts if and only if the directed network has a directed spanning tree. By Lemma 3.2, one only needs to prove that both $\mathcal{R}(\mu_i) > 0$ $(i = 2, 3, \dots, N)$ and (3.8) hold if and only if $\mathcal{R}(\lambda_{ij}) < 0$ $(i = 2, 3, \dots, N; j = 1, 2)$.

Let $\sqrt{\beta^2\mu_i^2 - 4\alpha\mu_i} = c + \mathbf{i}d$, where c and d are real, and $\mathbf{i} = \sqrt{-1}$. From (3.5), $\mathcal{R}(\lambda_{ij}) < 0$ $(i = 2, 3, \dots, N; j = 1, 2)$ if and only if $-\beta\mathcal{R}(\mu_i) < c < \beta\mathcal{R}(\mu_i)$, which is equivalent to $\mathcal{R}(\mu_i) > 0$ and $c^2 < \beta^2\mathcal{R}^2(\mu_i)$ $(i = 2, 3, \dots, N)$. Then, it suffices to prove that (3.8) holds if and only if $c^2 < \beta^2\mathcal{R}^2(\mu_i)$ $(i = 2, 3, \dots, N)$. It is easy to see that

$$\beta^2\mu_i^2 - 4\alpha\mu_i = (c + \mathbf{i}d)^2.$$

Separating the real and imaginary parts, one has

$$\begin{aligned} c^2 - d^2 &= \beta^2[\mathcal{R}^2(\mu_i) - \mathcal{I}^2(\mu_i)] - 4\alpha\mathcal{R}(\mu_i), \\ cd &= \beta^2\mathcal{R}(\mu_i)\mathcal{I}(\mu_i) - 2\alpha\mathcal{I}(\mu_i). \end{aligned}$$

By simple calculations, one obtains

$$\begin{aligned} &c^4 - \{\beta^2[\mathcal{R}^2(\mu_i) - \mathcal{I}^2(\mu_i)] - 4\alpha\mathcal{R}(\mu_i)\}c^2 \\ &-\mathcal{I}^2(\mu_i)[\beta^2\mathcal{R}(\mu_i) - 2\alpha]^2 = 0. \end{aligned} \tag{3.9}$$

It is easy to check that $c^2 < \beta^2\mathcal{R}^2(\mu_i)$ if and only if (3.8) holds. □

Remark 3.4 In Theorem 3.3, in addition to the condition that the network has a directed spanning tree, (3.8) should also be satisfied. It is easy to verify that if all the other eigenvalues of the Laplacian matrix L are real, then (3.8) holds. From (3.8), it is found that both real and imaginary parts of the eigenvalues of the Laplacian matrix play important roles in reaching second-order consensus. Let $\frac{\mathcal{I}^2(\mu_k)}{\mathcal{R}(\mu_k)[\mathcal{R}^2(\mu_k) + \mathcal{I}^2(\mu_k)]} = \max_{2\le i\le N} \frac{\mathcal{I}^2(\mu_i)}{\mathcal{R}(\mu_i)[\mathcal{R}^2(\mu_i) + \mathcal{I}^2(\mu_i)]}$, where $2 \le k \le N$. Then, one can see that in order to reach consensus, the critical value β^2/α increases as $|\mathcal{I}(\mu_k)|$ increases and decreases as $\mathcal{R}(\mu_k)$ increases.

Remark 3.5 If $\frac{\beta^2}{\alpha} > \max_{2\le i\le N} \frac{1}{\mathcal{R}(\mu_i)}$ holds, (3.8) is satisfied for sure. So, the sufficient condition for reaching consensus, $\frac{\beta^2}{\alpha} > \max_{2\le i\le N} \frac{2}{\| \mu_i \| \cos\left(\frac{\pi}{2} - \tan^{-1}\frac{\mathcal{R}(\mu_i)}{\mathcal{I}(\mu_i)}\right)} =$

$\max_{2\le i\le N} \frac{2}{\| \mu_i \| \sin\left(\tan^{-1}\frac{\mathcal{R}(\mu_i)}{\mathcal{I}(\mu_i)}\right)} = \max_{2\le i\le N} \frac{2}{\mathcal{R}(\mu_i)}$, given in [97], is more conservative.
Here, the sufficient condition depends only on the real parts of the eigenvalues of the Laplacian matrix L, but are independent of their imaginary parts. Moreover, when β^2/α is very small, consensus may still be achieved even if $\frac{\beta^2}{\alpha} > \max_{2\le i\le N} \frac{1}{\mathcal{R}(\mu_i)}$ is not satisfied.

3.1.3 Second-Order Consensus in Delayed Directed Networks

In this subsection, the following second-order consensus protocol with time delays is considered:

$$\begin{aligned} \dot{x}_i(t) &= v_i, \\ \dot{v}_i(t) &= -\alpha \sum_{j=1}^{N} L_{ij} x_j(t-\tau) - \beta \sum_{j=1}^{N} L_{ij} v_j(t-\tau), i = 1, 2, \cdots, N, \end{aligned} \tag{3.10}$$

where $\tau > 0$ is the time-delay constant.

Let $x = (x_1^T, x_2^T, \dots, x_N^T)^T$, $v = (v_1^T, v_2^T, \dots, v_N^T)^T$, and $y = (x^T, v^T)^T$. Then, network (3.10) can be rewritten in a compact matrix form, as follows:

$$\dot{y}(t) = (\widetilde{L}_1 \otimes I_n)y + (\widetilde{L}_2 \otimes I_n)y(t-\tau), \tag{3.11}$$

where $\widetilde{L}_1 = \begin{pmatrix} O_N & I_N \\ O_N & O_N \end{pmatrix}$ and $\widetilde{L}_2 = \begin{pmatrix} O_N & O_N \\ -\alpha L & -\beta L \end{pmatrix}$.

In [137, 138, 140], stability and Hopf bifurcation of delayed networks were studied, where the time delays are regarded as bifurcation parameters. It was found that Hopf bifurcation occurs when time delays pass through some critical values where the conditions for local asymptotic stability of the equilibrium are not satisfied. Similarly, this subsection aims to find the maximum time delay with which the consensus can be achieved in the multi-agent system (3.11).

The characteristic equation of system (3.11) is $\det(\lambda I_{2N} - \widetilde{L}_1 - e^{-\lambda\tau}\widetilde{L}_2) = 0$, i.e.,

$$\begin{aligned} \det(\lambda I_{2N} - \widetilde{L}_1 - e^{-\lambda\tau}\widetilde{L}_2) &= \det \begin{pmatrix} \lambda I_N & -I_N \\ \alpha e^{-\lambda\tau} L & \lambda I_N + \beta e^{-\lambda\tau} L \end{pmatrix} \\ &= \det(\lambda^2 I_N + (\alpha + \beta\lambda)e^{-\lambda\tau} L) = \prod_{i=1}^{N} (\lambda^2 + (\alpha + \beta\lambda)e^{-\lambda\tau}\mu_i) = 0. \end{aligned} \tag{3.12}$$

Let $g_i(\lambda) = \lambda^2 + (\alpha + \beta\lambda)e^{-\lambda\tau}\mu_i$ and $g(\lambda) = \prod_{i=1}^{N} g_i(\lambda)$. From (3.12), it is easy to see that L has a zero eigenvalue of multiplicity m if and only if $g(\lambda) = 0$ has a zero root of multiplicity $2m$.

Lemma 3.6 *Suppose that the network contains a directed spanning tree. Then, $g(\lambda) = 0$ has a purely imaginary root if and only if*

$$\tau \in \Psi = \left\{ \frac{1}{\omega_{i1}}(2k\pi + \theta_{i1}) | \ i = 2, \ldots, N; k = 0, 1, \ldots \right\}, \tag{3.13}$$

where $0 \leq \theta_{i1} < 2\pi$*, which satisfies* $\cos\theta_{i1} = [\mathscr{R}(\mu_i)\alpha - \mathscr{I}(\mu_i)\omega_{i1}\beta]/\omega_{i1}^2$ *and* $\sin\theta_{i1} = [\mathscr{R}(\mu_i)\omega_{i1}\beta + \mathscr{I}(\mu_i)\alpha]/\omega_{i1}^2$*, and* $\omega_{i1} = \sqrt{\frac{\| \mu_i\|^2\beta^2 + \sqrt{\| \mu_i\|^4\beta^4 + 4 \| \mu_i\|^2\alpha^2}}{2}}$, $i = 2, \ldots, N$.

Proof. (Necessity). Let $\lambda = \mathbf{i}\omega_i$ $(\omega_i \neq 0)$. From $g_i(\lambda) = 0$, one has

$$\omega_i^2 = (\alpha + \mathbf{i}\beta\omega_i)e^{-\mathbf{i}\omega_i\tau}\mu_i. \tag{3.14}$$

Taking modulus on both sides of (3.14), one obtains

$$\omega_i^4 - [\mathscr{R}(\mu_i)^2 + \mathscr{I}(\mu_i)^2]\beta^2\omega_i^2 - [\mathscr{R}(\mu_i)^2 + \mathscr{I}(\mu_i)^2]\alpha^2 = 0. \tag{3.15}$$

Then,

$$\omega_i^2 = \frac{\| \mu_i\|^2\beta^2 + \sqrt{\| \mu_i\|^4\beta^4 + 4 \| \mu_i\|^2\alpha^2}}{2}. \tag{3.16}$$

Separating the real and imaginary parts of (3.14) yields

$$\begin{aligned} \omega_i^2 &= [\mathscr{R}(\mu_i)\alpha - \mathscr{I}(\mu_i)\omega_i\beta]\cos(\omega_i\tau) + [\mathscr{R}(\mu_i)\omega_i\beta + \mathscr{I}(\mu_i)\alpha]\sin(\omega_i\tau), \\ 0 &= [\mathscr{R}(\mu_i)\omega_i\beta + \mathscr{I}(\mu_i)\alpha]\cos(\omega_i\tau) - [\mathscr{R}(\mu_i)\alpha - \mathscr{I}(\mu_i)\omega_i\beta]\sin(\omega_i\tau). \end{aligned} \tag{3.17}$$

By simple calculations, one obtains

$$\begin{aligned} \cos(\omega_i\tau) &= \frac{\omega_i^2[\mathscr{R}(\mu_i)\alpha - \mathscr{I}(\mu_i)\omega_i\beta]}{[\mathscr{R}(\mu_i)\alpha - \mathscr{I}(\mu_i)\omega_i\beta]^2 + [\mathscr{R}(\mu_i)\omega_i\beta + \mathscr{I}(\mu_i)\alpha]^2}, \\ \sin(\omega_i\tau) &= \frac{\omega_i^2[\mathscr{R}(\mu_i)\omega_i\beta + \mathscr{I}(\mu_i)\alpha]}{[\mathscr{R}(\mu_i)\alpha - \mathscr{I}(\mu_i)\omega_i\beta]^2 + [\mathscr{R}(\mu_i)\omega_i\beta + \mathscr{I}(\mu_i)\alpha]^2}. \end{aligned} \tag{3.18}$$

From (3.15), it follows that $\omega_i^4 = [\mathscr{R}(\mu_i)\alpha - \mathscr{I}(\mu_i)\omega_i\beta]^2 + [\mathscr{R}(\mu_i)\omega_i\beta + \mathscr{I}(\mu_i)\alpha]^2$. Thus, (3.18) can be written as

$$\begin{aligned} \cos(\omega_i\tau) &= \frac{[\mathscr{R}(\mu_i)\alpha - \mathscr{I}(\mu_i)\omega_i\beta]}{\omega_i^2}, \\ \sin(\omega_i\tau) &= \frac{[\mathscr{R}(\mu_i)\omega_i\beta + \mathscr{I}(\mu_i)\alpha]}{\omega_i^2}. \end{aligned} \tag{3.19}$$

Let $\omega_{i1,2} = \pm\sqrt{\frac{\| \mu_i\|^2\beta^2 + \sqrt{\| \mu_i\|^4\beta^4 + 4 \| \mu_i\|^2\alpha^2}}{2}}$ and $0 \le \theta_{ij} < 2\pi$, which satisfies $\cos\theta_{ij} = \frac{[\mathscr{R}(\mu_i)\alpha - \mathscr{I}(\mu_i)\omega_{ij}\beta]}{\omega_{ij}^2}$ and $\sin\theta_{ij} = \frac{[\mathscr{R}(\mu_i)\omega_{ij}\beta + \mathscr{I}(\mu_i)\alpha]}{\omega_{ij}^2}$, $j = 1, 2$. Since $\mathscr{R}(\mu_i) \pm \mathbf{i}\mathscr{I}(\mu_i)$ are two eigenvalues of L, if $\mathscr{I}(\mu_i) \neq 0$, then there exists an integer s, $2 \le s \le N$, such that

$$\begin{aligned}&\cos(\omega_{i1}\tau) = \cos(\omega_{s2}\tau), \sin(\omega_{i1}\tau) = -\sin(\omega_{s2}\tau),\\&\theta_{i1} = 2\pi - \theta_{s2}.\end{aligned}$$

If $\mathscr{I}(\mu_i) = 0$, then

$$\begin{aligned}&\cos(\omega_{i1}\tau) = \cos(\omega_{i2}\tau), \sin(\omega_{i1}\tau) = -\sin(\omega_{i2}\tau),\\&\theta_{i1} = 2\pi - \theta_{i2}.\end{aligned}$$

It follows that

$$\begin{aligned}\tau \in \Psi = &\left\{ \frac{1}{\omega_{i1}}(2k\pi + \theta_{i1}) \mid i = 2, \dots, N; k = 0, 1, \dots \right\}\\&\bigcup \left\{ \frac{1}{\omega_{i2}}(2k\pi + \theta_{i2}) \mid i = 2, \dots, N; k = -1, -2, \dots \right\}\\= &\left\{ \frac{1}{\omega_{i1}}(2k\pi + \theta_{i1}) \mid i = 2, \dots, N; k = 0, 1, \dots \right\}\\&\bigcup \left\{ \frac{1}{\omega_{i1}}(2k\pi - \theta_{i2}) \mid i = 2, \dots, N; k = 1, 2, \dots \right\}\\= &\left\{ \frac{1}{\omega_{i1}}(2k\pi + \theta_{i1}) \mid i = 2, \dots, N; k = 0, 1, \dots \right\}.\end{aligned}$$

(Sufficiency). If $\tau = \frac{1}{\omega_{i1}}(2k\pi + \theta_{i1}) \in \Phi$, by the same process, one obtains that $g_i(\lambda) = 0$ when $\lambda = \mathbf{i}\omega_{i1}$ since (3.14) is satisfied. The proof is completed. □

Lemma 3.7 *[137] Consider the exponential polynomial*

$$\begin{aligned}&P(\lambda, e^{-\lambda\tau_1}, \cdots, e^{-\lambda\tau_m})\\&= \lambda^n + p_1^{(0)}\lambda^{n-1} + \cdots + p_{n-1}^{(0)}\lambda + p_n^{(0)} + [p_1^{(1)}\lambda^{n-1} + \cdots + p_{n-1}^{(1)}\lambda + p_n^{(1)}]e^{-\lambda\tau_1}\\&\quad + \cdots + [p_1^{(m)}\lambda^{n-1} + \cdots + p_{n-1}^{(m)}\lambda + p_n^{(m)}]e^{-\lambda\tau_m},\end{aligned}$$

where $\tau_i \ge 0$ $(i = 1, 2, \cdots, m)$ and $p_j^{(i)}$ $(i = 0, 1, \cdots, m; j = 1, 2, \cdots, n)$ are constants. As $(\tau_1, \tau_2, \cdots, \tau_m)$ vary, the sum of the orders of the zeros of $P(\lambda, e^{-\lambda\tau_1}, \cdots, e^{-\lambda\tau_m})$

on the open right-half plane can change only if a zero appears on or crosses the imaginary axis.

Lemma 3.8 *Suppose that the network contains a directed spanning tree. Let λ be the solution of $g_i(\lambda) = 0$, $2 \le i \le N$. Then, $d\lambda/d\tau$ exists at the point $\tau \in \Psi$ and*

$$\mathscr{R}\left(\frac{d\lambda}{d\tau}\right)\Big|_{\tau\in\Psi} > 0. \tag{3.20}$$

Proof. Let $\widetilde{g}_i(\lambda,\tau) = \lambda^2 + (\alpha + \beta\lambda)e^{-\lambda\tau}\mu_i$. Since $\widetilde{g}_i(\mathbf{i}\omega_0, \tau_0) = 0$ if $\tau_0 \in \Phi$ and $\mathbf{i}\omega_0$ is the corresponding purely imaginary root, $\widetilde{g}_i(\lambda,\tau)$ is continuous around the point $(\mathbf{i}\omega_0, \tau_0)$, $\frac{\partial \widetilde{g}_i}{\partial\lambda}$ and $\frac{\partial \widetilde{g}_i}{\partial\tau}$ are continuous, and $\frac{\partial \widetilde{g}_i}{\partial\lambda}|_{(\mathbf{i}\omega_0,\tau_0)} \neq 0$, λ is differentiable with respect to τ around the point $(\mathbf{i}\omega_0, \tau_0)$ according to the implicit function theorem [102].

Taking the derivative of λ with respect to τ in $g_i(\lambda) = 0$, one obtains

$$2\lambda\frac{d\lambda}{d\tau} + e^{-\lambda\tau}\mu_i\left[\beta\frac{d\lambda}{d\tau} - (\alpha+\beta\lambda)(\lambda + \tau\frac{d\lambda}{d\tau})\right] = 0. \tag{3.21}$$

If $\tau \in \Psi$, then $\lambda = \mathbf{i}\omega_{ij}$ for some i and j, $2 \le i \le N, 1 \le j \le 2$. It follows that

$$\begin{aligned}
&\frac{d\lambda}{d\tau}\Big|_{\lambda=\mathbf{i}\omega_{ij}} \\
&= \frac{\mu_i\lambda(\alpha+\beta\lambda)}{2\lambda e^{\lambda\tau} + \mu_i[\beta - (\alpha+\beta\lambda)\tau]}\Big|_{\lambda=\mathbf{i}\omega_{ij}} \\
&= \frac{[\mathscr{R}(\mu_i) + \mathbf{i}\mathscr{I}(\mu_i)]\mathbf{i}\omega_{ij}[\alpha + \mathbf{i}\beta\omega_{ij}]}{2[\cos(\omega_{ij}\tau) + \mathbf{i}\,\sin(\omega_{ij}\tau)]\mathbf{i}\omega_{ij} + [\mathscr{R}(\mu_i) + \mathbf{i}\mathscr{I}(\mu_i)][(\beta - \alpha\tau) - \mathbf{i}\beta\omega_{ij}\tau]}.
\end{aligned} \tag{3.22}$$

Let

$$\begin{aligned}
q = {} & [-2\omega_{ij}\sin(\omega_{ij}\tau) + \mathscr{R}(\mu_i)(\beta - \alpha\tau) + \mathscr{I}(\mu_i)\beta\omega_{ij}\tau]^2 \\
& + [2\omega_{ij}\cos(\omega_{ij}\tau) - \mathscr{R}(\mu_i)\beta\omega_{ij}\tau + \mathscr{I}(\mu_i)(\beta - \alpha\tau)]^2.
\end{aligned}$$

By simple calculations, one obtains

$$\begin{aligned}
& q\mathscr{R}\left(\frac{d\lambda}{d\tau}\right)\Big|_{\tau\in\Psi} \\
& = -[\beta\omega_{ij}\mathscr{R}(\mu_i) + \alpha\mathscr{I}(\mu_i)][-2\omega_{ij}\sin(\omega_{ij}\tau) + \mathscr{R}(\mu_i)(\beta - \alpha\tau) \\
& \quad + \mathscr{I}(\mu_i)\beta\omega_{ij}\tau]\omega_{ij} + [\alpha\mathscr{R}(\mu_i) - \beta\omega_{ij}\mathscr{I}(\mu_i)] \\
& \quad \times[2\omega_{ij}\cos(\omega_{ij}\tau) - \mathscr{R}(\mu_i)\beta\omega_{ij}\tau + \mathscr{I}(\mu_i)(\beta - \alpha\tau)]\omega_{ij} \\
& = 2\omega_{ij}^2\{\sin(\omega_{ij}\tau)[\beta\omega_{ij}\mathscr{R}(\mu_i) + \alpha\mathscr{I}(\mu_i)] + \cos(\omega_{ij}\tau) \\
& \quad \times[\alpha\mathscr{R}(\mu_i) - \beta\omega_{ij}\mathscr{I}(\mu_i)]\} - \beta^2\omega_{ij}^2[\mathscr{R}^2(\mu_i) + \mathscr{I}^2(\mu_i)].
\end{aligned} \tag{3.23}$$

Substituting (3.15) and (3.19) into (3.23), one has

$$\begin{aligned} q\mathcal{R}\left(\frac{d\lambda}{d\tau}\right)\bigg|_{\tau\in\Psi} &= 2\omega_{ij}^4 - \beta^2\omega_{ij}^2[\mathcal{R}^2(\mu_i) + \mathcal{I}^2(\mu_i)] \\ &= 2[\mathcal{R}(\mu_i)^2 + \mathcal{I}(\mu_i)^2](\beta^2\omega_i^2 + \alpha^2) - \beta^2\omega_{ij}^2[\mathcal{R}^2(\mu_i) + \mathcal{I}^2(\mu_i)] \\ &= [\mathcal{R}(\mu_i)^2 + \mathcal{I}(\mu_i)^2](\beta^2\omega_i^2 + 2\alpha^2) > 0. \end{aligned} \tag{3.24}$$

This completes the proof. □

Theorem 3.9 *Suppose that the network contains a directed spanning tree and (3.8) is satisfied. Then, second-order consensus in system (3.11) is achieved if and only if*

$$\tau < \tau_0 = \min_{2\leq i\leq N}\left\{\frac{\theta_{i1}}{\omega_{i1}}\right\}, \tag{3.25}$$

where $0 \leq \theta_{i1} < 2\pi$, *which satisfies* $\cos\theta_{i1} = [\mathcal{R}(\mu_i)\alpha - \mathcal{I}(\mu_i)\omega_{i1}\beta]/\omega_{i1}^2$ *and* $\sin\theta_{i1} = [\mathcal{R}(\mu_i)\omega_{i1}\beta + \mathcal{I}(\mu_i)\alpha]/\omega_{i1}^2$, $\omega_{i1} = \sqrt{\dfrac{\| \mu_i\|^2\beta^2 + \sqrt{\| \mu_i\|^4\beta^4 + 4\| \mu_i\|^2\alpha^2}}{2}}$, *and* μ_i *are the nonzero eigenvalues of the Laplacian matrix L,* $i = 2, 3, \ldots, N$.

Proof. Since the network contains a directed spanning tree and (3.8) is satisfied, from Theorem 3.3 it follows that the second-order consensus can be achieved in system (3.11) when $\tau = 0$, where $g(\lambda) = 0$ has exactly a zero root of multiplicity two and all the other roots have negative real parts. When τ varies from 0 to τ_0, by Lemma 3.6, a purely imaginary root emerges. From Lemmas 3.7 and 3.8, one knows that $g(\lambda) = 0$ has exactly a zero root of multiplicity two and all the other roots have negative real parts when $0 \leq \tau < \tau_0$, and there is at least one root with positive real part $\tau > \tau_0$. Therefore, second-order consensus cannot be achieved when $\tau \geq \tau_0$. The proof is completed. □

Remark 3.10 In [137, 138, 140], stability and Hopf bifurcation were studied for delayed networks, where the time delays are regarded as bifurcation parameters. Similar ideas are used here. The result in Theorem 3.9 is important in that a necessary and sufficient condition is established by computing the critical value τ_0 for the maximum allowable time delay.

3.1.4 Simulation Examples

In this subsection, two simulation examples are given to verify the theoretical analysis.

1. Second-Order Consensus in Directed Networks

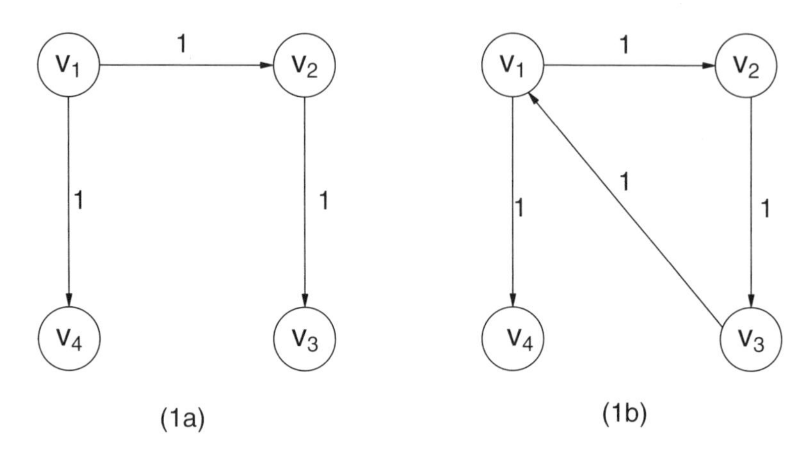

Figure 3.1 Network structure of a network with a directed spanning tree

Consider the network (3.3) with the topology shown in Fig. 3.1. In Fig. 3.1(a), the Laplacian matrix L has a simple zero eigenvalue and all the other eigenvalues are real. Consensus in (3.3) can be reached for any $\alpha > 0$ and $\beta > 0$. In Fig. 3.1(b), the Laplacian matrix L is $\begin{pmatrix} 1 & 0 & -1 & 0 \\ -1 & 1 & 0 & 0 \\ 0 & -1 & 1 & 0 \\ -1 & 0 & 0 & 1 \end{pmatrix}$ and its four eigenvalues are $0, 1, 1.5 + 0.866i, 1.5 - 0.866i$. Let $\alpha = 1$ and apply Theorem 3.3. Then, second-order consensus in the multi-agent system (3.3) can be achieved if and only if $\beta > 0.4082$. The position and velocity states of all the agents are shown in Fig. 3.2(a) and Fig. 3.2(b), where consensus cannot be achieved when $\beta = 0.4$ but it can be reached if $\beta = 0.415$. It is easy to see that by appropriately choosing some $\alpha > 0$ and $\beta > 0$, consensus can be achieved but then may fail if a connection between two agents is added.

2. Second-Order Consensus in Delayed Directed Networks

Consider the network (3.10) with a structure shown in Fig. 3.1(b) where $\alpha = \beta = 1$. When $\tau = 0$, from Theorem 3.3, one knows that second-order consensus can be achieved in the network. By simple calculations using Theorem 3.9, the second-order consensus can be reached if and only if $\tau < 0.29415$. The position and velocity states of all the agents are shown in Fig. 3.3(a) and Fig. 3.3(b), where consensus is achieved when $\tau = 0.29$ but it cannot be reached if $\tau = 0.30$.

3.2 Second-Order Consensus in Nonlinear Multi-agent Systems

3.2.1 *Preliminaries*

When the network reaches second-order consensus in (3.2), the velocities of all agents converge to $\sum_{j=1}^{N} \xi_j v_j(0)$, which depends only on the initial velocities of the

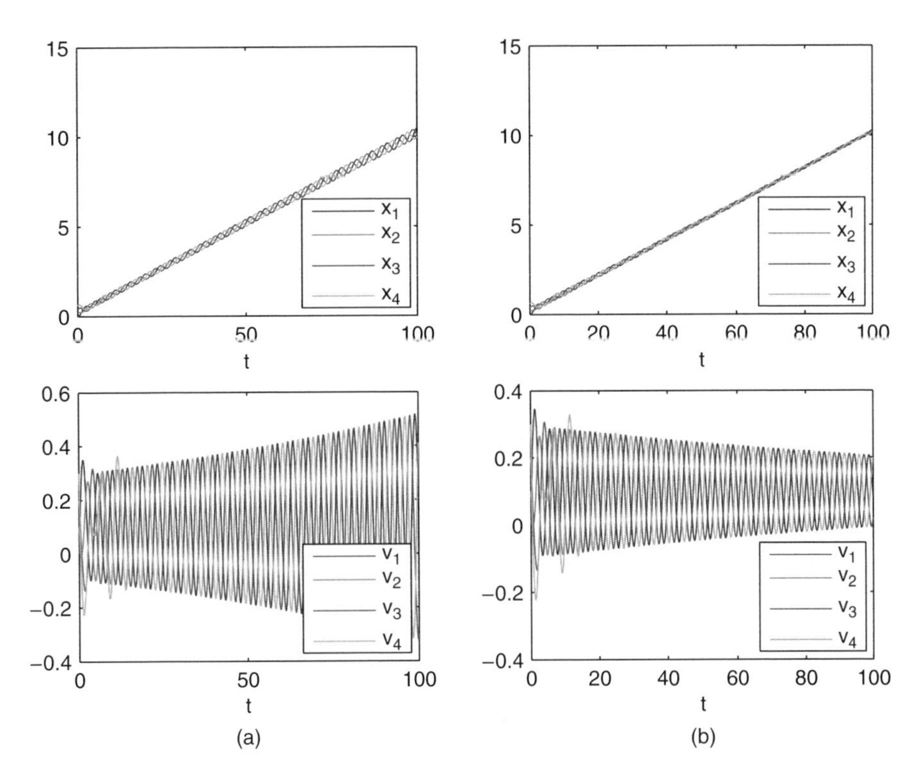

Figure 3.2 Position and velocity states of agents in a network, where $\beta = 0.4$ (a) and $\beta = 0.415$ (b)

agents, where $\xi = (\xi_1, \ldots, \xi_N)^T$ is the nonnegative left eigenvector of L associated with eigenvalue 0 satisfying $\xi^T 1_N = 1$ [93, 95]. However, in most of the applications of multi-agent formations, the velocity of each agent is generally not a constant, but a time-varying variable. Therefore, in this section, the following second-order consensus protocol with time-varying velocities is considered

$$\begin{aligned} \dot{x}_i(t) &= v_i(t), \\ \dot{v}_i(t) &= f(x_i(t), v_i(t), t) - \alpha \sum_{j=1}^{N} L_{ij} x_j(t) - \beta \sum_{j=1}^{N} L_{ij} v_j(t), i = 1, 2, \cdots, N, \end{aligned} \tag{3.26}$$

where $f : R^n \times R^n \times R^+ \to R^n$ is a continuously differentiable vector-valued function. Here, f can be taken as $f = -\nabla U(x, v)$, where $U(x, v)$ is a potential function, then the multi-agent system (3.26) includes many popular swarming and flocking models [44, 82] as special cases.

Clearly, since $\sum_{j=1}^{N} L_{ij} = 0$, if a consensus can be achieved, the solution $s(t) = (s_1(t), s_2(t)) \in R^{2n}$ of the system (3.26) must be a possible trajectory of an isolated

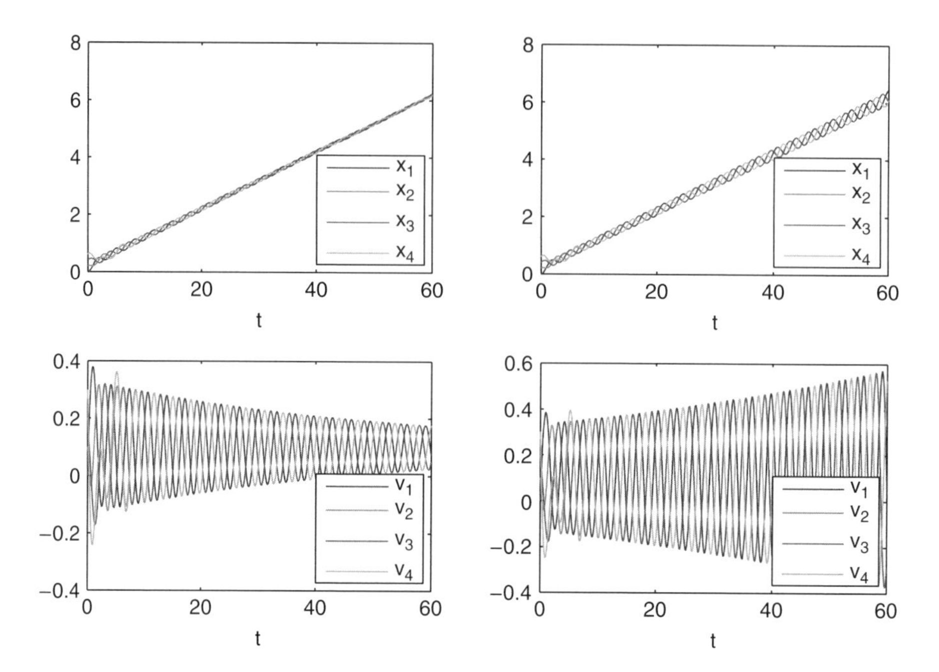

Figure 3.3 Position and velocity states of agents in a delayed network, where $\tau = 0.29$ (a) and $\tau = 0.30$ (b)

node satisfying

$$\dot{s}_1(t) = s_2(t),$$
$$\dot{s}_2(t) = f(s_1(t), s_2(t), t).$$

Here, $s(t)$ may be an isolated equilibrium point, a periodic orbit, or even a chaotic orbit.

For convenience, we say a scalar is an irreducible matrix of order one. Next, a lemma is given to show the relation between an irreducible matrix and the corresponding strong connectivity in the network.

Lemma 3.11 *(Theorem 3.2.1 in [13], Theorem 6.2.24 in [52]) A matrix G is irreducible if and only if its corresponding network $\mathcal{G}$ is strongly connected.*

Lemma 3.12 *(Schur Complement [12]) The following linear matrix inequality (LMI)*

$$\begin{pmatrix} \mathcal{Q}(x) & \mathcal{S}(x) \\ \mathcal{S}(x)^T & \mathcal{R}(x) \end{pmatrix} > 0,$$

where $\mathcal{Q}(x) = \mathcal{Q}(x)^T, \mathcal{R}(x) = \mathcal{R}(x)^T$, is equivalent to one of the following conditions:

(i) $\mathcal{Q}(x) > 0, \mathcal{R}(x) - \mathcal{S}(x)^T \mathcal{Q}(x)^{-1} \mathcal{S}(x) > 0;$
(ii) $\mathcal{R}(x) > 0, \mathcal{Q}(x) - \mathcal{S}(x) \mathcal{R}(x)^{-1} \mathcal{S}(x)^T > 0.$

Definition 3.13 The multi-agent system (3.26) is said to achieve second-order consensus if for any initial conditions,

$$\lim_{t\to\infty} \| x_i(t) - x_j(t) \| = 0, \lim_{t\to\infty} \| v_i(t) - v_j(t) \| = 0, \forall i,j = 1,2,\cdots,N.$$

3.2.2 Second-Order Consensus in Strongly Connected Networks

In this subsection, second-order consensus in strongly connected networks with nonlinear dynamics (3.26) is first investigated.

Assumption 3.14 *There exist nonnegative constants ρ_1 and ρ_2 such that*

$$\| f(x,v,t) - f(y,z,t) \| \leq \rho_1 \| x-y \| + \rho_2 \| v-z \|, \tag{3.27}$$

$\forall x,y,v,z \in R^n, \forall t \geq 0$.

Note that Assumption 3.14 is a Lipschitz-type condition, satisfied by many well-known systems.

Let $\hat{x}_i(t) = x_i(t) - \sum_{k=1}^N \xi_k x_k(t)$ and $\hat{v}_i(t) = v_i(t) - \sum_{k=1}^N \xi_k v_k(t)$ represent the position and velocity vectors relative to the average position and velocity of the agents in system (3.26), where $\xi = (\xi_1,\ldots,\xi_N)^T$ is the positive left eigenvector of L associated with eigenvalue 0 satisfying $\xi^T 1_N = 1$. Then, one obtains the following error dynamical system:

$$\begin{aligned}
\dot{\hat{x}}_i(t) =& \hat{v}_i(t),\\
\dot{\hat{v}}_i(t) =& f(x_i(t),v_i(t),t) - \sum_{k=1}^N \xi_k f(x_k(t),v_k(t),t) - \alpha \sum_{j=1}^N L_{ij}x_j(t)\\
&-\beta \sum_{j=1}^N L_{ij}v_j(t) - \alpha \sum_{k=1}^N \xi_k \sum_{j=1}^N L_{kj}x_j(t) - \beta \sum_{k=1}^N \xi_k \sum_{j=1}^N L_{kj}v_j(t),\\
& i = 1,2,\cdots,N.
\end{aligned} \tag{3.28}$$

Since $\xi^T L = 0$, one has $\sum_{k=1}^N \xi_k \sum_{j=1}^N L_{kj}x_j(t) = [(\xi^T L)\otimes I_n]x = 0$ and $\sum_{k=1}^N \xi_k \sum_{j=1}^N L_{kj}v_j(t) = [(\xi^T L)\otimes I_n]v = 0$. Note that $\sum_{j=1}^N L_{ij} = 0$, so (3.28) can be written as

$$\begin{aligned}
\dot{\hat{x}}_i(t) =& \hat{v}_i(t),\\
\dot{\hat{v}}_i(t) =& f(x_i(t),v_i(t),t) - \sum_{k=1}^N \xi_k f(x_k(t),v_k(t),t) - \alpha \sum_{j=1}^N L_{ij}\hat{x}_j(t)\\
&-\beta \sum_{j=1}^N L_{ij}\hat{v}_j(t), i = 1,2,\cdots,N.
\end{aligned} \tag{3.29}$$

Let $\widehat{x} = (\widehat{x}_1^T, \widehat{x}_2^T, \ldots, \widehat{x}_N^T)^T$, $v = (\widehat{v}_1^T, \widehat{v}_2^T, \ldots, \widehat{v}_N^T)^T$, $f(x, v, t) = (f^T(x_1(t), v_1(t), t), \ldots, f^T(x_N(t), v_N(t), t))^T$ and $\widehat{y} = (\widehat{x}^T, \widehat{v}^T)^T$. Then, system (3.29) can be recast in a compact matrix form, as follows:

$$\dot{\widehat{y}}(t) = F(x, v, t) + (\widetilde{L} \otimes I_n)\widehat{y}(t), \tag{3.30}$$

where $F(x, v, t) = \begin{pmatrix} 0_{Nn} \\ ((I_N - 1_N\xi^T) \otimes I_n)f(x, v, t) \end{pmatrix}$ and $\widetilde{L} = \begin{pmatrix} O_N & I_N \\ -\alpha L & -\beta L \end{pmatrix}$.

The algebraic graph theory, especially the notion of algebraic connectivity, has been well developed for undirected networks [37]. For directed graphs, however, it has not yet been fully developed. For example, there are no standard definitions for the algebraic connectivity and consensus convergence rate for directed graphs while their counterparts for undirected graphs have been widely used to study the consensus problem.

Let $a(L)$ and $b(\widetilde{L}_{ii})$ denote $a_\xi(L)$ defined in Definition 2.11 and $b_{\widetilde{\xi}_i}(\widetilde{L}_{ii})$ defined in Definition 2.26 for simplicity throughout this section where the chosen positive vectors satisfy $\xi^T L = 0$ and $\widetilde{\xi}_i^T \widetilde{A}^i = 0$.

Theorem 3.15 *Suppose that the network is strongly connected and Assumption 3.14 holds. Then, second-order consensus in system (3.26) is achieved if*

$$a(L) > \frac{1}{2}\left(\frac{\rho_1}{\alpha} + \frac{\alpha}{\beta^2} + \frac{\rho_1}{\beta} + \sqrt{\left(\frac{\rho_1}{\alpha} - \frac{\alpha}{\beta^2} - \frac{\rho_1}{\beta}\right)^2 + \frac{(\alpha+\beta)^2\rho_2^2}{\alpha^2\beta^2}}\right). \tag{3.31}$$

Proof. Consider the following Lyapunov function candidate

$$V(t) = \frac{1}{2}\widehat{y}^T(t)(\Omega \otimes I_n)\widehat{y}(t), \tag{3.32}$$

where $\Omega = \begin{pmatrix} 2\alpha\widehat{L} & \frac{\alpha\Xi}{\beta} \\ \frac{\alpha\Xi}{\beta} & \Xi \end{pmatrix}$. It will be shown that $V(t) \geq 0$ and $V(t) = 0$ if and only if $\widehat{y} = 0$. From the definition of $a(L)$, one has

$$\begin{aligned} V(t) &= \alpha\widehat{x}^T(t)(\widehat{L} \otimes I_n)\widehat{x}(t) + \frac{\alpha}{2\beta}\widehat{x}^T(t)(\Xi \otimes I_n)\widehat{v}(t) + \frac{\alpha}{2\beta}\widehat{v}^T(t)(\Xi \otimes I_n)\widehat{x}(t) \\ &\quad + \frac{1}{2}\widehat{v}^T(t)(\Xi \otimes I_n)\widehat{v}(t) \\ &\geq \frac{1}{2}\widehat{y}^T(t)(\widehat{Q} \otimes I_n)\widehat{y}(t), \end{aligned} \tag{3.33}$$

where $\widehat{Q} = \begin{pmatrix} 2a(L)\alpha\Xi & \frac{\alpha\Xi}{\beta} \\ \frac{\alpha\Xi}{\beta} & \Xi \end{pmatrix}$. By Lemma 3.12, $\widehat{Q} > 0$ is equivalent to that $\Xi > 0$ and $2a(L)\alpha\Xi - \frac{\alpha^2}{\beta^2}\Xi > 0$. From (3.31), one has $a(L) > \frac{1}{2}\left(\frac{\rho_1}{\alpha} + \frac{\alpha}{\beta^2} + \frac{\rho_1}{\beta} + \left|\frac{\rho_1}{\alpha} - \frac{\alpha}{\beta^2} - \frac{\rho_1}{\beta}\right|\right) = \max\left(\frac{\rho_1}{\alpha}, \frac{\alpha}{\beta^2} + \frac{\rho_1}{\beta}\right)$, and thus $\widehat{Q} > 0$. Consequently, $V(t) \geq 0$ and $V(t) = 0$ if and only if $\widehat{y} = 0$.

Let $\overline{x} = \sum_{j=1}^{N} \xi_j x_j$ and $\overline{v} = \sum_{j=1}^{N} \xi_j v_j$. Taking the derivative of $V(t)$ along the trajectories of (3.30) yields

$$\begin{aligned}
\dot{V}(t) =& \widehat{y}^T(t)(\Omega \otimes I_n)[F(x, v, t) + (\widetilde{L} \otimes I_n)\widehat{y}] \\
=& \frac{\alpha}{\beta}\widehat{x}^T(t)[(\Xi(I_N - 1_N\xi^T)) \otimes I_n]f(x, v, t) \\
&+\widehat{v}^T(t)[(\Xi(I_N - 1_N\xi^T)) \otimes I_n]f(x, v, t) + \widehat{y}^T(t)[(\Omega\widetilde{L}) \otimes I_n]\widehat{y} \\
=& \left[\frac{\alpha}{\beta}\widehat{x}^T(t) + \widehat{v}^T(t)\right](\Xi \otimes I_n)[f(x, v, t) - 1_N \otimes f(\overline{x}, \overline{v}, t)] \\
&+\frac{1}{2}\widehat{y}^T(t)[(\Omega\widetilde{L} + \widetilde{L}^T\Omega) \otimes I_n]\widehat{y}(t) \\
&- \left[\frac{\alpha}{\beta}\widehat{x}^T(t) + \widehat{v}^T(t)\right](\Xi \otimes I_n)[(1_N\xi^T) \otimes I_n]f(x, v, t) \\
&+ \left[\frac{\alpha}{\beta}\widehat{x}^T(t) + \widehat{v}^T(t)\right](\Xi \otimes I_n)[1_N \otimes f(\overline{x}, \overline{v}, t)].
\end{aligned} \tag{3.34}$$

Since $\widehat{x}(t) = [(I_N - 1_N\xi^T) \otimes I_n]x(t)$, $\widehat{v}(t) = [(I_N - 1_N\xi^T) \otimes I_n]v(t)$, and $\xi^T 1_N = 1$, one has

$$\begin{aligned}
&\widehat{x}^T(t)(\Xi \otimes I_n)[1_N \otimes f(\overline{x}, \overline{v}, t)] \\
&= [1_N^T \otimes f^T(\overline{x}, \overline{v}, t)](\Xi \otimes I_n)[(I_N - 1_N\xi^T) \otimes I_n]x(t) \\
&= \{[1_N^T\Xi(I_N - 1_N\xi^T)] \otimes f^T(\overline{x}, \overline{v}, t)\}x(t) \\
&= \{[\xi^T(I_N - 1_N\xi^T)] \otimes f^T(\overline{x})\}x(t) = 0,
\end{aligned} \tag{3.35}$$

and

$$\begin{aligned}
&\widehat{x}^T(t)(\Xi \otimes I_n)[(1_N\xi^T) \otimes I_n]f(x, v, t) \\
=& f^T(x, v, t)[(\xi 1_N^T) \otimes I_n](\Xi \otimes I_n)[(I_N - 1_N\xi^T) \otimes I_n]x(t) \\
=& f^T(x, v, t)\{[\xi 1_N^T\Xi(I_N - 1_N\xi^T)] \otimes I_n\}x(t) \\
=& f^T(x, v, t)\{[\xi\xi^T(I_N - 1_N\xi^T)] \otimes I_n\}x(t) = 0.
\end{aligned} \tag{3.36}$$

Similarly, one can obtain

$$\hat{v}^T(t)(\Xi \otimes I_n)[1_N \otimes f(\bar{x}, \bar{v}, t)] = 0, \tag{3.37}$$

and

$$\hat{v}^T(t)(\Xi \otimes I_n)[(1_N \xi^T) \otimes I_n]f(x, v, t) = 0. \tag{3.38}$$

In addition,

$$\Omega\widetilde{L} = \begin{pmatrix} 2\alpha\widehat{L} & \frac{\alpha\Xi}{\beta} \\ \frac{\alpha\Xi}{\beta} & \Xi \end{pmatrix} \begin{pmatrix} O_N & I_N \\ -\alpha L & -\beta L \end{pmatrix} = \begin{pmatrix} -\frac{\alpha^2}{\beta}\Xi L & 2\alpha\widehat{L} - \alpha\Xi L \\ -\alpha\Xi L & \frac{\alpha\Xi}{\beta} - \beta\Xi L \end{pmatrix}. \tag{3.39}$$

Note that $\widehat{L} = \frac{1}{2}(\Xi L + L^T\Xi)$, so

$$\frac{1}{2}(\Omega\widetilde{L} + \widetilde{L}^T\Omega) = \begin{pmatrix} -\frac{\alpha^2}{\beta}\widehat{L} & O_N \\ O_N & \frac{\alpha\Xi}{\beta} - \beta\widehat{L} \end{pmatrix}. \tag{3.40}$$

By Assumption 3.14, one obtains

$$\begin{aligned} &\hat{x}^T(t)(\Xi \otimes I_n)[f(x, v, t) - 1_N \otimes f(\bar{x}, \bar{v}, t)] \\ &= \sum_{i=1}^{N} (x_i - \bar{x})^T \xi_i [f(x_i, v_i, t) - f(\bar{x}, \bar{v}, t)] \\ &\leq \sum_{i=1}^{N} \| \hat{x}_i \| \xi_i(\rho_1 \| \hat{x}_i \| + \rho_2 \| \hat{v}_i \|) \\ &= \rho_1 \sum_{i=1}^{N} \xi_i \| \hat{x}_i\|^2 + \rho_2 \sum_{i=1}^{N} \xi_i \| \hat{x}_i \| \| \hat{v}_i \|, \end{aligned} \tag{3.41}$$

and

$$\begin{aligned} &\hat{v}^T(t)(\Xi \otimes I_n)[f(x, v, t) - 1_N \otimes f(\bar{x}, \bar{v}, t)] \\ &\leq \rho_1 \sum_{i=1}^{N} \xi_i \| \hat{v}_i\|^2 + \rho_2 \sum_{i=1}^{N} \xi_i \| \hat{x}_i \| \| \hat{v}_i \| . \end{aligned} \tag{3.42}$$

Combining (3.35)–(3.42), one has

$$\begin{aligned} \dot{V}(t) = &\left[\frac{\alpha}{\beta}\hat{x}^T(t) + \hat{v}^T(t)\right] (\Xi \otimes I_n)[f(x, v, t) - 1_N \otimes f(\bar{x}, \bar{v}, t)] \\ &+ \frac{1}{2}\hat{y}^T(t)[(\Omega\widetilde{L} + \widetilde{L}^T\Omega) \otimes I_n]\hat{y}(t) \end{aligned}$$

$$\leq \frac{\alpha}{\beta}\rho_1 \sum_{i=1}^{N} \xi_i \parallel \widehat{x}_i \parallel^2 + \left(\frac{\alpha}{\beta}+1\right)\rho_2 \sum_{i=1}^{N} \xi_i \parallel \widehat{x}_i \parallel \parallel \widehat{v}_i \parallel + \rho_1 \sum_{i=1}^{N} \xi_i \parallel \widehat{v}_i \parallel^2$$

$$-\frac{\alpha^2}{\beta}\widehat{x}^T(\widehat{L}\otimes I_n)\widehat{x} + \widehat{v}^T\left[\left(\frac{\alpha\Xi}{\beta} - \beta\widehat{L}\right)\otimes I_n\right]\widehat{v}$$

$$\leq \left[\frac{\alpha}{\beta}\rho_1 - \frac{\alpha^2}{\beta}a(L)\right]\sum_{i=1}^{N} \xi_i \parallel \widehat{x}_i \parallel^2 + \left(\frac{\alpha}{\beta}+1\right)\rho_2 \sum_{i=1}^{N} \xi_i \parallel \widehat{x}_i \parallel \parallel \widehat{v}_i \parallel$$

$$+\left[\rho_1 + \frac{\alpha}{\beta} - \beta a(L)\right]\sum_{i=1}^{N} \xi_i \parallel \widehat{v}_i \parallel^2$$

$$= \parallel \widehat{x} \parallel^T \overline{Q} \parallel \widehat{x} \parallel, \tag{3.43}$$

where $\parallel \widehat{x} \parallel = (\parallel \widehat{x}_1 \parallel, \ldots, \parallel \widehat{x}_N \parallel)^T$ and

$$\overline{Q} = \begin{pmatrix} \left[\frac{\alpha}{\beta}\rho_1 - \frac{\alpha^2}{\beta}a(L)\right]\Xi & \frac{1}{2}\left(\frac{\alpha}{\beta}+1\right)\rho_2\Xi \\ \frac{1}{2}\left(\frac{\alpha}{\beta}+1\right)\rho_2\Xi & \left[\rho_1 + \frac{\alpha}{\beta} - \beta a(L)\right]\Xi \end{pmatrix}.$$

By Lemma 3.12, $\overline{Q} < 0$ is equivalent to that

$$a(L) > \frac{\rho_1}{\alpha}$$

and

$$\left[a(L) - \frac{\rho_1}{\alpha}\right]\left[a(L) - \frac{\alpha}{\beta^2} - \frac{\rho_1}{\beta}\right] > \frac{(\alpha+\beta)^2\rho_2^2}{4\alpha^2\beta^2}.$$

By simple calculations, one obtains (3.31). Therefore, the second-order consensus is achieved in system (3.26) under condition (3.31). This completes the proof. □

Corollary 3.16 *Suppose that the network is undirected and Assumption 3.14 holds. Then, second-order consensus in system (3.26) is achieved if*

$$\lambda_2(L) > \frac{1}{2}\left(\frac{\rho_1}{\alpha} + \frac{\alpha}{\beta^2} + \frac{\rho_1}{\beta} + \sqrt{\left(\frac{\rho_1}{\alpha} - \frac{\alpha}{\beta^2} - \frac{\rho_1}{\beta}\right)^2 + \frac{(\alpha+\beta)^2\rho_2^2}{\alpha^2\beta^2}}\right).$$

Proof. If the network is undirected, then by Definition 2.11, $a(L) = \lambda_2(L)$. This completes the proof. □

Remark 3.17 In Definition 2.11, the general algebraic connectivity is defined for a strongly connected network, which is shown in Theorem 3.15 to be the key factor for reaching network consensus. The right-hand side of condition (3.31) depends on the coupling strengths α and β, and the nonlinear constants ρ_1 and ρ_2. Thus, $a(L)$ is a key factor concerning the network structure that can be used to describe the ability for reaching consensus.

If $f = 0$, then $\rho_1 = \rho_2 = 0$ and system (3.26) is reduced to the linear system (3.3).

Corollary 3.18 *Suppose that the network is strongly connected. Then, second-order consensus in system (3.26) is achieved if*

$$a(L) > \frac{\alpha}{\beta^2}. \tag{3.44}$$

Up to this point, it is still a challenging problem to compute $a(L)$ and to find the relationship between $a(L)$ and the eigenvalues of L. Fortunately, the following useful result can be obtained as a byproduct of the above results.

Corollary 3.19 *Suppose that the network is strongly connected. Then, the following statement holds:*

$$a(L) \leq \min_{2\leq i\leq N} \frac{\mathcal{R}(\mu_i)[\mathcal{R}^2(\mu_i) + \mathcal{I}^2(\mu_i)]}{\mathcal{I}^2(\mu_i)} = \min_{2\leq i\leq N} \left[\mathcal{R}(\mu_i) + \frac{\mathcal{R}^3(\mu_i)}{\mathcal{I}^2(\mu_i)}\right]. \tag{3.45}$$

In addition, if the network is undirected, then $a(L) = \lambda_2(L)$.

Proof. From the results in Section 3.1, one knows that second-order consensus in system (3.3) is achieved if and only if the network contains a directed spanning tree, and moreover $\min_{2\leq i\leq N} \frac{\mathcal{R}(\mu_i)[\mathcal{R}^2(\mu_i) + \mathcal{I}^2(\mu_i)]}{\mathcal{I}^2(\mu_i)} > \frac{\alpha}{\beta^2}$. By Corollary 3.18, one obtains a sufficient condition $a(L) > \frac{\alpha}{\beta^2}$ for the network consensus. Thus, (3.45) is satisfied. □

Remark 3.20 In general, it is very difficult to compute $a(L)$, not to mention finding the relationship between $a(L)$ and the eigenvalues of L. However, if the network is strongly connected, Corollary 3.19 yields that $a(L) \leq \min_{2\leq i\leq N} \left[\mathcal{R}(\mu_i) + \frac{\mathcal{R}^3(\mu_i)}{\mathcal{I}^2(\mu_i)}\right]$. This result is useful for both theoretical analysis in algebraic graph theory and relevant applications.

3.2.3 Second-Order Consensus in Rooted Networks

In this subsection, second-order consensus in networks containing a directed spanning tree with time-varying velocities is further investigated.

One can change the order of the node indices to obtain the Frobenius normal form (2.29). Without loss of generality, assume that the adjacency matrix G of $\mathscr{G}$ is in the Frobenius normal form. Suppose that the condition (3.31) in Theorem 3.15 holds in the first strongly connected component, so that the final states of the nodes in this component satisfy [70, 133]

$$\begin{aligned} \dot{x}_s(t) &= v_s(t) + \mathcal{O}(e^{\epsilon t}), \\ \dot{v}_s(t) &= f(x_s, v_s, t) + \mathcal{O}(e^{\epsilon t}), \end{aligned} \tag{3.46}$$

where $x_s \in R^n$, $v_s \in R^n$ and $\epsilon < 0$. Let $\widetilde{x}_i = x_i - x_s$, $\widetilde{v}_i = v_i - v_s$, $s_i = \sum_{j=1}^{i} q_j$, $e_i = (\widetilde{x}^T_{s_{i-1}}, \ldots, \widetilde{x}^T_{s_i}, \widetilde{v}^T_{s_{i-1}}, \ldots, \widetilde{v}^T_{s_i})^T$, $e = (e_2^T, \ldots, e_m^T)^T$, $\widetilde{L}^*_{ii} = \begin{pmatrix} O_{q_i} & I_{q_i} \\ -\alpha \widetilde{L}_{ii} & -\beta \widetilde{L}_{ii} \end{pmatrix}$, $\widetilde{L}^*_{ij} = \begin{pmatrix} O_{q_i \times q_j} & I_{q_i \times q_j} \\ -\alpha \widetilde{L}_{ij} & -\beta \widetilde{L}_{ij} \end{pmatrix}$, $f^*_i(x, v, t) = (f^T(x_{s_{i-1}}, v_{s_{i-1}}, t), \ldots, f^T(x_{s_i}, v_{s_i}, t))^T$, and $F^*_i(x, v, t) = \begin{pmatrix} 0_{q_i n} \\ f^*_i(x, v, t) - 1_{q_i} \otimes f(x_s, v_s, t) \end{pmatrix}$, $i = 1, 2, \ldots, m$. Then, the network (3.26) can be recast in a compact matrix form, as follows:

$$\dot{e}(t) = F^*(x, v, t) + (\widetilde{L}^* \otimes I_n) e(t) + \mathcal{O}(e^{\epsilon t}), \tag{3.47}$$

where $F^*(x, v, t) = (F_2^{*T}(x, v, t), \ldots, F_N^{*T}(x, v, t))^T$, and $\widetilde{L}^* = (\widetilde{L}^*_{ij})_{(m-1)\times(m-1)}$, $2 \le i$, $j \le m$.

Theorem 3.21 *Suppose that the network contains a directed spanning tree and Assumption 3.14 holds. Then, second-order consensus in system (3.26) is achieved if*

$$\begin{aligned} &\min_{2 \le j \le m} \{a(\widetilde{L}_{11}), b(\widetilde{L}_{jj})\} \\ &> \frac{1}{2}\left(\frac{\rho_1}{\alpha} + \frac{\alpha}{\beta^2} + \frac{\rho_1}{\beta} + \sqrt{\left(\frac{\rho_1}{\alpha} - \frac{\alpha}{\beta^2} - \frac{\rho_1}{\beta} \right)^2 + \frac{(\alpha+\beta)^2 \rho_2^2}{\alpha^2 \beta^2}} \right). \end{aligned} \tag{3.48}$$

Proof. From Theorem 3.15, one knows that under condition (3.48), second-order consensus can be achieved in the first strongly connected component. Thus, all the states of the nodes in this component satisfy (3.46).

Consider the Lyapunov functional candidate:

$$V(t) = \sum_{i=2}^{m} \Delta_i e_i^T(t) (Q_i^* \otimes I_n) e_i(t), \tag{3.49}$$

where $Q_i^* = \begin{pmatrix} 2\alpha\widetilde{L}_{jj} & \frac{\alpha\overline{\Xi}_i}{\beta} \\ \frac{\alpha\overline{\Xi}_i}{\beta} & \overline{\Xi}_i \end{pmatrix}$, $\overline{\Xi}_i = \text{diag}(\widetilde{\xi}_{i1}, \widetilde{\xi}_{i2}, \ldots, \widetilde{\xi}_{iq_i})$and Δ_i are positive constants, $i = 2, \ldots, m$.

Taking the derivative of $V(t)$along (3.47), and using (3.40)–(3.42), one gets

$$\begin{aligned}
\dot{V} &= 2\sum_{i=2}^{m} \Delta_i e_i^T(t)(Q_i^* \otimes I_n)\dot{e}_i(t) \\
&\leq 2\sum_{i=2}^{m} \Delta_i e_i^T(t)(Q_i^* \otimes I_n)\left[F_i^*(x, v, t) + \sum_{j=1}^{i}(\widetilde{L}_{ij}^* \otimes I_n)e_j(t) + \mathcal{O}(e^{\epsilon t})\right] \\
&\leq 2\sum_{i=2}^{m} \Delta_i \parallel e_i \parallel \begin{pmatrix} \frac{\alpha}{\beta}\rho_1\overline{\Xi}_i & \frac{1}{2}(\frac{\alpha}{\beta}+1)\rho_2\overline{\Xi}_i \\ \frac{1}{2}(\frac{\alpha}{\beta}+1)\rho_2\overline{\Xi}_i & \rho_1\overline{\Xi}_i \end{pmatrix} \parallel e_i \parallel \\
&\quad +2\sum_{i=2}^{m} \Delta_i e_i^T(t)(Q_i^*\widetilde{L}_{ii}^* \otimes I_n)e_i(t) \\
&\quad +2\sum_{i=2}^{m} \Delta_i \left\| e_i^T(t)(Q_i^* \otimes I_n)\left[\sum_{j=1}^{i-1}(\widetilde{L}_{ij}^* \otimes I_n)e_j(t) + \mathcal{O}(e^{\epsilon t})\right]\right\| \\
&\leq 2\sum_{i=2}^{m} \Delta_i \parallel e_i \parallel \Gamma_i \parallel e_i \parallel + 2\sum_{i=2}^{m}\sum_{j=1}^{i-1} \Delta_i \parallel e_i \parallel [\Gamma_{ij} \parallel e_j \parallel + \mathcal{O}(e^{\epsilon t})],
\end{aligned} \tag{3.50}$$

where $\parallel e_i \parallel = (\parallel \widetilde{x}_{s_{i-1}} \parallel, \ldots, \parallel \widetilde{x}_{s_i} \parallel, \parallel \widetilde{v}_{s_{i-1}}^T \parallel, \ldots, \parallel \widetilde{v}_{s_i} \parallel)^T$,
$\Gamma_i = \begin{pmatrix} \frac{\alpha}{\beta}\rho_1\overline{\Xi}_i - \frac{\alpha^2}{\beta}b(\overline{L}_i)\overline{\Xi}_i & \frac{1}{2}\left(\frac{\alpha}{\beta}+1\right)\rho_2\overline{\Xi}_i \\ \frac{1}{2}\left(\frac{\alpha}{\beta}+1\right)\rho_2\overline{\Xi}_i & \left(\rho_1 + \frac{\alpha}{\beta}\right)\overline{\Xi}_i - \beta b(\widetilde{L}_{ii})\overline{\Xi}_i \end{pmatrix}$, and Γ_{ij} are matrices with appropriate dimensions, $2 \leq i,j \leq m$. From Lemma 2.7, one knows that if $\Gamma_i < 0$ $(2 \leq i \leq m)$ and by choosing Δ_{i+1} sufficiently smaller than $\Delta_j$for $j < i+1$, then second-order consensus can be achieved in system (3.26). This completes the proof. □

Similarly, as in Corollary 3.19, one can prove the following general result.

Corollary 3.22 *Suppose that the network has a directed spanning tree. Then,*

$$\begin{aligned}\min_{2\leq j\leq m}\{a(\overline{L}_1), b(\overline{L}_j)\} &\leq \min_{2\leq i\leq N}\frac{\mathscr{R}(\mu_i)[\mathscr{R}^2(\mu_i)+\mathscr{I}^2(\mu_i)]}{\mathscr{I}^2(\mu_i)}\\ &= \min_{2\leq i\leq N}\left[\mathscr{R}(\mu_i)+\frac{\mathscr{R}^3(\mu_i)}{\mathscr{I}^2(\mu_i)}\right]. \end{aligned} \tag{3.51}$$

Remark 3.23 In addition to the general algebraic connectivity $a(\widetilde{L}_{11})$in strongly connected networks, the general algebraic connectivity $b(\widetilde{L}_{ii})(2 \leq i \leq m)$ in each strongly connected component of a directed network has also been defined here. It is shown in Theorem 3.21 that $\min_{2\leq j\leq m}\{a(\widetilde{L}_{11}), b(\widetilde{L}_{jj})\}$plays a key role in reaching consensus and can be used to describe the consensus ability in a network with fixed structure. As a byproduct, (3.51) is obtained, which is useful in algebraic graph theory in its own right.

Remark 3.24 Theorem 3.21 can also be used to study various leader-follower multi-agent systems. Suppose that the network has a directed spanning tree and the first strongly connected component has only one node that is a root of the directed network. Then, all the states of the followers converge to that of the leader if

$$\min_{2\leq j\leq m}\{b(\widetilde{L_{jj}})\} > \frac{1}{2}\left(\frac{\rho_1}{\alpha}+\frac{\alpha}{\beta^2}+\frac{\rho_1}{\beta}+\sqrt{\left(\frac{\rho_1}{\alpha}-\frac{\alpha}{\beta^2}-\frac{\rho_1}{\beta}\right)^2+\frac{(\alpha+\beta)^2\rho_2^2}{\alpha^2\beta^2}}\right).$$

3.2.4 Simulation Examples

In this subsection, a simulation example is given to demonstrate the potential of our theoretical analysis.

Consider the second-order consensus protocol with time-varying velocities in system (3.26), where the network structure is shown in Fig. 2.1 with the weights on the connections. The nonlinear function $f$is described by Chua's circuit [24]

$$f(x_i(t), v_i(t), t) = \begin{pmatrix} \varsigma(-v_{i1}+v_{i2}-l(v_{i1})),\\ v_{i1}-v_{i2}+v_{i3},\\ -\rho v_{i2}, \end{pmatrix}, \tag{3.52}$$

where $l(v_{i1}) = bv_{i1} + 0.5(a-b)(|v_{i1}+1| - |v_{i1}-1|)$. The isolated system (3.52) is chaotic when $\varsigma = 10$, $\rho = 18$, $a = -4/3$, and $b = -3/4$, as shown in [142]. In view of Assumption 3.14, by computation, one obtains $\rho_1 = 4.3871$ and $\rho_2 = 0$. Let $\alpha = 5$ and $\beta = 6$. From Fig. 2.1, it is easy to see that the network contains a directed spanning tree, where the nodes 1–4 and 5–7 belong to the first and second strongly connected components, respectively. By Lemma 2.21 and Definition 2.26, one has $a(\widetilde{L}_{11}) =$

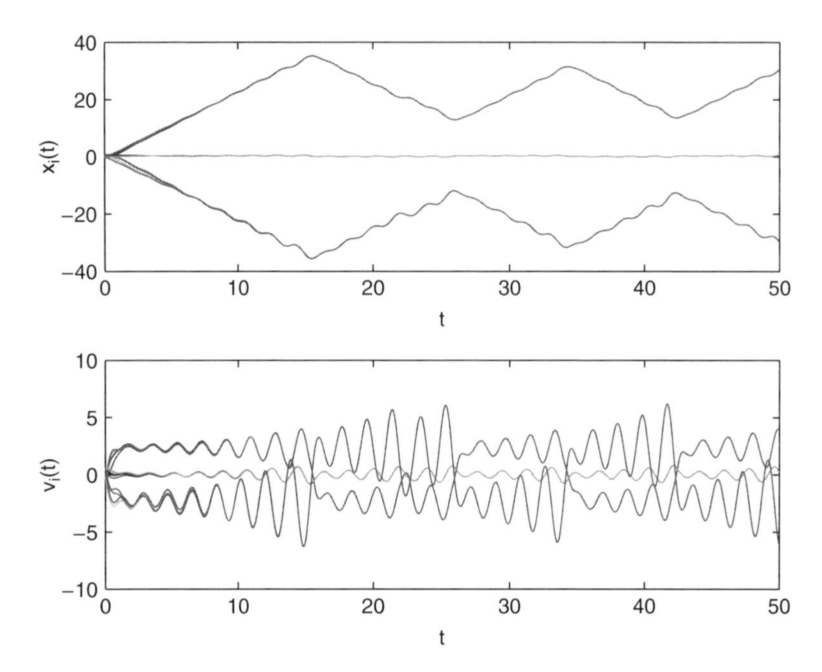

Figure 3.4 Position and velocity states of multiple agents in a network with time-varying velocities

1.8118 and $b(\widetilde{L_{22}}) = 1.0206$, where $\widetilde{\xi}_1 = (0.2727, 0.1818, 0.1364, 0.4091)^T$ and $\widetilde{\xi}_2 = (0.4615, 0.3077, 0.2308)^T$. By Theorem 3.21, one has that $\min\{a(\widetilde{L}_{11}), b(\widetilde{L}_{22})\} = 1.0206 > \frac{\rho_1}{\alpha} + \frac{\alpha}{\beta^2} + \frac{\rho_1}{\beta} + \left|\frac{\rho_1}{\alpha} + \frac{\alpha}{\beta^2} + \frac{\rho_1}{\beta}\right| = 0.8701$. Therefore, second-order consensus can be achieved in the multi-agent system (3.26). The position and velocity states of all the agents are shown in Fig. 3.4.

3.3 Notes

In Subsection 3.1, some second-order consensus algorithms for multi-agent dynamical systems with directed topologies have been studied. Detailed analysis has been performed for the case where the second-order dynamics of each agent are determined by both position and velocity terms. A necessary and sufficient condition has been derived to ensure second-order consensus in multi-agent systems where the network has a directed spanning tree. It has been found that both the real and imaginary parts of the eigenvalues of the Laplacian matrix play key roles in reaching consensus. Moreover, the scenario when communication delays are present in the network has been investigated. A necessary and sufficient condition has also been established, and it was shown that, in this case, the second-order consensus can be achieved in the multi-agent systems with a directed spanning tree if and only if the time delay is less than a critical value.

The study in this section on second-order consensus algorithms can serve as a stepping stone for introducing more complicated and realistic agent dynamics to groups of mobile agents. Moreover, the effects of more complicated inter-agent couplings on group behaviors are being investigated. For example, it is of great importance to generalize the results of this section to the case when the network topology evolves with time, or has certain hierarchical features.

In Subsection 3.2, some second-order consensus algorithms for multi-agent dynamical systems with directed topologies and nonlinear dynamics have been studied. Detailed analysis has been performed on the case in which the second-order dynamics of each agent are determined by both position and velocity terms. Two notions of generalized algebraic connectivity have been introduced to strongly connected network components so as to describe the ability of reaching consensus in a directed network. Some sufficient conditions have also been derived for reaching second-order consensus in multi-agent systems with time-varying velocities.

The study of second-order consensus protocols in multi-agent systems with directed network topologies is still a challenging problem, and this chapter can serve as a stepping stone to study more complicated agent dynamics by combining ideas in algebraic graph theory and control approach. Future works will be on the effects of group behaviors in more complicated networks, such as time-varying networks, stochastic networks, and switching networks, among others.

4

Higher-Order Consensus in Multi-agent Systems

This chapter studies general higher-order distributed consensus protocols in multi-agent dynamical systems [153]. First, network synchronization is investigated, with some necessary and sufficient conditions derived for higher-order consensus. It is found that consensus can be reached if and only if all subsystems are asymptotically stable. Based on this result, consensus regions are characterized. It is proved that for the mth-order consensus, there are at most $\left\lfloor \frac{m+1}{2} \right\rfloor$ (the integer part of the real number $\frac{m+1}{2}$) disconnected stable and unstable consensus regions. It is shown that consensus can be achieved if and only if all the nonzero eigenvalues of the Laplacian matrix lie in the stable consensus regions. Moreover, it shows that the ratio of the largest to the smallest nonzero eigenvalues of the Laplacian matrix plays a key role in reaching consensus, and a scheme for choosing the coupling strength is derived. Furthermore, a leader-follower control problem with full-state and partial-state observations in multi-agent dynamical systems is considered, respectively, which reveals that the agents with very small degrees must be informed.

4.1 Preliminaries

Consider the following mth-order dynamics in a multi-agent system:

$$\begin{aligned}
\dot{\xi}_i^{(1)}(t) &= \xi_i^{(2)}(t), \\
&\vdots \\
\dot{\xi}_i^{(m-1)}(t) &= \xi_i^{(m)} \\
\dot{\xi}_i^{(m)}(t) &= u_i, i = 1, 2, \ldots, N,
\end{aligned} \tag{4.1}$$

Distributed Cooperative Control of Multi-agent Systems, First Edition.
Wenwu Yu, Guanghui Wen, Guanrong Chen, and Jinde Cao.

where $\xi_i^{(k)} \in R^n$ are the states of the ith node, which denotes the $(k-1)$st derivative of $\xi_i^{(1)}$, $k = 1, \ldots, l$, and u_i is the control input. For notational simplicity, only $n = 1$ is considered throughout the chapter. One can easily generate the results for the case of $n > 1$ by using Kronecker products [53].

Definition 4.1 The multi-agent system (4.1) is said to achieve the mth-order consensus if, for any well-defined initial conditions,

$$\lim_{t\to\infty} \|\xi_i^{(k)}(t) - \xi_j^{(k)}(t)\| = 0, \forall k = 1, \ldots, m;\ i,j = 1, 2, \ldots, N.$$

In order to achieve the mth-order consensus, the following control input protocol is designed [100]:

$$u_i(t) = c \sum_{j=1,\ j\neq i}^{N} G_{ij} \sum_{k=1}^{m} \alpha_k \left(\xi_j^{(k)}(t) - \xi_i^{(k)}(t)\right), i = 1, 2, \ldots, N, \tag{4.2}$$

where $c > 0$ is the outer coupling strength, $\alpha_k > 0$ are the inner coupling strengths, $G = (G_{ij})_{N\times N}$ is the coupling configuration matrix representing the topological structure of the network and thus is the weighted adjacency matrix of the network, and the Laplacian matrix $L = (L_{ij})_{N\times N}$ is defined by

$$L_{ii} = -\sum_{j=1,\ j\neq i}^{N} L_{ij}\ , L_{ij} = -G_{ij}, i \neq j, \tag{4.3}$$

which ensures the diffusion property that $\sum_{j=1}^{N} L_{ij} = 0$.

Note that when $m = 1$, the consensus algorithm (4.1)–(4.2) reduces to the well-known first-order consensus protocol [57, 72, 81, 98, 117, 134, 141, 142, 151, 165, 166]:

$$\xi_i^{(1)}(t) = -c\alpha_1 \sum_{j=1}^{N} L_{ij}\xi_j^{(1)}(t), i = 1, 2, \ldots, N. \tag{4.4}$$

Note also that, the control input (4.2) can be rewritten in a simpler equivalent form:

$$u_i(t) = -c \sum_{j=1}^{N} L_{ij} \sum_{k=1}^{m} \alpha_k \xi_j^{(k)}(t), i = 1, 2, \ldots, N. \tag{4.5}$$

Throughout this chapter, let $0 = \lambda_1 \leq \lambda_2 \leq \cdots \leq \lambda_N$ be the N eigenvalues of the Laplacian matrix L, and $1_N \in R^N$ ($0_N \in R^N$) be a vector with all entries being 1 (0).

4.2 Higher-Order Consensus in a General Form

4.2.1 Synchronization in Complex Networks

Consider a complex dynamical network consisting of N identical nodes with linearly diffusive coupling [72, 117, 134, 141, 142, 151, 165, 166]

$$\dot{\eta}_i(t) = g(\eta_i) - c\sum_{j=1}^{N} L_{ij}D\eta_j(t), i = 1, 2, \dots, N, \tag{4.6}$$

where $\eta_i \in R^m$ is the state vector of the ith node, $g(\eta_i) = C\eta_i$ with $C \in R^{N\times N}$ being a matrix describing the dynamics of each single node, $D \in R^{m\times m}$ is the inner coupling matrix, and L_{ij} is the Laplacian matrix, $i, j = 1, \dots, N$. The goal of synchronization in complex networks is to achieve $\lim_{t\to\infty} \|\eta_i(t) - \eta_j(t)\| = 0, \forall i,$ $j = 1, \dots, N$.

Note that a solution of an isolated node satisfies

$$\dot{s}(t) = g(s(t)), \tag{4.7}$$

where $s(t) = (s_1, \dots, s_m)^T$ is the state vector. Here, only a linear model $g(s(t)) = Cs(t)$ is considered. For a general form of $g(s(t))$, $s(t)$ can be an equilibrium point, a periodic orbit, or even a chaotic orbit. For more details about synchronization in complex networks with nonlinear dynamics, see [72, 117, 134, 141, 142, 151, 165, 166].

Let $\eta = (\eta_1^T, \dots, \eta_N^T)^T$ and rewrite system (4.6) into a matrix form:

$$\dot{\eta}(t) = [(I_N \otimes C) - c(L \otimes D)]\eta(t). \tag{4.8}$$

Let $\Lambda = \text{diag}(\lambda_1, \dots, \lambda_N)$ be the diagonal matrix associated with L, i.e., there exists a unitary matrix P such that $P^T LP = \Lambda$. Then, one has

$$\begin{aligned}(P^T \otimes I_m)\dot{\eta}(t) &= [(P^T \otimes I_m)(I_N \otimes C) - c(\Lambda \otimes D)(P^T \otimes I_m)]\eta(t)\\ &= [(P^T \otimes C) - c(\Lambda \otimes D)(P^T \otimes I_m)]\eta(t)\\ &= [(I_N \otimes C) - c(\Lambda \otimes D)](P^T \otimes I_m)\eta(t).\end{aligned}$$

Let $x(t) = (P^T \otimes I_m)\eta(t) = (x_1^T, \dots, x_N^T)^T$. Then, the above complex network model can be written as

$$\dot{x}(t) = [(I_N \otimes C) - c(\Lambda \otimes D)]x(t),$$

or

$$\dot{x}_i(t) = (C - c\lambda_i D)x_i(t), i = 1, \dots, N. \tag{4.9}$$

Theorem 4.2 *Suppose that the network $\mathcal{G}$ is connected. Synchronization in network (4.8) can be reached if and only if the following $N-1$ linear systems are asymptotically stable:*

$$\dot{x}_i(t) = (C - c\lambda_i D)x_i(t), i = 2, \dots, N. \tag{4.10}$$

Proof. (Sufficiency). Since the network is connected, $1_N/\sqrt{N}$ is the unit eigenvector of the Laplacian matrix L associated with the simple zero eigenvalue. From (4.9), one has $\eta(t) = (P \otimes I_m)x(t)$, where $LP = P\Lambda$ and $P = (p_1, \dots, p_N)$. It is easy to verify that $p_1 = 1_N/\sqrt{N}$ is the unit eigenvector of the Laplacian matrix L associated with the eigenvalue 0, i.e., $Lp_1 = 0$. Since the $N-1$ systems in (4.10) are asymptotically stable, one has $\lim_{t\to\infty} \|x_i\| \to 0$ for $i = 2, \dots, N$. Therefore,

$$\lim_{t\to\infty} \left\| \eta(t) - \frac{1}{\sqrt{N}}(x_1(t)^T, \dots, x_1(t)^T)^T \right\| = 0,$$

where $\dot{x}_1(t) = Cx_1(t)$.

(Necessity). If synchronization in network (4.8) can be reached, then there exists a vector $\eta^*(t) \in R^m$ such that $\lim_{t\to\infty} \|\eta(t) - 1_N \otimes \eta^*(t)\| = 0$. Since $0_N^T = 1_N^T LP = 1_N^T P\Lambda = (\lambda_1 1_N^T p_1, \dots, \lambda_N 1_N^T p_N)$, one has $1_N^T p_i = 0$ for $i = 2, \dots, N$. Therefore, $x_i(t) = (p_i^T \otimes I_m)\eta(t) \to (p_i^T 1_N) \otimes (I_m \eta^*(t)) = 0$, as $t \to \infty$ for all $i = 2, \dots, N$. This completes the proof. □

4.2.2 Higher-Order Consensus in a General Form

If one chooses $\eta_i = (\xi_i^{(1)}, \dots, \xi_i^{(m)})^T$, $C = \begin{pmatrix} 0 & 1 & \cdots & 0 \\ \vdots & \ddots & \ddots & \vdots \\ 0 & 0 & \ddots & 1 \\ 0 & 0 & \cdots & 0 \end{pmatrix}_{m\times m}$, and $D = \begin{pmatrix} 0 & 0 & \cdots & 0 \\ \vdots & \ddots & \dots & 0 \\ \alpha_1 & \alpha_2 & \cdots & \alpha_m \end{pmatrix}_{m\times m}$, then the mth protocol in multi-agent systems (4.1) with the control input (4.5) is a special model of the complex network (4.6). Thus, the mth-order consensus problem in system (4.1) with the control input (4.5) can be transformed into the synchronization problem $\lim_{t\to\infty} \|\eta_i(t) - \eta_j(t)\| = 0, \forall i, j = 1, \dots, N$, in a network, which has just been discussed above.

Remark 4.3 In [97, 100], second-order and third-order consensus in multi-agent systems were studied by solving linear systems (4.8). It has been shown that second-order (third-order) consensus can be reached if and only if $(I_N \otimes C) - (L \otimes D)$ has exactly two (three) zero eigenvalues and all the other eigenvalues have negative real parts. Theorem 4.2 above generates the results in [97, 100] to any positive integer m by using a different method originated from synchronization in complex networks, i.e., employing the transverse stability to the synchronization manifold. In [36], the stability in cooperative networks with identical linear subsystems was considered where the equation with $i = 1$, which is not in condition (4.10) of Theorem 4.2 here, also needs to be checked.

Corollary 4.4 *Suppose that the network $\mathcal{G}$ is connected. The mth-order consensus in the multi-agent system (4.1) with protocol (4.5) can be reached if and only if the real* *parts of the roots in the following $N-1$ equations are all negative:*

$$\lambda^m + \alpha_m c\lambda_i \lambda^{m-1} + \cdots + \alpha_2 c\lambda_i \lambda + \alpha_1 c\lambda_i = 0, i = 2, \ldots, N, \tag{4.11}$$

i.e., polynomial (4.11) is stable.

Proof. Let λ be an eigenvalue of matrix $C - c\lambda_i D$. Then, one has $\det(\lambda I_m - C + c\lambda_i D) = 0$. Note that

$$\det(\lambda I_m - C + \lambda_i D) = \det \begin{pmatrix} \lambda & -1 & \cdots & 0 \\ \vdots & \ddots & \ddots & \vdots \\ 0 & 0 & \ddots & -1 \\ \alpha_1 c\lambda_i & \alpha_2 c\lambda_i & \cdots & \lambda + \alpha_m c\lambda_i \end{pmatrix}_{m\times m}$$

$$= \lambda^m + \alpha_m c\lambda_i \lambda^{m-1} + \cdots + \alpha_2 c\lambda_i \lambda + \alpha_1 c\lambda_i,$$

where the last equation is obtained by determinantal expansion in the last row. Eqs. (4.10) are asymptotically stable if and only if the real parts of the roots in (4.11) are all negative. The proof is completed. □

Remark 4.5 If $m = 1$ and $\lambda = -\alpha_1 c\lambda_2$, then first-order consensus can be achieved if and only if the network is connected [57, 98]. If $m = 2$, then the real parts of the roots in $\lambda^2 + \alpha_2 c\lambda_i \lambda + \alpha_1 c\lambda_i = 0$ are all negative if the network is connected and thus second-order consensus can be reached. However, for a general directed network, second-order consensus may fail even if the network contains a directed spanning tree [97]. In [146], some necessary and sufficient conditions for second-order consensus in multi-agent dynamical systems with directed topologies were derived. Similarly, consensus in multi-agent systems with specific higher-order dynamics was discussed in [131]. However, in this section, the results are based on a general framework, i.e., synchronization in complex networks, and consensus in multi-agent systems with general identical linear subsystems (4.8) are studied. It should also be noted that the necessary and sufficient conditions in Theorem 4.2 and Corollary 4.4 can be used to study consensus in general directed networks where λ_i are complex values, $i = 2, \ldots, N$, as discussed in [131].

4.2.3 Consensus Region in Higher-Order Consensus

It is intuitive to see that the Routh–Hurwitz criterion [9] can be applied to obtain a necessary and sufficient condition by solving the stability of Eq. (4.11) in Corollary 4.4. However, this approach could result in the calculation of many high-order polynomial inequalities, which are very complicated and cannot be implemented for a general system.

Notice that it is generally difficult to check if Eq. (4.11) is stable for a large-scale network. As an alternative, let $\sigma > 0$ be a variable and $\mathcal{S} = \{\sigma | \lambda^m + \alpha_m \sigma \lambda^{m-1} + \cdots + \alpha_2 \sigma \lambda + \alpha_1 \sigma = 0 \quad \text{is stable}\}$ be the stable consensus region. Then, the problem is transformed to find if all the nonzero eigenvalues of the Laplacian matrix scaled by a factor c lie in the stable consensus region $\mathcal{S}$. Consequently, the following criterion can be easily verified.

Corollary 4.6 *Suppose that the network $\mathcal{G}$ is connected. The mth-order consensus in the multi-agent system (4.1) with protocol (4.5) can be reached if and only if*

$$c\lambda_i \in \mathcal{S}, \quad i = 2, \ldots, N. \tag{4.12}$$

Let $\widetilde{P}(\lambda, \sigma) = \lambda^m + \alpha_m \sigma \lambda^{m-1} + \cdots + \alpha_2 \sigma \lambda + \alpha_1 \sigma$, where $\sigma \in [c\lambda_2, c\lambda_N]$. Therefore, if the polynomial $\widetilde{P}(\lambda, \sigma)$ is robustly stable for all $\sigma \in [c\lambda_2, c\lambda_N]$, the stability of all the polynomials in Eq. (4.11) can be obtained. Consequently, the following well-known Kharitonov's theorem in robust control can be applied.

Proposition 4.7 *(Kharitonov's theorem [9]) Suppose that the network $\mathcal{G}$ is connected. The mth-order consensus in the multi-agent system (4.1) with protocol (4.5) can be reached if the four Kharitonov polynomials of $\widetilde{P}(\lambda, \sigma)$ with parameters $\sigma \in [c\lambda_2, c\lambda_N]$ are stable.*

The computation of stability of four Kharitonov polynomials is very simple. However, the obtained condition in Proposition 4.7 is only sufficient since one needs to check if the polynomial $\widetilde{P}(\lambda, \sigma)$ is robustly stable for all $\sigma \in [c\lambda_2, c\lambda_N]$. Note that in Corollary 4.4, in order to reach mth-order consensus in the multi-agent system (4.1), one only needs to check the stability of the polynomial $\widetilde{P}(\lambda, \sigma)$ at some discrete points $\sigma = c\lambda_2, \ldots, c\lambda_N$. In some particular cases, if the stable consensus regions are disconnected, the four Kharitonov polynomials may be unstable; however, all the nonzero eigenvalues of the Laplacian matrix scaled by a factor c can still lie in these disconnected stable consensus regions.

In order to obtain a necessary and sufficient condition for reaching mth-order consensus in the multi-agent system (4.1), the next objective is to find the structure of the stable consensus region $\mathcal{S}$. In [31, 65], disconnected synchronization regions of complex networks were discussed. It was shown that there exist some disconnected synchronization regions in several particular complex networks when the synchronous state is an equilibrium point. In the following, a general mth-order dynamical multi-agent system (4.6) is considered, with corresponding criteria derived.

Lemma 4.8 *Consider the polynomial*

$$P(\lambda) = \lambda^m + \alpha_m \sigma \lambda^{m-1} + \cdots + \alpha_2 \sigma \lambda + \alpha_1 \sigma,$$

where α_i $(i = 1, 2, \ldots, m)$ are constants. As σ is varied, the sum of the orders of the zeros of $P(\lambda)$ on the open right-half plane can change only if a zero appears on or crosses the imaginary axis.

Proof. The zeros of $P(\lambda)$ continuously depend on the parameter σ. The result thus follows. □

Lemma 4.9 *$P(\lambda) = \lambda^m + \alpha_m \sigma \lambda^{m-1} + \cdots + \alpha_2 \sigma \lambda + \alpha_1 \sigma = 0$ has a purely imaginary root if and only if*

$$\sigma \in \begin{cases} \left\{ \dfrac{(-1)^{(m-2)/2}\theta^{m/2}}{\sum\limits_{k=0}^{(m-2)/2} (-1)^k \alpha_{2k+1}\theta^k} > 0 \,\middle|\, \theta \in \mathcal{Q}_1 \right\}; & m \text{ is even}, \\ \left\{ \dfrac{(-1)^{(m-3)/2}\theta^{(m-1)/2}}{\sum\limits_{k=0}^{(m-3)/2} (-1)^k \alpha_{2k+2}\theta^k} > 0 \,\middle|\, \theta \in \mathcal{Q}_2 \right\}; & m \text{ is odd}, \end{cases} \tag{4.13}$$

where $\mathcal{Q}_1 = \left\{ \theta > 0 \,\middle|\, \sum\limits_{k=0}^{(m-2)/2} (-1)^k \alpha_{2k+2}\theta^k = 0 \right\}$, $\mathcal{Q}_2 = \left\{ \theta > 0 \,\middle|\, \sum\limits_{k=0}^{(m-1)/2} (-1)^k \alpha_{2k+1}\theta^k = 0 \right\}$.

Proof. It is easy to see that $P(\lambda) = 0$ has m zero roots if and only if $\sigma = 0$. Let $\lambda = i\omega$ ($\omega \neq 0$). Then, one can show that $P(\lambda) = 0$ has a purely imaginary root if and only if

$$P(\lambda) = (i\omega)^m + \alpha_m \sigma (i\omega)^{m-1} + \cdots + \alpha_2 \sigma (i\omega) + \alpha_1 \sigma = 0. \tag{4.14}$$

If $m = 2l$, where l is a positive integer, then by separating the real and imaginary parts of (4.13), one obtains

$$\begin{cases} (-1)^{m/2}\omega^m + [(-1)^{(m-2)/2}\alpha_{m-1}\omega^{m-2} + \cdots - \alpha_3\omega^2 + \alpha_1]\sigma \\ = (-1)^{m/2}\omega^m + \sigma \sum\limits_{k=0}^{(m-2)/2} (-1)^k \alpha_{2k+1}\omega^{2k} = 0, \\ (-1)^{(m-2)/2}\alpha_m\omega^{m-2} + (-1)^{(m-4)/2}\alpha_{m-2}\omega^{m-4} + \cdots - \alpha_4\omega^2 + \alpha_2 \\ = \sum\limits_{k=0}^{(m-2)/2} (-1)^k \alpha_{2k+2}\omega^{2k} = 0. \end{cases} \tag{4.15}$$

Let $\mathcal{Q}_1 = \left\{ \theta > 0 \,\middle|\, \sum\limits_{k=0}^{(m-2)/2} (-1)^k \alpha_{2k+2}\theta^k = 0 \right\}$. Then, $\mathcal{Q}_1$ has at most $(m-2)/2$ elements. Thus, from (4.15), one obtains $\sigma = \dfrac{(-1)^{(m-2)/2}\omega^m}{\sum\limits_{k=0}^{(m-2)/2} (-1)^k \alpha_{2k+1}\omega^{2k}}$.

If $m = 2l - 1$, where l is a positive integer, one can similarly get

$$\begin{cases} (-1)^{(m-1)/2}\omega^m + [(-1)^{(m-3)/2}\alpha_{m-1}\omega^{m-2} + \cdots + \alpha_2\omega]\sigma \\ = (-1)^{(m-1)/2}\omega^m + \sigma\omega \sum_{k=0}^{(m-3)/2} (-1)^k \alpha_{2k+2}\omega^{2k} = 0, \\ (-1)^{(m-1)/2}\alpha_m\omega^{m-1} + (-1)^{(m-3)/2}\alpha_{m-3}\omega^{m-3} + \cdots - \alpha_3\omega^2 + \alpha_1 \\ = \sum_{k=0}^{(m-1)/2} (-1)^k \alpha_{2k+1}\omega^{2k} = 0. \end{cases} \tag{4.16}$$

The proof is thus completed. □

From Lemma 4.9, one can see that there exist at most $\left\lfloor \frac{m-1}{2} \right\rfloor$ different positive values of σ such that $P(\lambda) = 0$ has a purely imaginary root, where $\left\lfloor \frac{m-1}{2} \right\rfloor$ represents the integer part of the real number $\frac{m-1}{2}$. Without loss of generality, suppose that there are r different positive values, $0 = \sigma_0 < \sigma_1 < \cdots < \sigma_r < \sigma_{r+1} = \infty$, such that $P(\lambda) = 0$ has a purely imaginary root if and only if $\sigma = \sigma_i$ for some $1 \le i \le r$ ($r \le \left\lfloor \frac{m-1}{2} \right\rfloor$). Let $\mathcal{S}_i = (\sigma_{i-1}, \sigma_i)$, $i = 1, \ldots, r+1$.

Lemma 4.10 *Suppose that the network $\mathcal{G}$ is connected. If there exists a positive value $\overline{\sigma} \in S_i$ such that $\lambda^m + \alpha_m\overline{\sigma}\lambda^{m-1} + \cdots + \alpha_2\overline{\sigma}\lambda + \alpha_1\overline{\sigma} = 0$ is stable (unstable), then, for any $\sigma \in \mathcal{S}_i = (\sigma_{i-1}, \sigma_i)$, $\lambda^m + \alpha_m\sigma\lambda^{m-1} + \cdots + \alpha_2\sigma\lambda + \alpha_1\sigma = 0$ is stable (unstable), where $\lambda^m + \alpha_m\sigma\lambda^{m-1} + \cdots + \alpha_2\sigma\lambda + \alpha_1\sigma = 0$ has a purely imaginary root when $\sigma = \sigma_i$, $i = 1, \ldots, r$.*

Proof. If $\lambda^m + \alpha_m\sigma\lambda^{m-1} + \cdots + \alpha_2\sigma\lambda + \alpha_1\sigma = 0$ is stable (unstable), then all the roots of the polynomial have negative real parts (at least one root has a positive real part). By Lemma 4.9, one knows that when $\sigma = \sigma_{i-1}$ or $\sigma = \sigma_i$, $P(\lambda)$ has a purely imaginary root. In view of Lemma 4.8, the sum of the orders of the zeros of $P(\lambda)$ on the open right-half plane cannot change if $\sigma \in \mathcal{S}_i = (\sigma_{i-1}, \sigma_i)$. This completes the proof. □

The positive real axis is now partitioned into $r+1$ intervals and r points: $(0, \infty) = \mathcal{S}_1 \bigcup \{\sigma_1\} \bigcup \cdots \bigcup \{\sigma_r\} \bigcup \mathcal{S}_{r+1}$. $P(\lambda) = 0$ is not stable when $\sigma = \sigma_i$, $i = 1, \ldots, r$. From Lemma 4.10, in each interval $\mathcal{S}_i$, the sum of the orders of the zeros of $P(\lambda)$ on the open right-half plane or left-half plane remains the same for all $\sigma \in \mathcal{S}_i = (\sigma_{i-1}, \sigma_i)$, $i = 1, \ldots, r+1$.

Definition 4.11 $\mathcal{S}_i$ is called a *stable consensus region* if, for any $\sigma \in S_i$, $\lambda^m + \alpha_m\sigma\lambda^{m-1} + \cdots + \alpha_2\sigma\lambda + \alpha_1\sigma = 0$ is stable; otherwise, $\mathcal{S}_i$ is called an *unstable consensus region*.

From Lemma 4.10, it is easy to see that the positive real axis is partitioned into several stable and unstable consensus regions. In [59, 65], the synchronization regions of complex networks are classified into three types: unbounded region, bounded region, and empty region. Here, in the mth-order consensus problem, the stable consensus region can be any of these three types or even a union of them which may be composed of several disconnected stable consensus regions, as pointed out in [31].

Theorem 4.12 *Suppose that the network $\mathcal{G}$ is connected. The mth-order consensus in the multi-agent system (4.1) with protocol (4.5) can be reached if and only if*

$$c\lambda_i \in \mathcal{S}, \quad i = 2, \dots, N, \tag{4.17}$$

where $\mathcal{S} = \bigcup_{i\in\mathcal{N}} \mathcal{S}_i$ and $\mathcal{N} = \{i|\mathcal{S}_i$ is a stable consensus region, $i = 1, \dots, r+1\}$.

Remark 4.13 For $m = 1, 2$, one can easily check that all the roots of $P(\lambda) = 0$ have negative real parts, thus the stable consensus region is $(0, \infty)$. However, this may not be true for $m \geq 3$. By computing σ_i, where $P(\lambda) = \lambda^m + \alpha_m\sigma_i\lambda^{m-1} + \cdots + \alpha_2\sigma_i\lambda + \alpha_1\sigma_i = 0$ has a purely imaginary root as in Lemma 4.9, one can easily get $r+1$ disconnected regions, $\mathcal{S}_i = (\sigma_{i-1}, \sigma_i)$, $i = 1, \dots, r+1$. In view of Lemma 4.10, stable and unstable consensus regions can be derived.

Corollary 4.14 *Suppose that the network $\mathcal{G}$ is connected and $\mathcal{S}_i = (\sigma_{i-1}, \sigma_i)$ is a stable region for some i, $i = 1, \dots, r+1$. If $\frac{\lambda_N}{\lambda_2} < \frac{\sigma_i}{\sigma_{i-1}}$, then there exists a value $c > 0$, such that $c\lambda_j \in \mathcal{S}_i$ for all $j = 2, \dots, N$. Thus, the mth-order consensus in the multi-agent system (4.1) with protocol (4.5) can be reached.*

Proof. It suffices to prove that there exists a $c > 0$ such that $\sigma_{i-1} < c\lambda_2$ and $c\lambda_N < \sigma_i$. Let $\frac{\lambda_N}{\lambda_2} + \epsilon^* = \frac{\sigma_i}{\sigma_{i-1}}$, where ϵ^* is a positive constant. One can choose $c^* > 0$ such that $c^*\lambda_2 = \sigma_{i-1} + \epsilon$, where ϵ is a sufficiently small positive value. From the condition $\frac{\lambda_N}{\lambda_2} < \frac{\sigma_i}{\sigma_{i-1}}$, one has $c^*\lambda_N < \sigma_i - \epsilon^*\sigma_{i-1} + \epsilon(\sigma_i/\sigma_{i-1} - \epsilon^*)$. Letting ϵ be sufficiently small completes the proof. □

Remark 4.15 If $\sigma_{i-1} = 0$ or $\sigma_i = \infty$, then the condition $\frac{\lambda_N}{\lambda_2} < \frac{\sigma_i}{\sigma_{i-1}}$ is satisfied. The eigen-ratio $\frac{\lambda_N}{\lambda_2}$ can be considered as the consensus ability of the network. The smaller the $\frac{\lambda_N}{\lambda_2}$, the easier the network reaches consensus by choosing an appropriate value of c.

4.3 Leader-Follower Control in Multi-agent Systems

In this section, the leader-follower control problem in multi-agent systems is discussed. Assume that the leader evolves according to the following dynamics:

$$\begin{aligned}
\dot{\xi}_0^{(1)}(t) &= \xi_0^{(2)}(t),\\
&\vdots\\
\dot{\xi}_0^{(m-1)}(t) &= \xi_0^{(m)}(t),\\
\dot{\xi}_0^{(m)}(t) &= u_0,
\end{aligned} \tag{4.18}$$

where $\xi_0^{(k)} \in R^n$ are the states of the leader and u_0 is a control input.

4.3.1 Leader-Follower Control in Multi-agent Systems with Full-State Feedback

In this subsection, the dynamics of all the followers, labeled $1, \ldots, N$, are governed by the multi-agent system (4.1) with the control input

$$\begin{aligned}
u_i(t) = u_0 &+ c \sum_{j=1, j\neq i}^{N} G_{ij} \sum_{k=1}^{m} \alpha_k \left(\xi_j^{(k)}(t) - \xi_i^{(k)}(t)\right)\\
&+ c\beta_i \sum_{k=1}^{m} \alpha_k \left(\xi_0^{(k)}(t) - \xi_i^{(k)}(t)\right), i = 1, 2, \ldots, N,
\end{aligned} \tag{4.19}$$

where $\beta_i \geq 0$ are the coupling strengths, $i = 1, 2, \ldots, N$. If $\beta_i > 0$, then follower i can get the leader's information; if $\beta_i = 0$, the leader's states are not available for follower i. Therefore, only a small fraction of agents can sense the leader's information.

Let $e_i^{(j)} = \xi_i^{(j)} - \xi_0^{(j)}$ denote the relative error states to the leader, $e_i = (e_i^{(1)}, \ldots, e_i^{(m)})^T$, and $e = (e_1^T, \ldots, e_N^T)^T$. Then, the error dynamics can be written as

$$\dot{e}(t) = [(I_N \otimes C) - c(\widetilde{L} \otimes D)]e(t), \tag{4.20}$$

where $\widetilde{L} = L + A$ and $A = \text{diag}(\beta_1, \ldots, \beta_N)$.

From Lemma 2.32, one knows that $L + A$ is positive definite if the network $\mathcal{G}$ is connected and there is at least one positive diagonal element in A. Let $0 < \mu_1 \leq \cdots \leq \mu_N$ be the N eigenvalues of $\widetilde{L}$.

Corollary 4.16 *Suppose that the network $\mathcal{G}$ is connected. All agents in the multi-agent system (4.1) with the control input (4.19) can follow the leader in (4.18) asymptotically if and only if the real parts of the roots in the following N equations are all negative:*

$$\lambda^m + \alpha_m c\mu_i \lambda^{m-1} + \cdots + \alpha_2 c\mu_i \lambda + \alpha_1 c\mu_i = 0, i = 1, 2, \ldots, N. \tag{4.21}$$

Proof. By following the same arguments as in Theorem 4.2 and Corollary 4.4, the result can be easily proved. □

Similar to Lemmas 4.9 and 4.10, one can compute σ_i, where $\lambda^m + \alpha_m \sigma \lambda^{m-1} + \cdots + \alpha_2 \sigma \lambda + \alpha_1 \sigma = 0$ has a purely imaginary root when $\sigma = \sigma_i$, $i = 1, \ldots, r$.

Theorem 4.17 *Suppose that the network $\mathcal{G}$ is connected. All agents in the multi-agent system (4.1) with the control input (4.19) can follow the leader in (4.18) asymptotically if and only if*

$$c\mu_i \in \mathcal{S}, i = 1, \ldots, N, \tag{4.22}$$

where $\mathcal{S} = \bigcup_{i \in \mathcal{N}} \mathcal{S}_i$ and $\mathcal{N} = \{i | \mathcal{S}_i$ is a stable consensus region, $i = 1, \ldots, r+1\}$.

From Theorems 4.12 and 4.17, it is easy to see that the computation of stable consensus regions in Lemma 4.10 plays a key role in the mth-order consensus and leader-follower control of multi-agent systems. The eigenvalues of $\widetilde{L}$ may lie in different stable regions. Since all the eigenvalues of $\widetilde{L}$ can be changed by choosing different followers, i.e., with different i where $\beta_i > 0$, it is desirable if all these eigenvalues μ_i lie in a particular stable region.

Corollary 4.18 *Suppose that the network $\mathcal{G}$ is connected and $\mathcal{S}_i = (\sigma_{i-1}, \sigma_i)$ is a stable region for some i, $i = 1, \ldots, r+1$. If*

$$\sigma_{i-1} I_N < c(L + A) < \sigma_i I_N, \tag{4.23}$$

then all agents in the multi-agent system (4.1) with the control input (4.19) can follow the leader in (4.18) asymptotically.

Proof. If $\mathcal{S}_i$ is a stable region, then under condition (4.23), one has $c\mu_1 > \sigma_{i-1}$ and $c\mu_N < \sigma_i$, which indicates that $c\mu_j \in \mathcal{S}_i$ for all $j = 1, \ldots, N$. This completes the proof. □

Note that it is impossible for all agents to sense the leader in reality. To reduce the number of informed agents, some local feedback injections may be applied to a fraction of network nodes, which is known as pinning control [23, 151, 166]. It is still a challenging problem, even today, to know how to choose the minimum number of informed agents such that (4.23) can be satisfied.

Corollary 4.19 *Suppose that the network $\mathcal{G}$ is connected and $\mathcal{S}_i = (\sigma_{i-1}, \sigma_i)$ is a stable region for some i, $i = 1, \ldots, r+1$. Under condition (4.23), it is necessary that*

$$\sigma_{i-1} < cL_{jj} \tag{4.24}$$

holds for uninformed agents j ($\beta_j = 0$) and

$$cL_{kk} < \sigma_i \tag{4.25}$$

is satisfied for all agents, $k = 1, 2, \ldots, N$.

Proof. Under condition (4.23), one has $\sigma_{i-1} < c(L_{jj} + \beta_j) < \sigma_i$ since every diagonal element of a positive definite matrix is positive. For uninformed agents j with $\beta_j = 0$, condition (4.24) is satisfied. For all agents, $cL_{jj} \leq c(L_{jj} + \beta_j) < \sigma_i$ hold. This completes the proof. □

Remark 4.20 If one aims to choose some informed agents such that all the eigenvalues of $\widetilde{L}$ lie in a stable consensus region $\mathcal{S}_i = (\sigma_{i-1}, \sigma_i)$, (4.24) must hold for the uninformed agents, which shows that the agents with very small degrees L_{jj} must be informed. For all agents, the maximum degree must be lower than σ_i/c; otherwise, some eigenvalues of $\widetilde{L}$ cannot lie in $\mathcal{S}_i$. If $\mathcal{S}_{r+1} = (\sigma_r, \infty)$ is a stable consensus region, then by Corollary 4.18, the condition (4.23) can be written as $c(L + A) - \sigma_r I_N > 0$. If c is sufficiently large, the mth-order consensus can be reached by informing only one agent [23]. Some interesting schemes for choosing informed agents have been discussed in [151]. It was found that the nodes with low degrees should be informed first, which is contrary to the common view that the most highly connected nodes should be informed first. Furthermore, it has been shown that the derived pinning condition with leader's information given in a high-dimensional setting can be reduced to a low-dimensional condition without pinning controllers involved [151].

4.3.2 Leader-Follower Control with Observers

In many cases, it is impossible for an agent to measure all the states $\xi_j^{(k)}$ of its neighbors and the leader, $j = 0, 1, \ldots, N$; $k = 1, \ldots, m$. For some order k, $\alpha_k = 0$, which means that $\xi_j^{(k)}$ may be unavailable. In the second-order ($m = 2$) leader-follower control problem of multi-agent systems [50, 51], some observers were designed under the condition that the velocity states $\xi_j^{(2)}$ cannot be measured by the agents. Here, a general higher-order leader-follower control problem is considered. One natural question is whether Corollary 4.16 can still hold. Then, by designing an appropriate control input, the similar results can also be obtained.

Lemma 4.21 *Suppose that the network $\mathcal{G}$ is connected. If system (4.21) is stable, then $\alpha_k > 0$ for all $k = 1, \ldots, m$.*

Therefore, Corollary 4.16 cannot hold if there is a k such that $\alpha_k = 0$. Usually, higher-order states are not available, for example, the velocity states in second-order dynamical multi-agent systems [50, 51]. Here, for simplicity, assume that $\alpha_m = \alpha_{m-1} = \alpha_{m-l+1} = 0$, $1 \leq l \leq m - 1$. In this case, a control input is designed as follows:

$$u_i(t) = u_0 + c \sum_{j=1,\, j\neq i}^{N} G_{ij} \sum_{k=1}^{m-l} \alpha_k \left(\xi_j^{(k)}(t) - \xi_i^{(k)}(t) \right)$$

$$+ c\beta_i \sum_{k=1}^{m-l} \alpha_k \left(\xi_0^{(k)}(t) - \xi_i^{(k)}(t) \right) - k_1(\xi_i^{(m)} - \widetilde{\xi}_i^{(1)}),$$

$$
\begin{aligned}
\dot{\tilde{\xi}}_i^{(1)}(t) &= c\sum_{j=1,\ j\neq i}^{N} G_{ij}\gamma_1\left(\xi_j^{(1)}(t)-\xi_i^{(1)}(t)\right)+c\beta_i\gamma_1(\xi_0^{(1)}(t)\xi_i^{(1)})\\
&\quad -k_2(\xi_i^{(m)}-\tilde{\xi}_i^{(2)}),\\
&\ \ \vdots\\
\dot{\tilde{\xi}}_i^{(l-1)}(t) &= c\sum_{j=1,\ j\neq i}^{N} G_{ij}\gamma_{l-1}\left(\xi_j^{(1)}(t)-\xi_i^{(1)}(t)\right)+c\beta_i\gamma_{m+l-1}(\xi_0^{(1)}(t)-\xi_i^{(1)})\\
&\quad -k_l(\xi_i^{(m)}-\tilde{\xi}_i^{(l)}),\\
\dot{\tilde{\xi}}_i^{(l)}(t) &= c\sum_{j=1,\ j\neq i}^{N} G_{ij}\gamma_{l}\left(\xi_j^{(1)}(t)-\xi_i^{(1)}(t)\right)+c\beta_i\gamma_{m+l}(\xi_0^{(1)}(t)-\xi_i^{(1)}),\\
&\quad i=1,2,\ldots,N,
\end{aligned}
\tag{4.26}
$$

where $\tilde{\xi}_i^{(j)}$ are slack variables denoting the estimations of $\xi_0^{(m)}(t)$, and $\gamma_j>0$ and $k_j>0$ are the coupling strengths, $1\leq j\leq l$. Let $y_i=(\xi_i^{(1)}-\xi_0^{(1)},\ldots,\xi_i^{(m)}-\xi_0^{(m)},\tilde{\xi}_i^{(1)}-\xi_0^{(m)},\ldots,\tilde{\xi}_i^{(l)}-\xi_0^{(m)})^T\in R^{m+l}$ and $y=(y_1^T,\ldots,y_N^T)^T$. Then, one has

$$
\dot{y}_i(t)=\hat{C}y_i(t)-c\sum_{j=1}^{N}\tilde{L}_{ij}\hat{D}y_j(t), i=1,2,\ldots,N,
\tag{4.27}
$$

where $\hat{C}=\begin{pmatrix} 0 & 1 & \cdots & 0 & 0 & \cdots & 0\\ \vdots & \ddots & \ddots & \vdots & 0 & \cdots & 0\\ 0 & 0 & \ddots & 1 & 0 & \cdots & 0\\ 0 & 0 & \cdots & -k_1 & k_1 & \cdots & 0\\ \vdots & \vdots & \cdots & \vdots & \ddots & \ddots & \ddots\\ 0 & 0 & \cdots & -k_l & 0 & 0 & k_l\\ 0 & 0 & \cdots & 0 & 0 & 0 & 0 \end{pmatrix}_{(m+l)\times(m+l)}$ and $\hat{D}=\begin{pmatrix} 0 & 0 & \cdots & 0 & \cdots & 0\\ \vdots & \ddots & \ldots & 0 & \vdots & 0\\ \alpha_1 & \cdots & \alpha_{m-l} & 0 & \cdots & 0\\ \gamma_1 & 0 & \cdots & 0 & \cdots & 0\\ \vdots & 0 & \cdots & 0 & \vdots & 0\\ \gamma_l & 0 & \cdots & 0 & \cdots & 0 \end{pmatrix}_{(m+l)\times(m+l)}$.

The following result is obtained based on Theorem 4.2.

Theorem 4.22 *Suppose that the network $\mathcal{G}$ is connected. All agents in the multi-agent system (4.1) with the control input (4.26) can follow the leader in (4.18) asymptotically if and only if the following N linear systems are asymptotically stable:*

$$
\dot{x}_i(t)=(\hat{C}-c\mu_i\hat{D})x_i(t), i=1,\ldots,N.
\tag{4.28}
$$

Corollary 4.23 *Suppose that the network $\mathcal{G}$ is connected. All agents in the multi-agent system (4.1) with the control input (4.26) can follow the leader in (4.18) asymptotically if and only if the real parts of the roots in the following N equations are all negative:*

$$\lambda^{m+l} + \sum_{j=1}^{m+l} \theta_{i,\,j}\lambda^{j-1} = 0, i = 1, \ldots, N, \tag{4.29}$$

where

$$\theta_{i,\,j+1} = \begin{cases} c\mu_i\gamma_{l-j}\prod_{s=1}^{l-j} k_s, & 0 \le j \le l-1, \\ c\mu_i\alpha_{j+1-l}, & l \le j \le m-1, \\ \prod_{s=1}^{m+l-j} k_s, & m \le j \le m+l-1. \end{cases}$$

Proof. Let λ be an eigenvalue of matrix $\hat{C} - c\mu_i\hat{D}$. Then, one has

$$\begin{aligned} &\det(\lambda I_{m+l} - \hat{C} + c\mu_i\hat{D}) \\ &= \lambda^{m+l} + \lambda^m \sum_{j=0}^{l-1}\prod_{s=1}^{l-j} k_s\lambda^j + +c\mu_i \sum_{j=1}^{m-l} \alpha_j\lambda^{l+j-1} + c\mu_i \sum_{j=0}^{l-1} \gamma_{l-j} \prod_{s=1}^{l-j} k_s\lambda^j = 0. \end{aligned}$$

This completes the proof. □

Remark 4.24 By designing the control input as in (4.26) and introducing some feedback coupling strengths γ_s and k_s, $s = 1, \ldots, l$, one has $\theta_{i,j} > 0$ for all $i = 1, \ldots, N$; $j = 1, \ldots, m + l$. Similarly, the consensus region can also be studied as in the above sections, thus it is omitted here.

4.4 Simulation Examples

In this section, some simulation examples are given to illustrate the theoretical results.

4.4.1 Consensus Regions

Consider the multi-agent system (4.8) with $m = 5$, $\alpha_1 = 1$, $\alpha_2 = 1.5$, $\alpha_3 = 4.16$, $\alpha_4 = 4$, and $\alpha_5 = 4$. From Lemma 4.9, one has $\sigma_1 = 0.3816$ and $\sigma_2 = 16.3784$. By simple calculations, it is easy to verify that $\mathcal{S}_2 = (0.3816, 16.3784)$ is a stable consensus region. On regions $\mathcal{S}_1 = (0, 0.3816)$ and $\mathcal{S}_3 = (16.3784, \infty)$, $\lambda^m + \alpha_m\sigma\lambda^{m-1} + \ldots + \alpha_2\sigma\lambda + \alpha_1\sigma = 0$, there are two eigenvalues with positive real parts and thus are unstable. It follows from Theorem 4.12 that the mth-order consensus can be reached in system (4.8) if and only if $c\lambda_i \in (0.3816, 16.3784)$ for all $i = 2, \ldots, N$.

A scale-free network is performed in the simulation, where $N = 1000$, the number of initial nodes is five, and at each time step a new node is introduced and connected to five existing nodes in the network [8]. By computation, one obtains that $\lambda_2 = 2.8674$ and $\lambda_N = 120.25$. From Corollary 4.14, one knows that $41.9369 = \frac{\lambda_N}{\lambda_2} < \frac{\sigma_2}{\sigma_1} = 42.9203$. Therefore, by choosing $c = 0.135$, one has $0.3871 = c\lambda_2 \le c\lambda_i \le c\lambda_N = 16.2338 \in (0.3816, 16.3784)$, $i = 2, \ldots, N$. Thus, the mth-order consensus of this network example in the form of the multi-agent system (4.8) can be reached. The locations of $c\lambda_i$ for the simulated scale-free network are illustrated in Fig. 4.1.

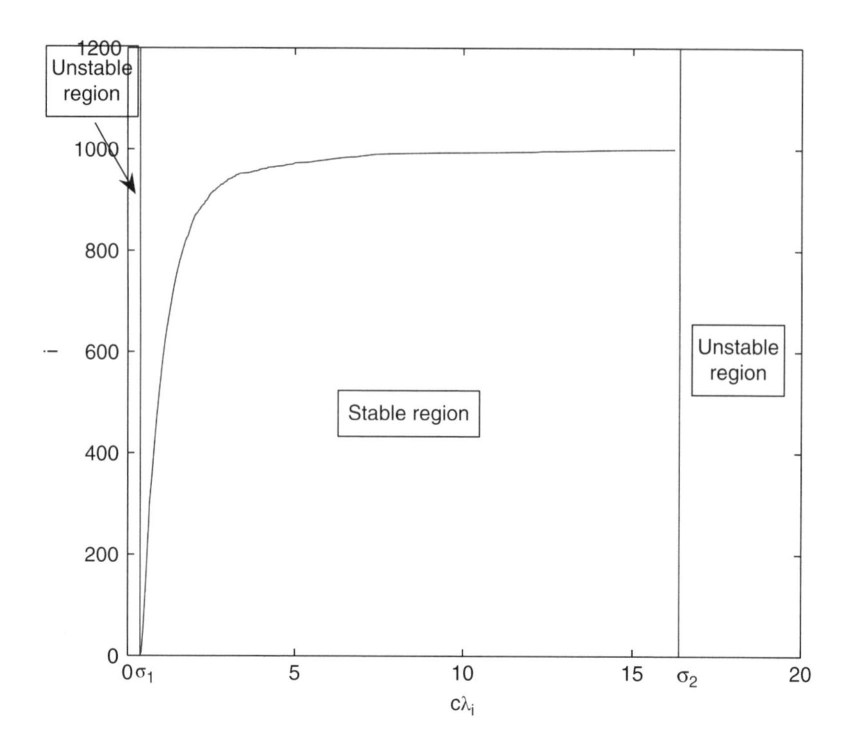

Figure 4.1 Locations of $c\lambda_i$, $i = 2, \dots, N$

4.4.2 Leader-Follower Control with Full-State Feedback

Consider the multi-agent system (4.19) with $m = 5$, $c = 0.8$, $\beta_i = 30$ $\alpha_1 = 1$, $\alpha_2 = 2$, $\alpha_3 = 8$, $\alpha_4 = 4$, and $\alpha_5 = 5$. From Lemma 4.9, one has $\sigma_1 = 0.5557$. It is easy to verify that $\mathcal{S}_2 = (0.5557, \infty)$ is a stable consensus region while $\mathcal{S}_1 = (0, 0.5557)$ is unstable, where $\lambda^m + \alpha_m \sigma \lambda^{m-1} + \cdots + \alpha_2 \sigma \lambda + \alpha_1 \sigma = 0$ has two eigenvalues with positive real parts. The same scale-free network is simulated as above, assuming that there are 50 informed agents with the largest degrees, which can measure the information of the leader. In this case, one has $c\mu_1 = 0.6259 > \sigma_1 = 0.5557$. Therefore, $c\mu_i \in \mathcal{S}_2$ for all $i = 1, \dots, N$. Thus, all agents in the multi-agent system (4.19) can follow the leader in (4.18) asymptotically. The locations of $c\mu_i$ for the simulated scale-free network are illustrated in Fig. 4.2.

4.4.3 Leader-Follower Control with Observers

Consider the multi-agent system (4.26), where $\xi_i^{(3)}$ are not available for informed agents and a new variable $\xi_i^{(4)}$ is designed to estimate $\xi_i^{(3)}$. The parameters in (4.26) are $m = 3$, $c = 0.8$, $\beta_i = 30$ $\alpha_1 = 1$, $\alpha_2 = 3$, $k_1 = 0.5$, and $\gamma_1 = 2$. Eq. (4.29) is rewritten as follows:

$$\lambda^4 + k_1 \lambda^3 + c\mu_i \alpha_2 \lambda^2 + c\mu_i \alpha_1 \lambda + c\mu_i \gamma_1 k_1 = 0, i = 1, \dots, N.$$

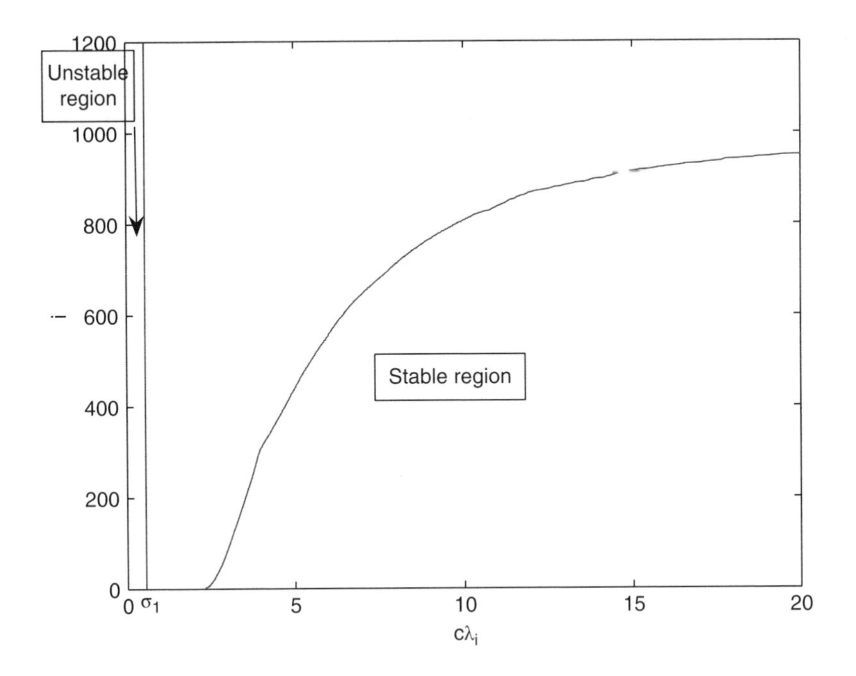

Figure 4.2 Locations of $c\mu_i$, $i = 1, \dots, N$

From the well-known Routh–Hurwitz criteria, if $k_1\alpha_2 > \alpha_1$ and $c\mu_i > \dfrac{k_1^3\gamma_1}{\alpha_1(k_1\alpha_2 - \alpha_1)} = 0.5$, then the real parts of the solutions in the above N equations are all negative. Therefore, $\mathcal{S}_2 = (0.5, \infty)$ is a stable consensus region. One can perform the same scale-free network as above and also choose the same informed agents. In doing so, similarly one has $c\mu_1 = 0.6259 > \sigma_1 = 0.5$. Therefore, $c\mu_i \in \mathcal{S}_2$ for all $i = 1, \dots, N$. Thus, all agents in the multi-agent system (4.26) can follow the leader in (4.18) asymptotically, as has also been verified by simulations (omitted for brevity).

4.5 Notes

In this chapter, general higher-order distributed consensus protocols in multi-agent dynamical systems have been studied. Some necessary and sufficient conditions have been derived for ensuring higher-order consensus, and it has been found that consensus can be reached if and only if all subsystems are asymptotically stable. Based on this result, consensus regions have been characterized, showing that consensus can be achieved if and only if all the nonzero eigenvalues of the Laplacian matrix lie in the stable consensus regions. It has also been found that the ratio of the largest to the smallest nonzero eigenvalues of the Laplacian matrix plays a key role in reaching consensus, and a scheme for choosing an appropriate coupling strength has been derived. Finally, a leader-follower control problem, with full-state and partial state observations in multi-agent dynamical systems, has been studied, which reveals that the agents with very small degrees must be informed.

The distributed consensus protocols developed in this chapter are very helpful for the design of cooperative control in multi-agent dynamical systems, which could involve more complicated and realistic dynamics of autonomous mobile agents, for such as swarming and flocking of agents with higher-order dynamics, switching topologies, and non-identical dynamics, leaving an interesting topic for future research.

5

Stability Analysis of Swarming Behaviors

Swarm behavior cohesion can be achieved in a distributed fashion, where each agent may only have local information about its nearest neighbors. In the classical swarming scheme, in addition to the consensus protocol [57], a cohesion function responsible for attraction and repulsion among agents was designed. The aim for reaching swarm is that all agents can achieve cohesion. Synchronous distributed coordination rules for swarming groups in one- or two-dimensional spaces were studied in [58], where convergence and stability analyses were performed. In [44, 45], stability properties of a continuous-time model for swarm aggregation in the n-dimensional space were discussed, and an asymptotic bound for the spatial size of the swarm was computed using the parameters in the swarm model.

The stability of a continuous-time swarm model with nonlinear profiles is investigated in this chapter [148]. It is shown that, under mild conditions, all agents in a swarm can reach cohesion within a finite time, where upper bounds of the cohesion are derived in terms of the parameters in the swarm model. The results are then generalized by considering stochastic noise and switching between nonlinear profiles. Furthermore, swarm models with changing communication topologies and unbounded repulsive interactions between agents are studied via nonsmooth analysis, where the sensing range of each agent is limited but the possibility of collision among nearby agents can be high.

5.1 Preliminaries

The swarm model considered in [44] is first reviewed. In a swarm of N agents in the n-dimensional Euclidean space, the motion dynamics of the agent i, $1 \leq i \leq N$, are described by

$$\dot{x}_i(t) = \sum_{j=1,\ j\neq i}^{N} g(x_i(t) - x_j(t)), \tag{5.1}$$

Distributed Cooperative Control of Multi-agent Systems, First Edition.
Wenwu Yu, Guanghui Wen, Guanrong Chen, and Jinde Cao.

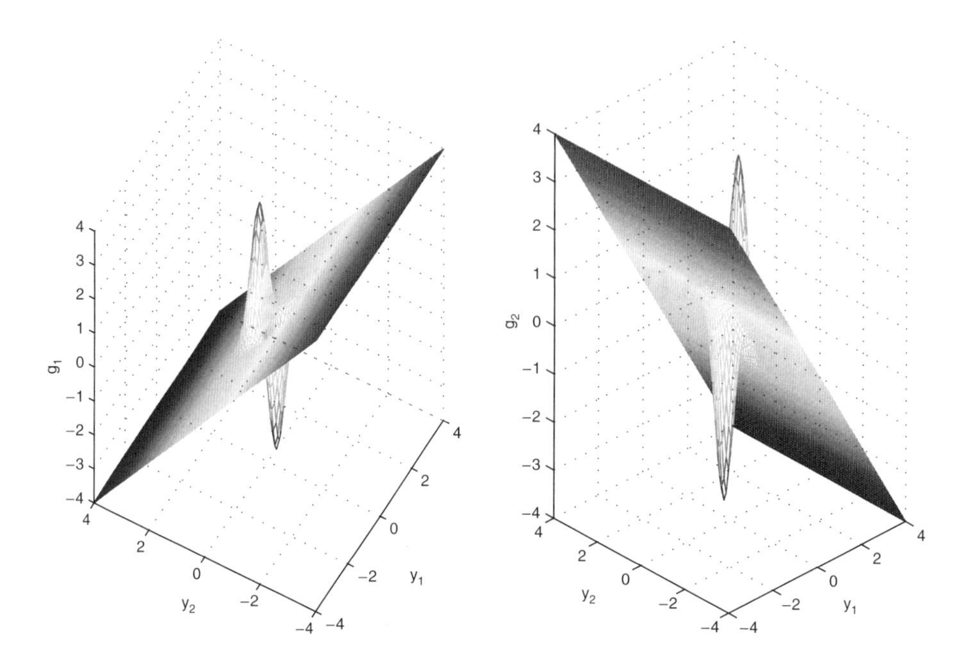

Figure 5.1 Attraction/repulsion function $g(\cdot)$, with $a = 1, b = 20, c = 0.2$, and $y \in R^2$

where $x_i \in R^n$ is the position of agent i, and $g(\cdot)$ represents the interaction force between the corresponding agents in the form of repulsion and attraction given by, in particular (Fig. 5.1):

$$g(y) = -y\left(a - be^{-\frac{\|y\|^2}{c}}\right), \tag{5.2}$$

where $y \in R^n$, a, b, and c are positive constants satisfying $b > a$. Here, the terms $-ay$ and $bye^{-\frac{\|y\|^2}{c}}$ represent the attraction and repulsion between agents, respectively, and the correspondingly $g(\cdot)$ has two equilibria, $y = 0$ and $\| y \| = \delta = \sqrt{c\ \ln(b/a)}$. Note that the attraction dominates when the two agents are far away from each other, and the repulsion dominates when they are close to each other. Because of this particular property, $g(\cdot)$ is widely used to describe the interactions among agents in swarming biological systems. However, as it will become apparent later in this chapter, there are some drawbacks in using attraction/repulsion function (5.2), and this motivates the attempt to improve the swarm model (5.1) and (5.2) in this chapter.

First, observe that the attraction/repulsion force between any chosen pair of agents i and j is anti-symmetric in i and j, namely, $g(x_i - x_j) = -g(x_j - x_i)$. As a result, if one examines the average position of the swarm, $\bar{x} = \frac{1}{N}\sum_{i=1}^{N} x_i$, it is easy to see that

$$\dot{\bar{x}} = -\frac{1}{N}\sum_{i=1}^{N}\sum_{j=1,\ j\neq i}^{N}(x_i - x_j)g(x_i - x_j) = 0, \tag{5.3}$$

which means that $\bar{x}$ is a constant and will not change with time. In real biological systems, however, each agent's motion dynamics are not only determined by inter-agent interactions, but also by each agent's intrinsic dynamics. For example, in a social foraging swarm, each agent tends to move towards a region with higher nutrient concentration. Consequently, in a biological swarm, the average position of all agents is not a constant, but more likely a dynamical variable as the whole group of agents are in motion. In this chapter, therefore, the following generalized swarm model with a nonlinear profile is considered:

$$\dot{x}_i(t) = f(x_i(t)) + \sum_{j=1,\ j\neq i}^{N} g(x_i(t) - x_j(t)), \tag{5.4}$$

where $f(x_i) = (f_1(x_i), f_2(x_i), \ldots, f_n(x_i))^T$ is a nonlinear function describing the intrinsic dynamics of each agent.

Assumption 5.1 *For all $x, y \in R^n$, there exists a constant θ such that*

$$\| f(x) - f(y)) \| \leq \theta \| x - y \| . \tag{5.5}$$

Here, $\theta > 0$ means that each agent needs energy from outside to become stable, while $\theta < 0$ means that each agent itself is already stable and may provide extra energy to other agents. Note that the Lipschitz condition (5.5) is very mild: if $\partial f_j / \partial x_{ij}$, $i = 1, 2, \ldots, N, j = 1, 2, \ldots, n$, are uniformly bounded, including in particular all linear time-invariant systems, then this condition is automatically satisfied.

Lemma 5.2 *Let $A \in R^{N\times N}$. $A_{ij} = 1$ $(i \neq j)$ and $A_{ii} = N - 1$, $i, j = 1, 2, \ldots, N$, namely,*

$$A = \begin{pmatrix} N-1 & 1 & \cdots & 1 \\ 1 & N-1 & \vdots & 1 \\ \vdots & \vdots & \ddots & 1 \\ 1 & 1 & \cdots & N-1 \end{pmatrix}. \tag{5.6}$$

Then, A has an eigenvalue $2(N - 1)$ with multiplicity 1, and an eigenvalue $N - 2$ with multiplicity $N - 1$.

Proof. Let λ be an eigenvalue of matrix A. Then, one has

$$|\lambda I_N - A| = \begin{vmatrix} \lambda - N + 1 & -1 & \cdots & -1 \\ -1 & \lambda - N + 1 & \vdots & -1 \\ \vdots & \vdots & \ddots & -1 \\ -1 & -1 & \cdots & \lambda - N + 1 \end{vmatrix}$$

$$= \begin{vmatrix} \lambda - N + 1 & -1 & \cdots & -1 \\ N - 2 - \lambda & \lambda - N + 2 & \vdots & 0 \\ \vdots & \vdots & \ddots & 0 \\ N - 2 - \lambda & 0 & \cdots & \lambda - N + 2 \end{vmatrix}$$

$$= \begin{vmatrix} \lambda - 2N + 2 & -1 & \cdots & -1 \\ 0 & \lambda - N + 2 & \vdots & 0 \\ \vdots & \vdots & \ddots & 0 \\ 0 & 0 & \cdots & \lambda - N + 2 \end{vmatrix}$$

$$= (\lambda - 2N + 2)(\lambda - N + 2)^{N-1},$$

where the second equality is obtained by subtracting the first row from the other rows, and the third equality is obtained by adding all the columns into the first column. This completes the proof. □

Now, define the error vectors $e_i(t) = x_i - \bar{x}(t)$. Then, one has the following error dynamical system:

$$\begin{aligned}
\dot{e}_i(t) =& f(x_i(t)) - \frac{1}{N}\sum_{j=1}^{N} f(x_j(t)) + \sum_{j=1,\ j\neq i}^{N} g(x_i(t) - x_j(t)) \\
=& f(x_i(t)) - \frac{1}{N}\sum_{j=1}^{N} f(x_j(t)) + \sum_{j=1,\ j\neq i}^{N} [-a(x_i(t) - x_j(t)) \\
&+ b(x_i(t) - x_j(t))e^{-\|x_i(t)-x_j(t)\|^2/c}] \\
=& f(x_i(t)) - \frac{1}{N}\sum_{j=1}^{N} f(x_j(t)) - aNe_i(t) \\
&+ \sum_{j=1, j\neq i}^{N} b(x_i(t) - x_j(t))e^{-\|x_i(t)-x_j(t)\|^2/c}.
\end{aligned} \tag{5.7}$$

5.2 Analysis of Swarm Cohesion

As a first step in the analysis of the generalized swarm model with a nonlinear profile, stability analysis of swarm cohesion is investigated in this section.

Theorem 5.3 *Suppose that Assumption 5.1 holds. Consider the generalized swarm model (5.4) with an attraction/replusion function (5.2). If*

$$a > \frac{2(N-1)\theta}{N^2}, \tag{5.8}$$

then all the agents of the swarm will converge to a hyperball centered at $\bar{x}(t)$,

$$B_\epsilon = \left\{ (x_1, \ldots, x_N) \left| \frac{1}{N} \sum_{i=1}^{N} \| x_i(t) - \bar{x}(t) \|^2 \leq \epsilon \right. \right\}, \tag{5.9}$$

where $\epsilon = \frac{b^2 c}{2e\left(a - \frac{2(N-1)\theta}{N^2}\right)^2}$. *Furthermore, all agents will move into the hyperball* B_ϵ *in a finite time specified by*

$$t = -\frac{1}{2\left(a - \frac{2(N-1)\theta}{N^2}\right)} \ln\left(\frac{N\epsilon}{2V(0)}\right). \tag{5.10}$$

Proof. Consider the following Lyapunov function candidate:

$$V(t) = \frac{1}{2} \sum_{i=1}^{N} e_i^T(t) e_i(t). \tag{5.11}$$

Taking the derivative of $V(t)$ along the trajectories of (5.7) gives

$$\begin{aligned} \dot{V} &= \sum_{i=1}^{N} e_i^T(t) \dot{e}_i(t) \\ &= \sum_{i=1}^{N} e_i^T(t) \left[f(x_i(t)) - \frac{1}{N} \sum_{j=1}^{N} f(x_j(t)) - aNe_i(t) \right. \\ &\quad \left. + \sum_{j=1,\, j\neq i}^{N} b(x_i(t) - x_j(t)) e^{-\|x_i(t)-x_j(t)\|^2/c} \right]. \end{aligned} \tag{5.12}$$

By Assumption 5.1, one has

$$\begin{aligned} &\sum_{i=1}^{N} e_i^T(t) \left[f(x_i(t)) - \frac{1}{N} \sum_{j=1}^{N} f(x_j(t)) \right] \\ &\leq \frac{\theta}{N} \sum_{i=1}^{N} \sum_{j=1,\, j\neq i}^{N} \| e_i(t) \| \| x_i(t) - x_j(t) \| \\ &\leq \frac{\theta}{N} \sum_{i=1}^{N} \sum_{j=1,\, j\neq i}^{N} \| e_i(t) \| (\| e_i(t) \| + \| e_j(t) \|). \end{aligned} \tag{5.13}$$

Note that the function $\| x_i(t) - x_j(t) \| e^{-\|x_i(t)-x_j(t)\|^2/c}$ is a bounded function with maximum value $\sqrt{\frac{c}{2}} e^{-\frac{1}{2}}$ attained when $\| x_i(t) - x_j(t) \| = \sqrt{\frac{c}{2}}$. Then, it follows that

$$\| e_i(t) \| \sum_{j=1,\ j\neq i}^{N} b \| x_i(t) - x_j(t) \| e^{-\|x_i(t)-x_j(t)\|^2/c}$$
$$\leq b(N-1) \| e_i(t) \| \sqrt{\frac{c}{2}}\, e^{-\frac{1}{2}}. \tag{5.14}$$

Substituting (5.13) and (5.14) into (5.12), one has

$$\dot{V} \leq \left(-aN + \frac{N-1}{N}\theta\right) \sum_{i=1}^{N} \| e_i(t)\|^2 + \frac{\theta}{N} \sum_{i=1}^{N} \sum_{j=1,\ j\neq i}^{N} \| e_i(t) \|\| e_j(t) \|$$
$$+b(N-1)\sqrt{\frac{c}{2}}\, e^{-\frac{1}{2}} \sum_{i=1}^{N} \| e_i(t) \|$$
$$\leq -N \| e(t) \| \Omega \| e(t) \| + b(N-1)\sqrt{\frac{c}{2}}\, e^{-\frac{1}{2}} \sum_{i=1}^{N} \| e_i(t) \|$$
$$\leq -N\lambda_{\min}(\Omega) \sum_{i=1}^{N} \| e_i(t)\|^2 + b(N-1)\sqrt{\frac{c}{2}}\, e^{-\frac{1}{2}} \sum_{i=1}^{N} \| e_i(t) \|, \tag{5.15}$$

where $\| e(t) \| = (\| e_1(t) \|, \| e_2(t) \|, \ldots, \| e_N(t) \|)^T$, and

$$\Omega = \begin{pmatrix} a - \frac{N-1}{N^2}\theta & -\frac{\theta}{N^2} & \cdots & -\frac{\theta}{N^2} \\ -\frac{\theta}{N^2} & a - \frac{N-1}{N^2}\theta & \vdots & -\frac{\theta}{N^2} \\ \vdots & \vdots & \ddots & -\frac{\theta}{N^2} \\ -\frac{\theta}{N^2} & -\frac{\theta}{N^2} & \cdots & a - \frac{N-1}{N^2}\theta \end{pmatrix}.$$

By Lemma 5.2 and (5.8), one has $\lambda_{\min}(\Omega) = a - \frac{2(N-1)\theta}{N^2}$. Note that

$$\left(\sum_{i=1}^{N} \| e_i(t) \|\right)^2 = \sum_{i=1}^{N} \sum_{j=1}^{N} \| e_i(t) \|\| e_j(t) \|$$
$$\leq \frac{1}{2} \sum_{i=1}^{N} \sum_{j=1}^{N} (\| e_i(t)\|^2 + \| e_j(t)\|^2) = N \sum_{i=1}^{N} \| e_i(t)\|^2.$$

Then, it follows that

$$
\begin{aligned}
\dot{V} \leq & -N\left(a-\frac{2(N-1)\theta}{N^2}\right)\sum_{i=1}^{N}\| e_i(t)\|^2+b(N-1)\sqrt{\frac{Nc}{2}}\,e^{-\frac{1}{2}}\sqrt{\sum_{i=1}^{N}\| e_i(t)\|^2} \\
= & -\left(a-\frac{2(N-1)\theta}{N^2}\right)\sum_{i=1}^{N}\| e_i(t)\|^2-\left(a-\frac{2(N-1)\theta}{N^2}\right)(N-1) \\
& \times\sqrt{\sum_{i=1}^{N}\| e_i(t)\|^2}\left(\sqrt{\sum_{i=1}^{N}\| e_i(t)\|^2}-\frac{b\sqrt{\frac{Nc}{2}}\,e^{-\frac{1}{2}}}{a-\frac{2(N-1)\theta}{N^2}}\right).
\end{aligned}
\tag{5.16}
$$

If $\frac{1}{N}\sum_{i=1}^{N}\| x_i(t)-\bar{x}\|^2 \geq \frac{b^2c}{2e(a-\frac{2(N-1)\theta}{N^2})^2}$, then one has

$$
\dot{V} \leq -\left(a-\frac{2(N-1)\theta}{N^2}\right)\sum_{i=1}^{N}\| e_i(t)\|^2 = -2\left(a-\frac{2(N-1)\theta}{N^2}\right)V(t). \tag{5.17}
$$

Therefore, the solutions of $V(t)$ satisfy

$$
V(t) \leq V_0 e^{-2\left(a-\frac{2(N-1)\theta}{N^2}\right)t}.
$$

Now, it is easy to see that the trajectories enter the boundary $\frac{1}{N}\sum_{i=1}^{N}\| x_i(t)-\bar{x}\|^2 = \epsilon$ in a finite time

$$
t \leq -\frac{1}{2\left(a-\frac{2(N-1)\theta}{N^2}\right)}\ln\left(\frac{N\epsilon}{2V(0)}\right).
$$

This completes the proof. □

Remark 5.4 The bound $\epsilon = \frac{b^2c}{2e\left(a-\frac{2(N-1)\theta}{N^2}\right)^2}$ increases as the parameters b and c increase, while it decreases as the parameter a increases. This is consistent with the balance $\delta = \sqrt{c\ln(b/a)}$ between the attraction and repulsion. In [44], $\theta = 0$ is considered, and the bound is a constant for given a, b, and c, and is independent of the size N. In this chapter, the bound ϵ of cohesion increases as $\theta > 0$ increases, which is closer to biological reality.

Remark 5.5 Note that the bound ϵ of the swarm depends on the size N. If N increases, the bound ϵ decreases, which means that the density of the swarm increases. Furthermore, for swarms with a very large number of agents, $\epsilon \to \frac{b^2c}{2ea^2}$. This, however, is inconsistent with the biological phenomena and is due to the fact that the attraction/repulsion function $g(\cdot)$ in (5.2) taken from [44] has an infinitely long effective range for any chosen pair of agents. This function $g(\cdot)$ will be modified in Section 5.5 later, so that it has only a limited effective range. In other words, there will be no interaction between a pair of agents that are out of a predetermined sensing range r.

Remark 5.6 In [45], a function f was investigated for stability analysis of social foraging swarms, where the assumptions on f are restrictive, such as requiring f to be bounded or linear. However, in this chapter, Assumption 5.1 is mild and applies to many well-known nonlinear systems.

5.3 Swarm Cohesion in a Noisy Environment

In this section, the cohesion of a swarm in a noisy environment is investigated. Consider the following stochastic swarm model:

$$dx_i(t) = \left[f(x_i(t)) + \sum_{j=1,\ j\neq i}^{N} g(x_i(t) - x_j(t)) \right] dt + v_i(t) d\nu_i, \tag{5.18}$$

where $v_i(t) \in R^n$ is an external noise intensity function of agent i, and $\nu_i(t)$ is an independent one-dimensional Brownian motion with expectation $\mathbf{E}\{d\nu_i(t)\} = 0$ and variance $\mathbf{D}\{d\nu_i(t)\} = 1$. The model is defined in a complete probability space $(\Omega, \mathscr{F}, \mathbb{P})$ with a natural filtration $\{\mathscr{F}_t\}_{t\geq 0}$ generated by $\{\nu_i(s) : 0 \leq s \leq t\}$, where Ω is associated with the canonical space generated by $\nu_i(t)$ and $\mathscr{F}$ is the associated σ-algebra generated by $\{\nu_i(t)\}$ with probability measure $\mathbb{P}$.

Assumption 5.7 *$v_i(t) \in R^n$ belongs to $L_\infty[0, \infty)$, i.e., $v_i(t)$ is a bounded vector function satisfying*

$$v_i^T(t) v_i(t) \leq \alpha_i, \forall t \in R, \tag{5.19}$$

where α_i is a positive constant, $i = 1, 2, \ldots, N$.

Theorem 5.8 *Suppose that Assumptions 5.1 and 5.7 hold. Consider the swarm model (5.18) with the attraction/replusion function $g(\cdot)$ defined by (5.2). If*

$$a > \frac{2(N-1)\theta}{N^2}, \tag{5.20}$$

then all the agents of the swarm will converge to a hyperball centered at $\bar{x}(t)$ in mean-square:

$$B_\eta = \left\{ (x_1, \ldots, x_N) \,\middle|\, \frac{1}{N} \sum_{i=1}^{N} \mathbf{E}(\| x_i(t) - \bar{x}(t) \|^2) \leq \eta \right\}, \tag{5.21}$$

where

$$\eta = \frac{\left(b\sqrt{\frac{Nc}{2e}} + \sqrt{\frac{b^2Nc}{2e} + \frac{4\alpha\left(a - \frac{2(N-1)\theta}{N^2}\right)}{N-1}}\right)^2}{4N\left(a - \frac{2(N-1)\theta}{N^2}\right)^2}, \text{ and } \alpha = \frac{1}{2}\sum_{i=1}^{N}\alpha_i.$$

Furthermore, cohesion will be achieved within the bound η in a finite time

$$t = -\frac{1}{2\left(a - \frac{2(N-1)\theta}{N^2}\right)} \ln\left(\frac{N\eta}{2\mathbf{E}V(0)}\right). \tag{5.22}$$

Proof. Consider the same Lyapunov function candidate as in (5.11). From the Itô formula [105], one obtains the following stochastic differential:

$$dV(t) = \mathscr{L}V(t)dt + \sum_{i=1}^{N} e_i^T(t)[v_i(t)dv_i(t)]. \tag{5.23}$$

By Assumption 5.7, the weak infinitesimal operator $\mathscr{L}$ of the stochastic process yields

$$\mathscr{L}V(t) = \sum_{i=1}^{N}\left[e_i^T(t)\dot{e}_i(t) + \frac{1}{2}v_i^T(t)v_i(t)\right] \le \sum_{i=1}^{N} e_i^T(t)\dot{e}_i(t) + \alpha, \tag{5.24}$$

where $\alpha = \frac{1}{2}\sum_{i=1}^{N}\alpha_i$. By following the same steps as in the proof of Theorem 5.3, one obtains the following expression which is similar to (5.16):

$$\begin{aligned}
\mathscr{L}V(t) &\le -N\left(a - \frac{2(N-1)\theta}{N^2}\right)\sum_{i=1}^{N} \| e_i(t)\|^2 \\
&\quad + b(N-1)\sqrt{\frac{Nc}{2}}\, e^{-\frac{1}{2}}\sqrt{\sum_{i=1}^{N} \| e_i(t)\|^2} + \alpha \\
&= -\left(a - \frac{2(N-1)\theta}{N^2}\right)\sum_{i=1}^{N} \| e_i(t)\|^2 - \left(a - \frac{2(N-1)\theta}{N^2}\right)(N-1) \\
&\quad \times \left(\sum_{i=1}^{N} \| e_i(t)\|^2 - \frac{b\sqrt{\frac{Nc}{2}}e^{-\frac{1}{2}}}{a - \frac{2(N-1)\theta}{N^2}}\sqrt{\sum_{i=1}^{N} \| e_i(t)\|^2} - \frac{\alpha}{(a - \frac{2(N-1)\theta}{N^2})(N-1)}\right).
\end{aligned} \tag{5.25}$$

Let $z = \sqrt{\sum_{i=1}^{N} \| e_i(t)\|^2}$ and $g(z) = z^2 - \frac{b\sqrt{\frac{Nc}{2}e^{-\frac{1}{2}}}}{a-\frac{2(N-1)\theta}{N^2}} z - \frac{\alpha}{\left(a-\frac{2(N-1)\theta}{N^2}\right)(N-1)}$. It is easy to see that $g(z) = 0$ has two solutions:

$$z_{1,2} = \frac{b\sqrt{\frac{Nc}{2e}} \pm \sqrt{\frac{b^2Nc}{2e} + \frac{4\alpha\left(a - \frac{2(N-1)\theta}{N^2}\right)}{N-1}}}{2\left(a - \frac{2(N-1)\theta}{N^2}\right)}, \tag{5.26}$$

where $z_1 > 0$ and $z_2 < 0$. If $z(t) \geq z_1$, then one has $g(z) \geq 0$, and it follows that

$$\mathscr{L}V(t) \leq -\left(a - \frac{2(N-1)\theta}{N^2}\right) \sum_{i=1}^{N} \| e_i(t)\|^2. \tag{5.27}$$

From the Itô formula, it follows that

$$\mathbf{E}V(t) - \mathbf{E}V(0) = \mathbf{E}\int_0^t \mathscr{L}V(s)ds \leq -2\left(a - \frac{2(N-1)\theta}{N^2}\right)\int_0^t \mathbf{E}V(s)ds.$$

Therefore, the solutions of $V(t)$ satisfy

$$\mathbf{E}V(t) \leq \mathbf{E}V(0)e^{-2\left(a-\frac{2(N-1)\theta}{N^2}\right)t}.$$

Thus, it is easy to show that the trajectories cross the boundary $\frac{1}{N}\mathbf{E}\left(\sum_{i=1}^{N} \|x_i(t) - \bar{x}\|^2\right) = \eta$ in a finite time:

$$t \leq -\frac{1}{2\left(a - \frac{2(N-1)\theta}{N^2}\right)} \ln\left(\frac{N\eta}{2\mathbf{E}V(0)}\right).$$

This completes the proof. □

5.4 Cohesion in Swarms with Switched Topologies

In a swarm system with a nonlinear profile, it is easy to check that the average position $\bar{x}(t)$ evolves according to

$$\dot{\bar{x}}(t) = \frac{1}{N}\sum_{i=1}^{N} f(x_i(t)). \tag{5.28}$$

Now let $f(y) = -\sigma(y - \rho)$, where $\sigma > 0$, $y, \rho \in R^n$. Then, $\dot{\bar{x}} = -\sigma(\bar{x} - \rho)$. Hence, one may check that $\bar{x}(t) \to \rho$ as $t \to \infty$. Here, ρ can be interpreted as the target average position of the swarming group whose average velocity is determined by σ. Since in biological swarms, each agent's velocity may change from time to time and the target position may also move, this leads to switched systems. The phenomenon that the profiles of agents may change at particular times is also of special interest to applications since such behaviors appear naturally in automatic control systems, neural networks, and communications. In this section, therefore, the following switched swarming system is considered:

$$dx_i(t) = \left[f_\gamma(x_i(t)) + \sum_{j=1,\ j\neq i}^{N} g_\gamma(x_i(t) - x_j(t)) \right] dt + v_i(t)dv_i, \tag{5.29}$$

where γ is a switching signal which takes values from the finite set $\mathscr{I} = \{1, 2, \ldots, \overline{N}\}$, and

$$g_\gamma(y) = -y\left(a_\gamma - b_\gamma e^{-\frac{\|y\|^2}{c_\gamma}} \right), \tag{5.30}$$

which means that the positive parameter values $(a_\gamma, b_\gamma, c_\gamma)$ are allowed to take values, at particular times, from the finite set $\left\{(a_1, b_1, c_1), \ldots, (a_{\overline{N}}, b_{\overline{N}}, c_{\overline{N}})\right\}$ [144].

Assumption 5.9 *There exist constants θ_γ such that*

$$\| f_\gamma(x) - f_\gamma(y)) \| \leq \theta_\gamma \| x - y \|, \forall x, y \in R^n, \gamma = 1, 2, \ldots, \overline{N}. \tag{5.31}$$

Theorem 5.10 *Suppose that Assumptions 5.1 and 5.9 hold. In the swarm model (5.29) with an attraction/replusion function (5.2), if*

$$a_\gamma > \frac{2(N-1)\theta_\gamma}{N^2}, \gamma = 1, 2, \ldots, \overline{N}, \tag{5.32}$$

then all the agents of the swarm will converge to a hyperball centered at $\overline{x}(t)$ in mean-square,

$$B_\eta = \max_\gamma \left\{ (x_1, \ldots, x_N) \left| \frac{1}{N} \sum_{i=1}^{N} \mathbf{E}(\| x_i(t) - \overline{x}(t)\|^2) \leq \eta_\gamma \right. \right\}, \tag{5.33}$$

where

$$\eta_\gamma = \frac{\left(b_\gamma \sqrt{\frac{Nc_\gamma}{2e}} + \sqrt{\frac{b_\gamma^2 N c_\gamma}{2e} + \frac{4\alpha\left(a_\gamma - \frac{2(N-1)\theta_\gamma}{N^2}\right)}{N-1}} \right)^2}{4N\left(a_\gamma - \frac{2(N-1)\theta_\gamma}{N^2}\right)^2}, \text{ and } \alpha = \frac{1}{2}\sum_{i=1}^{N} \alpha_i.$$

Furthermore, cohesion will be achieved within the bound η in a finite time

$$t = \max_{\gamma}\left(-\frac{1}{2\left(a_{\gamma} - \frac{2(N-1)\theta_{\gamma}}{N^2}\right)} \ln\left(\frac{N\eta}{2\mathbf{E}V(0)}\right)\right). \tag{5.34}$$

Proof. Choose the same Lyapunov candidate as in (5.11) to be the common Lyapunov function. Then, the proof can be completed in the same way as in the proof of Theorem 5.8. □

5.5 Cohesion in Swarms with Changing Topologies

As discussed in Remark 5.5 above, the attraction function $-ay$ in (5.2) has an infinite sensing range, which is not realistic for biological systems [44, 45]. In this section, therefore, a swarm model with an attraction function having a limited sensing range r is considered:

$$\dot{x}_i(t) = f(x_i(t)) + \sum_{j=1,\, j\neq i}^{N} h(x_i(t) - x_j(t)), \tag{5.35}$$

in which

$$h(x_i(t) - x_j(t)) = -\left(aa_{ij}(t) - be^{-\frac{\|x_i(t)-x_j(t)\|^2}{c}}\right)\left(x_i(t) - x_j(t)\right), \tag{5.36}$$

where $a_{ij} = a_{ji} = 1$ if $\| x_i(t) - x_j(t) \| \leq r$, otherwise, $a_{ij} = a_{ji} = 0$ for $i \neq j$; $a_{ii} = 0$ for $i = 1, 2, \ldots, N$. The attraction and repulsion function $h(y)$ is shown in Fig. 5.2. Let $l_{ij} = -a_{ij}$ for $i \neq j$, and $l_{ii} = \sum_{j=1,\, j\neq i}^{N} a_{ij}$. Then, system (5.35) can be written as

$$\begin{aligned}\dot{x}_i(t) = {} & f(x_i(t)) - \sum_{j=1}^{N} al_{ij}(t)x_j(t) \\ & + \sum_{j=1,\, j\neq i}^{N} be^{-\frac{\|x_i(t)-x_j(t)\|^2}{c}}(x_i(t) - x_j(t)).\end{aligned} \tag{5.37}$$

Again, let $e_i(t) = x_i - \overline{x}(t)$. Then, one obtains the following error dynamical system:

$$\begin{aligned}\dot{e}_i(t) = {} & f(x_i(t)) - \frac{1}{N}\sum_{j=1}^{N} f(x_j(t)) - a\sum_{j=1}^{N} l_{ij}(t)e_j(t) \\ & + \sum_{j=1,\, j\neq i}^{N} b(x_i(t) - x_j(t))e^{-\|x_i(t)-x_j(t)\|^2/c}.\end{aligned} \tag{5.38}$$

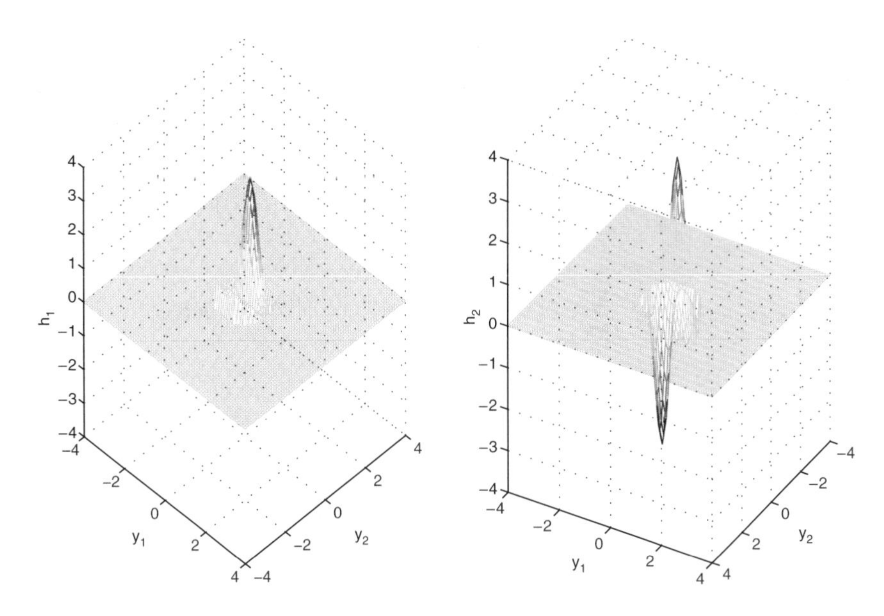

Figure 5.2 Attraction/repulsion function $h(\cdot)$, with $a = 1, b = 20, c = 0.2, r = 1$, and $y \in R^2$

Since the right-hand side of (5.38) is discontinuous, one cannot study it by using ordinary differential equations with classical solutions (continuously differential). Here, the nonsmooth analysis is applied [38].

Definition 5.11 [6] Suppose $E \subset R^n$. Map $x \to F(x)$ is called a set-value map from $E \to R^n$, if each point x of a set $E \subset R^n$ corresponds to a non-empty set $F(x) \subset R^n$.

Let $\dot{e}(t) = \phi(x)$, where $e(t) = \left(e_1^T, e_2^T, \ldots, e_N^T\right)^T$, and $x(t) = \left(x_1^T, x_2^T, \ldots, x_N^T\right)^T$.

Definition 5.12 [38, 69] (Filippov solution) A set-valued map is defined as

$$\psi(x) = \bigcap_{\delta>0} \bigcap_{\mu(\overline{N})=0} \mathbb{K}[\phi(B(x, \delta) - \overline{N})],$$

where $\mathbb{K}(E)$ is the closure of the convex hull of set E, $B(x, \delta) = \{y : \| y - x \| \leq \delta\}$, and $\mu(\overline{N})$ is the Lebesgue measure of set $\overline{N}$. A solution of the Cauchy problem for (5.38) with initial condition $x(0) = x_0$ is an absolutely continuous function $x(t)$, which satisfies $x(0) = x_0$ and the differential inclusion:

$$\dot{e}(t) \in \psi(x), a.e.t.$$

where a.e.t. means almost every time.

The concept of a Filippov solution is very important in engineering applications. All sets of measure zero are disregarded, which allows solutions to be defined at points even where the vector $\psi(x)$ is discontinuous. In addition, an arbitrary set of measure zero in $B(x, \delta)$ is excluded when evaluating x such that the result is the same for any two vector fields that differ on a set of measure zero. Following [87], the concept of Filippov solution is extended to:

$$\dot{e}_i(t) \in f(x_i(t)) - \frac{1}{N}\sum_{j=1}^{N} f(x_j(t)) - a\sum_{j=1}^{N} \mathbb{K}[l_{ij}(t)]e_j(t) + \sum_{j=1,\ j\neq i}^{N} b(x_i(t) - x_j(t))e^{-\|x_i(t)-x_j(t)\|^2/c}, a.e.t, \tag{5.39}$$

or equivalently,

$$\dot{e}_i(t) = f(x_i(t)) - \frac{1}{N}\sum_{j=1}^{N} f(x_j(t)) - a\sum_{j=1}^{N} d_{ij}(t)e_j(t) + \sum_{j=1, j\neq i}^{N} b(x_i(t) - x_j(t))e^{-\|x_i(t)-x_j(t)\|^2/c}, a.e.t, \tag{5.40}$$

where $d_{ij}(t) = \begin{cases} 1 & \| x_i(t) - x_j(t) \| < r \\ 0 & \| x_i(t) - x_j(t) \| > r \\ \xi_{ij} & \| x_i(t) - x_j(t) \| = r \end{cases}$ for $i \neq j$, $d_{ii} = -\sum_{j=1}^{N} d_{ij}$, and $\xi_{ij} \in [0, 1]$. Let $D = (d_{ij})_{N\times N}$.

Theorem 5.13 *Suppose that Assumption 5.1 holds, and the network with the Laplacian matrix L is connected at all times. Consider the swarm model (5.35) with an attraction/replusion function (5.36). If*

$$a\phi_t > \frac{2(N-1)\theta}{N}, \forall t > 0, \tag{5.41}$$

then all the agents of the swarm will converge to a hyperball centered at $\bar{x}(t)$*,*

$$B_{\epsilon_t} = \left\{ (x_1, \dots, x_N) \,\middle|\, \frac{1}{N}\sum_{i=1}^{N} \| x_i(t) - \bar{x}(t)\|^2 \leq \epsilon_t \right\}, \tag{5.42}$$

where

$$\phi_t = \min_{\tau\in[0,t]} \lambda_2(D(\tau)), \epsilon_t = \frac{b^2cN^2}{2e\left(a\phi_t - \frac{2(N-1)\theta}{N}\right)^2}.$$

Furthermore, cohesion will be achieved within the bound ϵ_t *in a finite time*

$$t = -\frac{N}{2\left(a\phi_t - \frac{2(N-1)\theta}{N}\right)} \ln\left(\frac{N\epsilon_t}{2V(0)}\right). \tag{5.43}$$

Proof. Consider the following Lyapunov function candidate:

$$V(t) = \frac{1}{2}\sum_{i=1}^{N} e_i^T(t)e_i(t). \tag{5.44}$$

Taking the derivative of $V(t)$ along the trajectories of (5.39), and using (5.13)–(5.16) and Lemma 2.9, one has

$$\begin{aligned}
\dot{V} &= \sum_{i=1}^{N} e_i^T(t)\dot{e}_i(t) \\
&= \sum_{i=1}^{N} e_i^T(t)\left[f(x_i(t)) - \frac{1}{N}\sum_{j=1}^{N} f(x_j(t)) - a\sum_{j=1}^{N} d_{ij}(t)e_j(t)\right. \\
&\quad \left. + \sum_{j=1,\, j\neq i}^{N} b(x_i(t) - x_j(t))e^{-\|x_i(t)-x_j(t)\|^2/c}\right] \\
&\leq -a\sum_{i=1}^{N}\sum_{j=1}^{N} d_{ij}(t)e_i^T(t)e_j(t) + \frac{2(N-1)}{N}\theta\sum_{i=1}^{N} \| e_i(t)\|^2 \\
&\quad + b(N-1)\sqrt{\frac{Nc}{2}}\, e^{-\frac{1}{2}}\sqrt{\sum_{i=1}^{N} \| e_i(t)\|^2} \\
&= -ae^T(t)(D(t)\otimes I_n)e(t) + \frac{2(N-1)}{N}\theta\sum_{i=1}^{N} \| e_i(t)\|^2 \\
&\quad + b(N-1)\sqrt{\frac{Nc}{2}}\, e^{-\frac{1}{2}}\sqrt{\sum_{i=1}^{N} \| e_i(t)\|^2} \\
&\leq \left[-a(\lambda_2(D(t)) + \frac{2(N-1)}{N}\theta\right]\sum_{i=1}^{N} \| e_i(t)\|^2 \\
&\quad + b(N-1)\sqrt{\frac{Nc}{2}}\, e^{-\frac{1}{2}}\sqrt{\sum_{i=1}^{N} \| e_i(t)\|^2},
\end{aligned} \tag{5.45}$$

By using a similar argument as in the proof Theorem 5.3, the claim can be proved. □

Remark 5.14 Note that the bound of the swarm, $\epsilon_t = \frac{b^2cN^2}{2e\left(a\phi_t - \frac{2(N-1)\theta}{N}\right)^2}$, depends on the size N. If N increases, the bound ϵ increases, which means that the boundary of cohesion increases due to the large number of agents. It is consistent with the biological phenomena, thanks to the limited sensing range r in the attraction/repulsion function $h(\cdot)$ in (5.36).

Generally, suppose that the network structure $L(t)$ changes very slowly:

Assumption 5.15 *There is at most one connection that changes in each time, for example, the connection between agents i and j, $i \neq j$.*

Lemma 5.16 *[32] For any given graph $\mathcal{G}$ of size N, its nonzero eigenvalues grow monotonically with the number of added edges, i.e., for any added edge $\widetilde{e}$, $\lambda_2(G+\widetilde{e}) \geq \lambda_2(G)$.*

Corollary 5.17 *Suppose that Assumptions 5.1 and 5.16 hold, and the network with the Laplacian matrix L is connected at all times. Consider the swarm model (5.35) with an attraction/replusion function (5.36). If*

$$a\phi_t > \frac{2(N-1)\theta}{N}, \forall t > 0,$$

then all the agents of the swarm will converge to a hyperball centered at $\overline{x}(t)$,

$$B_{\epsilon_t} = \left\{ (x_1, \ldots, x_N) \,\middle|\, \frac{1}{N} \sum_{i=1}^{N} \| x_i(t) - \overline{x}(t) \|^2 \leq \epsilon_t \right\},$$

where

$$\phi_t = \min_{\tau \in [0,t]} \lambda_2(L(\tau)), \epsilon_t = \frac{b^2 c N^2}{2e\left(a\phi_t - \frac{2(N-1)\theta}{N}\right)^2}.$$

Furthermore, cohesion will be achieved within the bound ϵ_t in a finite time

$$t = -\frac{N}{2\left(a\phi_t - \frac{2(N-1)\theta}{N}\right)} \ln\left(\frac{N\epsilon_t}{2V(0)}\right).$$

Proof. By Assumption 5.15 and Lemma 5.16, one knows that

$$\min\{\lambda_2(L(\tau-)), \lambda_2(L(\tau+))\} \leq \lambda_2(D(\tau)) \leq \max\{\lambda_2(L(\tau-)), \lambda_2(L(\tau+))\}.$$

The proof is completed. □

Remark 5.18 The difference between Theorem 5.13 and Corollary 5.17 is that under Assumption 5.15, the term $D(\tau)$ in Theorem 5.13 is replaced by $L(\tau)$ in Corollary 5.17. Therefore, when the network changes very slowly, the condition can be simplified by using L instead. In the above theorems, only D is used which can also be replaced by L under Assumption 5.15. Detailed analysis is omitted.

It is still not easy to verify whether or not the condition in (5.41) is satisfied for all $t \in R$. If the passivity degree $\theta < 0$, then it is possible that $\phi_t = 0$; if $\theta = 0$, then system (5.8) is more likely a linear model; if $\theta > 0$, then $\phi_t > 0$ must be satisfied. From condition (5.41), the cohesion can be reached even if the network is disconnected when $\theta < 0$. Since for periodic and chaotic nodes, $\theta > 0$, one may be interested in the condition under which $\phi_t > 0$. In the following, some conditions are given to ensure $\phi_t > 0$ for all $t \in R$.

Lemma 5.19 *[75, 76] For a connected graph $\mathcal{G}$ of order N, its second Laplacian eigenvalue λ_2 imposes upper bounds on the diameter diam(G) and the mean distance $\rho(G)$ of G as follows:*

$$(i)\ \lambda_2 \geq \frac{4}{N diam(G)},\ \ and\ \ (ii)\ \lambda_2 \geq \frac{2}{(N-1)\rho(G)-(N-2)/2}.$$

Theorem 5.20 *Suppose that Assumption 5.1 holds, and the network with the Laplacian matrix L is connected at all times. Consider the swarm model (5.35) with an attraction/replusion function (5.36). If*

$$a\delta_t > \frac{2(N-1)\theta}{N}, \forall t > 0, \tag{5.46}$$

then all the agents of the swarm will converge to a hyperball centered at $\overline{x}(t)$,

$$B_{\epsilon_t} = \left\{ (x_1, \ldots, x_N) \left| \frac{1}{N} \sum_{i=1}^{N} \| x_i(t) - \overline{x}(t)\|^2 \leq \epsilon_t \right. \right\}, \tag{5.47}$$

where $\delta_t = \min_{\tau\in[0,t]} \max \left\{ \frac{4}{N\ \mathrm{diam}(D(\tau))}, \frac{2}{(N-1)\rho(D(\tau))-(N-2)/2} \right\}$, *and* $\epsilon_t = \frac{b^2cN^2}{2e\left(a\delta_t - \frac{2(N-1)\theta}{N}\right)^2}$.
Furthermore, cohesion will be achieved within the bound ϵ_t in a finite time

$$t = -\frac{N}{2\left(a\delta_t - \frac{2(N-1)\theta}{N}\right)} \ln\left(\frac{N\epsilon_t}{2V(0)}\right). \tag{5.48}$$

Proof. By Lemma 5.19, it is easy to see that $\delta_t \geq \phi_t$. The proof can be completed by using the same method as in the proof of Theorem 5.13. □

Next, consider the stochastic switched swarm model (5.29), where

$$g_\gamma(x_i(t) - x_j(t)) = -\left(a_\gamma a_{ij}(t) - b_\gamma e^{-\frac{\|x_i(t)-x_j(t)\|^2}{c_\gamma}} \right)(x_i(t) - x_j(t)). \tag{5.49}$$

Theorem 5.21 *Suppose that Assumptions 5.1 and 5.9 hold, and the network with the Laplacian matrix L is connected at all times. Consider the swarm model (5.29) with an attraction/replusion function (5.49). If*

$$a_\gamma \phi_t > \frac{2(N-1)\theta_\gamma}{N}, \forall t > 0, \gamma = 1, 2, \ldots, \overline{N}, \tag{5.50}$$

then all the agents of the swarm will converge to a hyperball centered at $\overline{x}(t)$ in mean-square,

$$B_{\eta_t} = \max_\gamma \left\{ (x_1, \ldots, x_N) \,\middle|\, \frac{1}{N} \sum_{i=1}^{N} \mathbf{E}(\| x_i(t) - \overline{x}(t) \|^2) \leq \eta_{t,\gamma} \right\}, \tag{5.51}$$

where $\phi_t = \min_{\tau \in [0,t]} \lambda_2(D(\tau)), \eta_{t,\gamma} = \frac{\left(b_\gamma \sqrt{\frac{Nc_\gamma}{2e}} + \sqrt{\frac{b_\gamma^2 Nc_\gamma}{2e} + \frac{4\alpha(a_\gamma \phi_t - \frac{2(N-1)\theta_\gamma}{N})}{N(N-1)}}\right)^2 N^2}{4N(a_\gamma \phi_t - \frac{2(N-1)\theta_\gamma}{N})^2}$, *and* $\alpha = \frac{1}{2} \sum_{i=1}^{N} \alpha_i$. *Furthermore, cohesion will be achieved within the bound η in a finite time*

$$t = \max_\gamma \left(-\frac{N}{2\left(a_\gamma \phi_t - \frac{2(N-1)\theta_\gamma}{N}\right)} \ln \left(\frac{N\eta_t}{2\mathbf{E}V(0)} \right) \right). \tag{5.52}$$

In order to avoid possible collision when two agents are moving close to each other, the repulsive function should be sufficiently large around the origin. In what follows, an unbounded repulsive function is adopted. Consider the following swarm model:

$$\begin{aligned} \dot{x}_i(t) = & f(x_i(t)) - \sum_{j=1}^{N} a l_{ij}(t) x_j(t) \\ & + \sum_{j=1,\, j \neq i}^{N} g_r(\| x_i(t) - x_j(t) \|)(x_i(t) - x_j(t)), \end{aligned} \tag{5.53}$$

where $g_r(\cdot)$ is the repulsive function to be further described below.

Assumption 5.22 *For all $x, y \in R^n$, there exists a constant b such that*

$$|g_r(\| y \|)| \leq \frac{b}{\| y \|^2}. \tag{5.54}$$

Theorem 5.23 *Suppose that Assumptions 5.1 and 5.22 hold. Consider the swarm model (5.53). If*

$$a\phi_t > \frac{2(N-1)\theta}{N}, \forall t > 0, \tag{5.55}$$

then all the agents of the swarm will converge to a hyperball centered at $\bar{x}(t)$*,*

$$B_{\epsilon_t} = \left\{ (x_1, \ldots, x_N) \middle| \frac{1}{N} \sum_{i-1}^{N} \| x_i(t) - \bar{x}(t) \|^2 \leq \epsilon_t \right\}, \tag{5.56}$$

where

$$\phi_t = \min_{\tau \in [0,t]} \lambda_2(D(\tau)), \epsilon_t = \frac{bN}{2\left(a\phi_t - \frac{2(N-1)\theta}{N}\right)}.$$

Furthermore, cohesion will be achieved within the bound ϵ_t *in a finite time*

$$t = -\frac{N}{2\left(a\phi_t - \frac{2(N-1)\theta}{N}\right)} \ln\left(\frac{N\epsilon_t}{2V(0)}\right). \tag{5.57}$$

Proof. Consider the same Lyapunov function candidate as in (5.11). By Theorems 5.3 and 5.13, one obtains

$$\begin{aligned} \dot{V}(t) \leq & -\left(a\lambda_2(D(t)) - \frac{2(N-1)\theta}{N}\right) \sum_{i=1}^{N} \| e_i(t) \|^2 \\ & + \sum_{i=1}^{N} \sum_{j=1,\, j \neq i}^{N} g_r(\| x_i(t) - x_j(t) \|) e_i^T(t)(x_i(t) - x_j(t)). \end{aligned} \tag{5.58}$$

By Assumption 5.22 and based on the fact that $e_i(t) - e_j(t) = x_i(t) - x_j(t)$, one has

$$\begin{aligned} & \sum_{i=1}^{N} \sum_{j=1,\, j \neq i}^{N} g_r(\| x_i(t) - x_j(t) \|) e_i^T(t)(x_i(t) - x_j(t)) \\ = & \sum_{i=1}^{N} \sum_{j=1,\, j>i}^{N} [g_r(\| x_i(t) - x_j(t) \|) e_i^T(t)(x_i(t) - x_j(t)) \\ & + g_r(\| x_j(t) - x_i(t) \|) e_j^T(t)(x_j(t) - x_i(t))] \\ = & \sum_{i=1}^{N} \sum_{j=1,\, j>i}^{N} g_r(\| x_i(t) - x_j(t) \|)(e_i(t) - e_j(t))^T (x_i(t) - x_j(t)) \\ = & \frac{1}{2} \sum_{i=1}^{N} \sum_{j=1,\, j \neq i}^{N} g_r(\| x_i(t) - x_j(t) \|) \| x_i(t) - x_j(t) \|^2 \\ \leq & \frac{b}{2} N(N-1). \end{aligned} \tag{5.59}$$

Then, it follows that

$$\begin{aligned}\dot{V}(t) \leq & -\left(a\phi_t - \frac{2(N-1)\theta}{N}\right)\sum_{i=1}^{N} \| e_i(t)\|^2 + \frac{b}{2}N(N-1) \\ = & -\frac{1}{N}\left(a\phi_t - \frac{2(N-1)\theta}{N}\right)\sum_{i=1}^{N} \| e_i(t)\|^2 \\ & -(N-1)\left[\left(a\phi_t - \frac{2(N-1)\theta}{N}\right)\frac{1}{N}\sum_{i=1}^{N} \| e_i(t)\|^2 - \frac{b}{2}N\right]. \end{aligned} \quad (5.60)$$

If $\frac{1}{N}\sum_{i=1}^{N} \| e_i(t)\|^2 \geq \epsilon_t$, then one has

$$\dot{V}(t) \leq \frac{1}{N}\left(a\phi_t - \frac{2(N-1)\theta}{N}\right)\sum_{i=1}^{N} \| e_i(t)\|^2. \quad (5.61)$$

Consequently, it is easy to verify that the trajectories enter the boundary

$$\frac{1}{N}\mathbf{E}\left(\sum_{i=1}^{N} \| x_i(t) - \bar{x}\|^2\right) = \epsilon_t$$

in a finite time

$$t \leq -\frac{N}{2\left(a - \frac{2(N-1)\theta}{N}\right)} \ln\left(\frac{N\epsilon_t}{2V(0)}\right).$$

This completes the proof. □

Remark 5.24 The bound of the swarm, $\epsilon_t = \frac{bN}{2\left(a\phi_t - \frac{2(N-1)\theta}{N}\right)}$, depends on the size N. If N increases, the bound ϵ increases, which means that the boundary of cohesion increases due to the large number of agents. This is consistent with the biological phenomena, thanks to the particular attraction/repulsion function $g_r(\cdot)$ in (5.54).

Remark 5.25 The above analysis can still be useful even if the network is not connected. Suppose that the graph has k connected components of orders $n_1, n_2, \ldots, n_k$, where $n_1 + n_2 + \ldots + n_k = N$. Let L_i be the Laplacian matrix of component i of order n_i, $i = 1, 2, \ldots, k$. All the results can be used to study the swarm cohesion of the component i by replacing the Laplacian matrix L with L_i, for each $i = 1, 2, \ldots, k$. Detailed analysis is omitted here.

5.6 Simulation Examples

Example 1. Consider the following stochastic switched swarm model:

$$dx_i(t) = \left[f_\gamma(x_i(t)) + \sum_{j=1,\ j\neq i}^{N} g_\gamma(x_i(t) - x_j(t)) \right] dt + v_i(t)dv_i,$$

where $i = 1, 2, \ldots, 20$, $\gamma = 1, 2$, g_γ is shown in Fig. 5.1 with $a = 1$, $b = 20$, and $c = 0.2$, $v_i(t) = 0.02$, $f_1(y) = 4(1, 1)^T$, and $f_2(y) = 6(1, -1)^T$.

When $\gamma = 1$, $\mathbf{E}d\overline{x} = 4(1, 1)^T dt$, and when $\gamma = 2$, $\mathbf{E}d\overline{x} = 6(1, -1)^T dt$. Suppose that there is a triangular obstacle between the starting point and the target point. Simulation shows that the 20 agents first move in the direction of the 45 degree line ($\gamma = 1$), and after reaching a middle point move in the direction of the −45 degree line ($\gamma = 2$). Finally, they reach the target as shown in Fig. 5.3.

Hence, the proposed switched swarm model (5.29) is effective as the agents can change their directions at the middle point. By Theorem 5.10, cohesion of the swarm can be reached, as verified in Fig. 5.3. Note that $\theta = 0$, and the estimated bound of cohesion is $\sqrt{\epsilon} = 3.836$, which is much larger than the actual value.

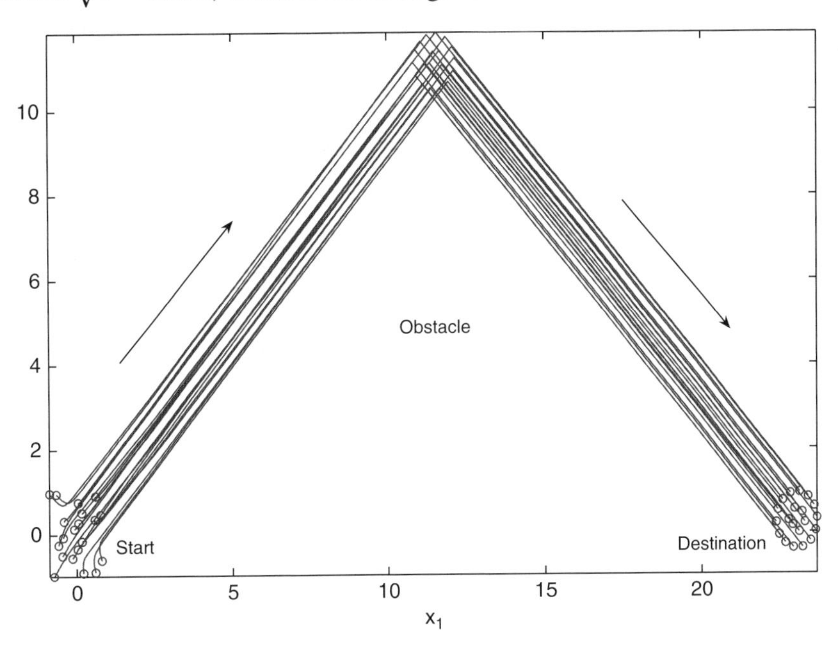

Figure 5.3 Agents' trajectories on the $x - y$ plan in the stochastic switched swarm model of Example 1

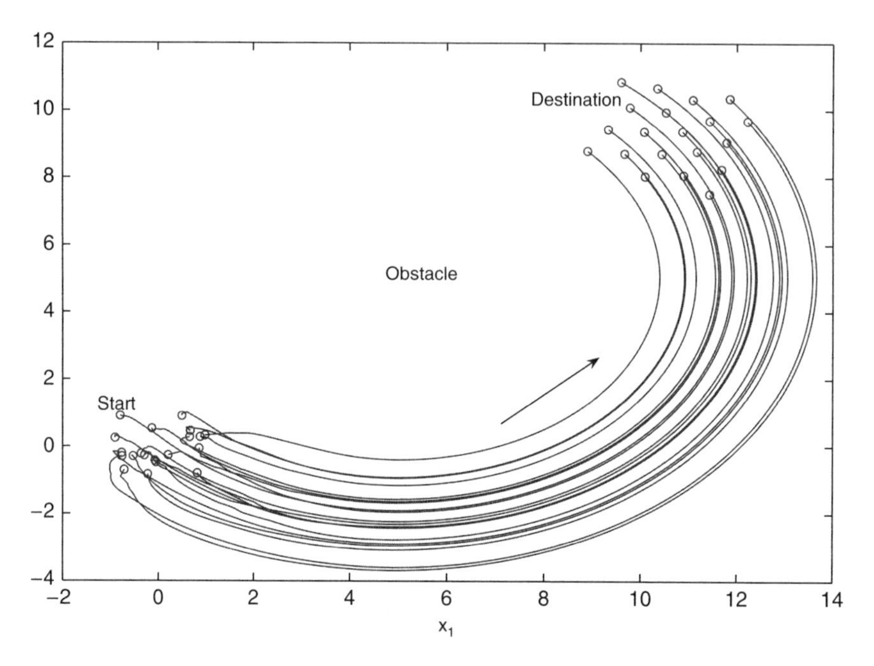

Figure 5.4 Agents' trajectories on the $x-y$ plan in the stochastic swarm model of Example 2 with a limited sensing range

Example 2. Consider the following stochastic swarm model with a limited sensing range:

$$dx_i(t) = \left[f(x_i(t)) + \sum_{j=1,\ j\neq i}^{N} h(x_i(t) - x_j(t)) \right] dt + v_i(t) dv_i,$$

where $i = 1, 2, \ldots, 20$, h is shown in Fig. 5.2 with $a = 1$, $b = 20$, $c = 0.2$ and $r = 1$, $v_i(t) = 0.01$, and $f(y) = 0.3 \begin{pmatrix} 0 & -1 \\ 1 & 0 \end{pmatrix} \left[y - \begin{pmatrix} 5 \\ 5 \end{pmatrix} \right]$.

Here, $\mathbf{E} d\overline{x} = 0.3 \begin{pmatrix} 0 & -1 \\ 1 & 0 \end{pmatrix} \left[\overline{x} - \begin{pmatrix} 5 \\ 5 \end{pmatrix} \right] dt$, which is a periodic orbit. Suppose that there is a semicircular obstacle between the starting point and the target, and an agent can only move to the target along that orbit.

By Theorem 5.21, cohesion of the swarm can be reached, as verified by the simulation result shown in Fig. 5.4. Note that $\theta = 0.3$, and $\phi_t = 0.66135$, so condition (5.46) is satisfied.

5.7 Notes

As evidenced by the fast-growing literature on complex swarming behaviors, models and corresponding analysis are in urgent need for gaining insight into biological collective behaviors in order to guide novel design of distributed coordination rules for engineering multi-agent systems. In this chapter, the stability of a continuous-time swarm model has been investigated. It was shown that, under some mild conditions, all agents of the swarm can reach cohesion in a finite time. In addition, by incorporating stochastic noise and switched topologies, more realistic swarm models have been studied; and the bounds of cohesion have been derived based on the parameters of the model. Furthermore, swarms with changing topologies and unbounded repulsions have been studied by nonsmooth analysis, where the sensing range of each agent is limited but the possibility of collision between nearby agents can be high. In future work, different types of attraction/repulsion interactions and the effects of different communication topologies on the cohesion of swarms should be further studied.

6

Distributed Leader-Follower Flocking Control

Using tools from algebraic graph theory and nonsmooth analysis in combination with ideas of collective potential functions, velocity consensus and navigation feedback, a distributed leader-follower algorithm for multi-agent dynamical systems with time-varying velocities is developed in this chapter, where each agent is governed by second-order dynamics [150]. The distributed leader-follower algorithm is applicable to the situation where the group has one virtual leader with time-varying velocity. For each agent, this algorithm consists of four terms: the first term is the self nonlinear dynamics which determines the final time-varying velocity; the second term is determined by the gradient of the collective potential between this agent and all of its neighbors; the third term is the velocity consensus term; and the fourth term is the navigation feedback term of the leader. To avoid an unpractical assumption that the informed agents can sense all the states of the leader, the new distributed algorithm is developed by making use of observer-based pinning navigation feedback. In this case, each informed agent has only partial information about the leader, yet the velocity of the whole group can still converge to that of the leader and the centroid of those informed agents.

6.1 Preliminaries

To keep this chapter self-sustained, some basic concepts and results in potential function, consensus, navigation feedback control, and nonsmooth analysis, are briefly introduced.

Distributed Cooperative Control of Multi-agent Systems, First Edition.
Wenwu Yu, Guanghui Wen, Guanrong Chen, and Jinde Cao.

6.1.1 Model Formulation

Suppose that there is a leader, which contains the motion information that the whole group of N agents need to follow. The leader moves according to the following dynamical model:

$$\begin{aligned} \dot{r}_o(t) &= v_o, \\ \dot{v}_o(t) &= f(r_o, v_o), \end{aligned} \tag{6.1}$$

where $r_o \in R^n$ and $v_o \in R^n$ are its position vector and velocity vector, respectively, and $f(r_o, v_o)$ is the input vector which governs the dynamics of the leader.

Consider the dynamics of a multi-agent system:

$$\begin{aligned} \dot{r}_i(t) &= v_i, \\ \dot{v}_i(t) &= u_i, \end{aligned} \tag{6.2}$$

where $r_i \in R^n$ is the position vector, $v_i \in R^n$ is the corresponding velocity vector, and $u_i \in R^n$ is the control input, for $i = 1, 2, \ldots, N$. Let $r = (r_1^T, r_2^T, \ldots, r_N^T)^T$, $v = (v_1^T, v_2^T, \ldots, v_N^T)^T$, and $u = (u_1^T, u_2^T, \ldots, u_N^T)^T$.

In distributed flocking algorithms, the following control input is of particular interest [82, 108, 110, 111, 112]:

$$u_i(t) = f(r_i, v_i) - \nabla_{r_i} V(r) + c \sum_{j \in \mathcal{N}_i} a_{ij}(\| r_j - r_i \|)(v_j - v_i) + \widetilde{f}_i, \tag{6.3}$$

where the first term is a nonlinear dynamical term, the second term is a gradient-based term, V is the collective potential function to be defined, the third term is the velocity consensus term, $\widetilde{f}_i$ is the navigation feedback based on information about the leader, $\mathcal{N}_i = \{j \in \mathcal{V} : \| r_j - r_i \| \leq r_s, j \neq i\}$ denotes the neighbors of agent i, r_s is the interaction range, and c is the coupling strength of velocity. Here, if $j \in \mathcal{N}_i$, then $a_{ij}(\| r_j - r_i \|) > 0$; otherwise, $a_{ij}(\| r_j - r_i \|) = 0$.

The collective potential function $V(r)$ is a nonnegative function with additional properties that are related to the overall geometric shape and graphical connectivity of system (6.2). In [82], a smooth collective potential function is given. It was pointed out [82] that the local minimum of $V(r)$ is an α-lattice and vice versa, which is responsible for collision avoidance and cohesion in the group. However, we will show that an α-lattice is a local minimum of $V(r)$ but the local minimum is not necessarily an α-lattice. In this chapter, the following potential function is considered.

Definition 6.1 The potential function is defined to be

$$V(r) = \frac{1}{2} \sum_{i} \sum_{j \neq i} \psi(\| r_{ij} \|), \tag{6.4}$$

where the function ψ is a continuously differentiable nonnegative function of the distance $\| r_{ij} \|$ between nodes i and j except at $\| r_{ij} \| = r_s$, such that

1. $\psi(\| r_{ij} \|)$ reaches its maximum at $\| r_{ij} \| = 0$ and attains its unique minimum at d;
2. ψ is nonincreasing when $\| r_{ij} \| \in [0, d]$, nondecreasing when $\| r_{ij} \| \in [d, r_s]$, and constant when $\| r_{ij} \| \in [r_s, \infty]$;
3. ψ is increasing and continuous at $\| r_{ij} \| = r_s$.

In this chapter, the following gradient-based term is considered:

$$-\nabla_{r_i} V(r) = \sum_{j \in \mathcal{N}_i} \phi(\| r_j - r_i \|) \frac{r_j - r_i}{\| r_j - r_i \|}, \tag{6.5}$$

where

$$\phi(\| r_j - r_i \|) = \begin{cases} \psi'((\| r_j - r_i \|)) & \| r_j - r_i \| < r_s, \\ \lim\limits_{\| r_j - r_i \| \to r_s^-} \psi'((\| r_j - r_i \|)) & \| r_j - r_i \| = r_s, \\ 0 & \| r_j - r_i \| > r_s. \end{cases}$$

Let $l_{ij}(\| r_j - r_i \|) = -a_{ij}(\| r_j - r_i \|)$, for $i \neq j$, $l_{ii}(r) = \sum_{j=1,\, j \neq i} a_{ij}(\| r_j - r_i \|)$, $A \otimes B$ denote the Kronecker product [52] of matrices A and B, I_n be the n-dimensional identity matrix, $\mathbf{1}_N$ be the N-dimensional column vector with all entries being 1, and $\bar{x} = \frac{1}{N} \sum\limits_{j=1}^{N} x_j$ and $\bar{v} = \frac{1}{N} \sum\limits_{j=1}^{N} v_j$ be the center position and the average velocity of the group, respectively.

In many cases, it is literally impossible for all agents to obtain the information of the leader. To reduce the number of informed agents, some local feedback injections are applied to only a small percentage of agents, which is known as pinning feedback [23, 151]. Here, the pinning strategy is applied to a small fraction δ $(0 < \delta < 1)$ of the agents. Without loss of generality, assume the first $l = \lfloor \delta N \rfloor$ agents can sense the leader and get its information. Moreover, in the real situations, it is impractical to assume that an informed agent can observe all the state components of the leader. Thus, assume that the first $l = \lfloor \delta N \rfloor$ agents can only measure partial state components of the leader by

$$\begin{aligned} x_i(t) &= H_i r_o(t), \ i = 1, 2, \ldots, l, \\ \widetilde{x}_i(t) &= \widetilde{H}_i v_o(t), \ i = 1, 2, \ldots, l, \end{aligned} \tag{6.6}$$

where $x_i(t) \in R^m$ and $\widetilde{x}_i(t) \in R^m$ are the measurements of agent i by observing the leader to get the information about $r_o(t)$ and $v_o(t)$ with $H_i \in R^{m \times n}$ and $\widetilde{H}_i \in R^{m \times n}$ for $i = 1, 2, \ldots, l$. The designed navigation feedback is described by

$$\begin{aligned} \widetilde{f}_i &= -D_i(H_i r_i - x_i) - \widetilde{D}_i(\widetilde{H}_i v_i - \widetilde{x}_i), & i &= 1, 2, \ldots, l, \\ \widetilde{f}_i &= 0, & i &= l+1, l+2, \ldots, N, \end{aligned} \tag{6.7}$$

where $D_i \in R^{n \times m}$ and $\widetilde{D}_i \in R^{n \times m}$ are the feedback control gain matrices.

In order to derive the main results, the following assumption is needed.

Assumption 6.2 *For all $x, y, r, s \in R^n$, there exist constants θ_1 and θ_2 such that*

$$(y-s)^T(f(x,y)-f(r,s)) \leq \theta_1(y-s)^T(x-r)+\theta_2\ \|y-s\|^2.$$

6.1.2 Nonsmooth Analysis

Due to the gradient-based term and the velocity consensus term in (6.3), the right-hand side of system (6.2) is discontinuous at r_s, so one cannot proceed by using classical methods (continuously differential solutions). In addition, the collective potential function in (6.4) is not differentiable. Therefore, nonsmmoth analysis and differential inclusion [6, 25, 38, 69] will be applied, which were studied in Chapter 5. For more detail, see the recent tutorial article [26] and the references therein.

Consider the following differential equation, in which the right-hand side may be discontinuous:

$$\dot{x} = F(x), \tag{6.8}$$

where $F : R^m \to R^m$ is measurable. For a discontinuous vector field $F(x)$, the existence of a continuously differentiable solution is not guaranteed. Therefore, a generalized solution, i.e., the Filippov solution is introduced in Definition 5.12.

A vector function $x(\cdot)$ is a Filippov solution of system (6.8) on $[t_0, t_1]$, which is absolutely continuous on $[t_0, t_1]$ and satisfies the following differential inclusion:

$$\dot{x}(t) \in \mathcal{F}\,(f(x)), a.e.t \in [t_0, t_1].$$

Here, $\mathcal{F}\,(f(x))$ is a set-valued map defined by

$$\mathcal{F}\,(f(x)) \equiv \bigcap_{\delta>0}\bigcap_{\mu(\overline{N})=0} \overline{co}[\phi(B(x,\delta)-\overline{N})],$$

where $\overline{co}(E)$ is the closure of the convex hull of set E, $B(x,\delta) = \{y : \|y-x\| \leq \delta\}$, and $\mu(\overline{N})$ is the Lebesgue measure of set $\overline{N}$. A equivalent definition is given by [87]: there exists N_f satisfying $\mu N_f = 0$, such that $\forall N_g \subset R^m$ with $\mu(N_g) = 0$,

$$\mathcal{F}\,(f(x)) \equiv \overline{co}\{\ \lim F(x_i)|x_i \to x, x \notin N_f \cup N_g\}.$$

By the calculus for computing Filippov's differential inclusion [87], the concept of Filippov solution in velocity consensus term is extended to the following:

$$c\sum_{j\in\mathcal{N}_i} a_{ij}(\|r_j-r_i\|)(v_j-v_i) \in c\sum_{j\in\mathcal{N}_i} \mathcal{F}\,(a_{ij}(\|r_{ij}(t)\|))(v_j-v_i), a.e.t, \tag{6.9}$$

where $\mathcal{F}\,(a_{ij}(\|r_{ij}(t)\|)) = \begin{cases} 1 & \|r_{ij}(t)\| < r_s \\ 0 & \|r_{ij}(t)\| > r_s \\ [0,1] & \|r_i(t)-r_j(t)\| = r_s \end{cases}$ for $i \neq j$.

Since the collective potential function in (6.4) is not differentiable, a generalized derivative is defined as follows.

Definition 6.3 (Generalized directional derivative [25]) Let g be Lipschitz near a given point $x \in Y$ and let w be any vector in X. The generalized directional derivative of g at x in the direction w, denoted by $g^0(x; w)$, is defined as

$$g^0(x; w) \equiv \lim_{y \to x, t \downarrow 0} \sup \frac{g(y + tw) - g(y)}{t}.$$

Note that this definition does not presuppose the existence of any limit, and it reduces to the classical directional derivative in the case where the function is differentiable.

Definition 6.4 (Generalized gradient [25]) The generalized gradient of g at x, denoted by $\partial g(x)$, is the subset of X^* given by

$$\partial g(x) = \{\zeta \in X^* | g^0(x; w) \geq < \zeta, w >, \forall w \in X\}.$$

The generalized gradient has the following basic properties:

Lemma 6.5 *[25] Let g be Lipschitz near x. Then, for every w in X, one has*

$$g^0(x; w) = \max\{ < \zeta, w > \mid \zeta \in \partial g(x)\}.$$

An additional condition to turn the inclusions to equalities is given below.

Definition 6.6 [25] A function g is said to be regular at x, if

(i) for all w, the classical one-sided directional derivative $g'(x; w)$ exists;
(ii) for all w, $g'(x; w) = g^0(x; w)$.

Lemma 6.7 *[25] (Calculus for generalized gradients of regular functions)*

(i) Let g_i $(i = 1, 2, \dots, m)$ be a finite family of regular functions, each of which is Lipschitz near x, and assume that for any nonnegative scalars s_i,

$$\partial\left(\sum_{i=1}^{m} s_i g_i\right)(x) = \sum_{i=1}^{m} s_i \partial g_i(x).$$

(ii) If g is strictly differentiable at x, then g is regular at x.
(iii) If g_i $(i = 1, 2, \dots, m)$ is a finite family of regular functions, each of which is regular at x, then for any nonnegative scalars s_i, $\sum_{i=1}^{m} s_i g_i(x)$ is regular.

Next, let

$$V(r) = \frac{1}{2} \sum_i \sum_{j \neq i} V_{ij}(\| r_{ij} \|), \tag{6.10}$$

where $V_{ij}(\| r_{ij} \|) = \psi(\| r_{ij} \|)$.

Lemma 6.8 *The function $-V_{ij}$ is regular everywhere in its domain.*

Proof. From the definition of V in Definition 6.1, it is clear that $-V_{ij}$ is strictly differentiable at $\| r_{ij} \| \neq r_s$. By Lemma 6.7, the function $-V_{ij}$ is regular at $\| r_{ij} \| \neq r_s$. In order to prove the regularity of $-V_{ij}$ in its domain, one only needs to show that $-V_{ij}$ is regular at r_s. In view of Definition 6.6, we need to establish the equality between the generalized directional derivative and the classical one-sided directional derivative of $-V_{ij}$ at r_s for any direction w.

The classical directional derivative of $-V_{ij}$ at r_s is given by

$$-V'_{ij}(r_s; w) = \lim_{t \downarrow 0} \frac{-V_{ij}(r_s + tw) - (-V_{ij}(r_s))}{t}.$$

If $w \geq 0$, then

$$-V'_{ij}(r_s; w) = \lim_{t \downarrow 0} \frac{V_{ij}(r_s) - V_{ij}(r_s + tw)}{t} = 0.$$

If $w < 0$, then

$$-V'_{ij}(r_s; w) = \lim_{t \downarrow 0} \frac{V_{ij}(r_s) - V_{ij}(r_s + tw)}{t} \equiv \kappa > 0.$$

The above inequalities are obtained based on the property of V in Definition 6.1. For the generalized directional derivative, one has the same two cases as follows.

If $w \geq 0$, then

$$-V^0_{ij}(r_s; w) = \lim_{y \to r_s, t \downarrow 0} \sup \frac{V_{ij}(y) - V_{ij}(y + tw)}{t} = 0.$$

If $w < 0$, then

$$-V^0_{ij}(r_s; w) = \lim_{y \to r_s, t \downarrow 0} \sup \frac{V_{ij}(y) - V_{ij}(y + tw)}{t} = \kappa > 0.$$

Therefore, for any direction w, one has $-V'_{ij}(r_s; w) = -V^0_{ij}(r_s; w)$, so $-V_{ij}$ is regular at r_s. The proof is completed. □

Remark 6.9 In [110, 111], it was claimed that V_{ij} is regular. However, it is shown above that $-V_{ij}$ should be regular instead.

To proceed further, a modified chain rule is first described in the following.

Lemma 6.10 *Let $x(\cdot)$ be a Filippov solution of $\dot{x} = F(x)$ on an interval containing t, and let $-V : R^m \to R$ be a Lipschitz and regular function. Then, $V(x(t))$ is absolutely continuous and, in addition, $\frac{d}{dt}V(x(t))$ exists almost everywhere:*

$$\frac{d}{dt}V(x(t)) \in \dot{\tilde{V}}(x), a.e.t, \tag{6.11}$$

where

$$\dot{\tilde{V}}(x) = \bigcap_{\xi \in \partial V(x(t))} \xi^T \mathscr{F}(f(x)). \tag{6.12}$$

Proof. By Lemmas 6.7 and 6.8, the function $-V$ is regular everywhere in its domain. Since V is Lipschitz and $x(\cdot)$ is absolutely continuous, $V(x(t))$ is absolutely continuous [106]. From the argument of [78], it follows that $V(x(t))$ is differentiable almost everywhere. Since V is Lipschitz and at a point where $V(x(t))$ and $x(t)$ are both differentiable, one has

$$\frac{d}{dt}V(x(t)) = \lim_{h\to 0}\frac{V(x(t+h)) - V(x(t))}{h} = \lim_{h\to 0}\frac{V(x(t) + h\dot{x}(t)) - V(x(t))}{h}.$$

Because of the regularity of $-V$ and Lemmas 6.8, by letting h tend to 0 from the right, one obtains

$$\begin{aligned}
\frac{d}{dt}V(x(t)) &= \lim_{h\to 0^+}\frac{V(x(t) + h\dot{x}(t)) - V(x(t))}{h}\\
&= -\lim_{h\to 0^+}\frac{(-V(x(t) + h\dot{x}(t))) - (-V(x(t)))}{h}\\
&= -(-V'(x(t);\dot{x}(t))) = -(-V^0(x(t);\dot{x}(t)))\\
&= -\max\{<\zeta, \dot{x}(t)> \mid \zeta \in \partial(-V)(x)\}\\
&= -\max\{<-\xi, \dot{x}(t)> \mid \xi \in \partial V(x)\}\\
&= \min\{<\xi, \dot{x}(t)> \mid \xi \in \partial V(x)\}.
\end{aligned}$$

Similarly, by letting h tend to 0 from the left, one obtains

$$\begin{aligned}
\frac{d}{dt}V(x(t)) &= \lim_{h\to 0^-}\frac{V(x(t) + h\dot{x}(t)) - V(x(t))}{h}\\
&= \lim_{h1\to 0^+}\frac{-V(x(t) - h1\dot{x}(t)) - (-V(x(t)))}{h1}\\
&= -V'(x(t);-\dot{x}(t)) = -V^0(x(t);-\dot{x}(t))\\
&= \max\{<\zeta, -\dot{x}(t)> \mid \zeta \in \partial(-V)(x)\}\\
&= \max\{<\xi, \dot{x}(t)> \mid \xi \in \partial V(x)\}.
\end{aligned}$$

Thus, one has

$$\frac{d}{dt}V(x(t)) = \{<\xi, \dot{x}(t)> \mid \forall \xi \in \partial V(x)\}.$$

Since $x(\cdot)$ is a Filippov solution satisfying

$$\dot{x}(t) \in \mathscr{F}(f(x)), a.e.t.$$

It follows that V is almost differentiable everywhere, and $\frac{d}{dt}V(x(t)) = \xi^T\eta$ for all $\xi \in \partial V(x(t))$ and some $\eta \in \mathscr{F}(f(x))$, equivalently,

$$\frac{d}{dt}V(x(t)) \in \bigcap_{\xi \in \partial V(x(t))} \xi^T \mathscr{F}(f(x)), a.e.t,$$

This completes the proof. □

Remark 6.11 In [106], only if V is regular, the chain rule is obtained. In this chapter, however, based on the regularity of $-V$, a modified chain rule can still be derived as shown above, and moreover the Lyapunov and LaSalle theorems can also be proved by a similar approach to that in [106].

6.2 Distributed Leader-Follower Control with Pinning Observers

In this section, the leader-follower control protocol of the multi-agent system (6.2) with pinning observer-based navigation feedback (6.7) is studied.

Let $\hat{r}_i = r_i - r_o$ and $\hat{v}_i = v_i - v_o$ represent the relative position and velocity vectors to the leader, $\hat{r}_{ij} = \hat{r}_i - \hat{r}_j = r_{ij}$, $\hat{v}_{ij} = \hat{v}_i - \hat{v}_j = v_{ij}$, $\hat{r} = (\hat{r}_1^T, \hat{r}_2^T, \ldots, \hat{r}_N^T)^T$, and $\hat{v} = (\hat{v}_1^T, \hat{v}_2^T, \ldots, \hat{v}_N^T)^T$. By Definition 5.12 and the calculation of Filippov's differential inclusion [87], the Filippov solution is extended to the following:

$$\begin{aligned}
\dot{\hat{r}}_i(t) &= \hat{v}_i,\\
\dot{\hat{v}}_i(t) &\in f(r_i, v_i) - f(r_o, v_o) - \hat{D}_i H_i \hat{r}_i - \hat{\tilde{D}}_i \tilde{H}_i \hat{v}_i + \sum_{j \in \mathcal{N}_i} \mathscr{F}(\phi(\|\hat{r}_{ji}\|)) \frac{\hat{r}_{ji}}{\|\hat{r}_{ji}\|}\\
&\quad + c \sum_{j \in \mathcal{N}_i} \mathscr{F}(a_{ij}(\|\hat{r}_{ji}\|))(\hat{v}_{ji}), a.e.t,
\end{aligned} \tag{6.13}$$

or equivalently,

$$\begin{aligned}
\dot{\hat{v}}_i(t) &= f(r_i, v_i) - f(r_o, v_o) - \hat{D}_i H_i \hat{r}_i - \hat{\tilde{D}}_i \tilde{H}_i \hat{v}_i + \sum_{j \in \mathcal{N}_i} \hat{\phi}_{ij} \frac{\hat{r}_{ji}}{\|\hat{r}_{ji}\|}\\
&\quad + c \sum_{j \in \mathcal{N}_i} \hat{a}_{ij} \hat{v}_{ji}, a.e.t,
\end{aligned} \tag{6.14}$$

where $\hat{D}_i = D_i$ and $\hat{\tilde{D}}_i = \tilde{D}_i$ for $i = 1, 2, \ldots, l$, $\hat{D}_i = \hat{\tilde{D}}_i = 0$ for $i = l+1, l+2, \ldots, N$,

$$\hat{\phi}_{ij} = \begin{cases} \phi(\|\hat{r}_{ij}(t)\|) & \|\hat{r}_{ij}(t)\| \neq r_s \\ \in [0, \phi(\|r_s^-\|)] & \|\hat{r}_{ij}(t)\| = r_s \end{cases} \text{ and } \hat{a}_{ij} = \begin{cases} 1 & \|\hat{r}_{ij}(t)\| < r_s \\ 0 & \|\hat{r}_{ij}(t)\| > r_s \\ \in [0, 1] & \|\hat{r}_{ij}(t)\| = r_s \end{cases} \text{ for } i \neq j.$$

Let $(B)^\eta = \frac{B+B^T}{2}$.

Definition 6.12 The position state component j $(1 \le j \le n)$ of the leader is said to be observable by an agent i $(1 \le i \le l)$, if there is a gain matrix D_i and a positive constant ϵ_j, such that

$$-s(D_iH_i)^{\eta}\widetilde{s} \le -\epsilon_j s_j \widetilde{s}_j,$$

for all $s = (s_1, s_2, \dots, s_n)^T \in R^n$ and $\widetilde{s} = (\widetilde{s}_1, \widetilde{s}_2, \dots, \widetilde{s}_n)^T \in R^n$.

Definition 6.13 The velocity state component j $(1 \le j \le n)$ of the leader is said to be observable by an agent i $(1 \le i \le l)$, if there is a gain matrix $\widetilde{D}_i$ and a positive constant $\widetilde{\epsilon}_j$, such that

$$-\widetilde{s}^T(\widetilde{D}_i\widetilde{H}_i)^{\eta}\widetilde{s} \le -\widetilde{\epsilon}_j\widetilde{s}_j^2,$$

for all $\widetilde{s} = (\widetilde{s}_1, \widetilde{s}_2, \dots, \widetilde{s}_n)^T \in R^n$.

Definition 6.14 The position (velocity) state component j $(1 \le j \le n)$ of the leader is said to be observable by a group of N agents if each of its position (velocity) state component is observable by at least one agent k $(1 \le k \le l)$.

Definition 6.15 The position (velocity) state of the leader is said to be observable by a group of N agents, if each of its position (velocity) state component is observable by the group of N agents.

Assumption 6.16 *Suppose that each network in $\mathscr{F}(A)$ is connected at all times.*

Assumption 6.17 *Suppose that the position and velocity of the leader are both observable by a group of N agents and, without loss of generality, suppose that this position is observable by agents $i_1, i_2, \dots, i_p$ $(i_p \le l)$ and this velocity is observable by agents $j_1, j_2, \dots, j_q$ $(j_q \le l)$. Then, there exist matrices D_{i_k} and $\widetilde{D}_{j_{\widetilde{k}}}$, $k = 1, 2, \dots, p$, $\widetilde{k} = 1, 2, \dots, q$, such that*

$$-\sum_{k=1}^{p} \widehat{r}_{i_k}^T(t)(D_{i_k}H_{i_k})^{\eta}\widehat{v}_{i_k} \le -\sum_{m=1}^{n}\sum_{k\in\mathscr{M}_m} \epsilon_{mk}\widehat{r}_{i_k m}\widehat{v}_{i_k m}, \tag{6.15}$$

and

$$-\sum_{\widetilde{k}=1}^{q} \widehat{v}_{j_{\widetilde{k}}}^T(t)(\widetilde{D}_{j_{\widetilde{k}}}\widetilde{H}_{j_{\widetilde{k}}})^{\eta}\widehat{v}_{j_{\widetilde{k}}} \le -\sum_{\widetilde{m}=1}^{n}\sum_{\widetilde{k}\in\widetilde{\mathscr{M}}_{\widetilde{m}}} \widetilde{\epsilon}_{\widetilde{m}\widetilde{k}}\widehat{v}_{j_{\widetilde{k}}\widetilde{m}}^2, \tag{6.16}$$

where the position state component m is observable by agents i_k $(1 \le k \le p)$ and the velocity state component $\widetilde{m}$ is observable by agents $j_{\widetilde{k}}$ $(1 \le \widetilde{k} \le q)$, respectively. Here, $\mathscr{M}_m$ and $\widetilde{\mathscr{M}}_{\widetilde{m}}$ denote the sets of all agents that can observe the position state component m and the velocity state component $\widetilde{m}$. If any position or velocity state component *of the leader cannot be observed by agent k, then one can simply let* $D_k = 0$ *or* $\widetilde{D}_k = 0$ *for* $1 \leq k \leq l$.

Define the sum of the total artificial potential function and the total relative kinetic energy function as follows:

$$\overline{U}(\widehat{r}, \widehat{v}) = V(\widehat{r}) + W(\widehat{v}) + \overline{W}(\widehat{r}), \tag{6.17}$$

where $V(\widehat{r}) = \frac{1}{2}\sum_i \sum_{j \neq i} \psi(\|\widehat{r}_{ij}\|)$, $W(\widehat{v}) = \frac{1}{2}\widehat{v}^T\widehat{v}$, and $\overline{W}(\widehat{r}) = \frac{1}{2}\sum_{m=1}^{n}\sum_{k \in \mathscr{M}_m} \epsilon_{mk}\widehat{r}^2_{i_k m}$. Let $U(\widehat{r}, \widehat{v}) = V(\widehat{r}) + W(\widehat{v})$.

Theorem 6.18 *Consider a group of N agents, which satisfies Assumptions 6.2 and 6.17. If*

$$\theta_2 I_N - \Xi_{\widetilde{m}} - c\mathscr{F}(\widehat{L}) < 0, \tag{6.18}$$

for all $\widehat{L} \in \mathscr{F}(L)$ *and* $1 \leq \widetilde{m} \leq n$, *where* $\Xi_{\widetilde{m}} = \mathrm{diag}(0, \ldots, 0, \underbrace{\epsilon_{\widetilde{m}\widetilde{k}_1}}_{j_{\widetilde{k}_1}}, 0, \ldots, 0, \underbrace{\epsilon_{\widetilde{m}\widetilde{k}_2}}_{j_{\widetilde{k}_2}}, 0, \ldots, 0, \underbrace{\epsilon_{\widetilde{m}\widetilde{k}_r}}_{j_{\widetilde{k}_r}}, 0) \in R^{N\times N}$, $\widetilde{k}_1, \ldots, \widetilde{k}_r \in \widetilde{\mathscr{M}_{\widetilde{m}}}$, *then*

- *(i) The states of dynamical system (6.13) converge to an equilibrium* $(r^*, 0)$, *where* r^* *is a local minimum of* $\overline{U}(\widehat{r}, 0)$ *in (6.17).*
- *(ii) All the agents move with the same velocity as the leader asymptotically.*
- *(iii) If the initial artificial potential energy is less than* $(k+1)\psi(0)$ *for some* $k \geq 0$, *then at most k agents may collide.*
- *(iv) If f is independent of the position term,*

$$\sum_{k=1}^{p} D_{i_k} H_{i_k} \widehat{r}_{i_k} \to 0, t \to \infty. \tag{6.19}$$

In addition, if for any $k \in \mathscr{M}_m$, $D_{i_k m} H_{i_k m}$ *are all equal, then*

$$\frac{1}{\mu(m)} \sum_{k \in \mathscr{M}_m} r_{i_k m}(t) \to r_{om}(t), t \to \infty,$$

where $\mu(m)$ *is the number of agents in* $\mathscr{M}_m$.

Proof. Since $-W$ and $-V$ are regular functions in Lemma 6.8, one knows that $-(W + V)$ is regular. Noting the following fact:

$$\frac{\partial \psi(\|\widehat{r}_{ij}\|)}{\partial \widehat{r}_{ij}} = \frac{\partial \psi(\|\widehat{r}_{ij}\|)}{\partial \widehat{r}_i} = -\frac{\partial \psi(\|\widehat{r}_{ij}\|)}{\partial \widehat{r}_j}, \tag{6.20}$$

and using Chain Rule II in [25], one has

$$\frac{\partial\psi(\|\hat{r}_{ij}\|)}{\partial\hat{r}_i} \subset \frac{\partial\psi(\|\hat{r}_{ij}\|)}{\partial\,\|\hat{r}_{ij}\|}\frac{\partial\,\|\hat{r}_{ij}\|}{\partial\hat{r}_i} = \begin{cases} \phi(\|\hat{r}_{ij}(t)\|)\frac{\hat{r}_{ij}}{\|\hat{r}_{ij}\|}, & \|\hat{r}_{ij}(t)\| \neq r_s, \\ \left\{\chi\frac{\hat{r}_{ij}}{\|\hat{r}_{ij}\|} \;:\; \chi \in [0, \phi(\|r_s^-\|)]\right\}, & \|\hat{r}_{ij}(t)\| = r_s. \end{cases} \tag{6.21}$$

Then

$$\frac{\partial U(\hat{r},\hat{v})}{\partial\hat{r}_i} = \sum_{j\in\mathcal{N}_i}\frac{\partial\psi(\|\hat{r}_{ij}\|)}{\partial\hat{r}_i}, \tag{6.22}$$

and

$$\frac{\partial U(\hat{r},\hat{v})}{\partial\hat{v}_i} = \hat{v}_i. \tag{6.23}$$

It follows that

$$\partial U(\hat{r},\hat{v}) = \left(\left(\frac{\partial U(\hat{r},\hat{v})}{\partial\hat{r}_1}\right)^T, \dots, \left(\frac{\partial U(\hat{r},\hat{v})}{\partial\hat{r}_N}\right)^T, \hat{v}_1^T, \dots, \hat{v}_N^T\right)^T. \tag{6.24}$$

Let

$$\mathcal{F}(\dot{\hat{v}}_i(t)) \equiv f(r_i, v_i) - f(r_o, v_o) - \hat{D}_i H_i \hat{r}_i - \hat{\tilde{D}}_i \tilde{H}_i \hat{v}_i + \sum_{j\in\mathcal{N}_i}\mathcal{F}(\phi(\|\hat{r}_{ji}\|))\frac{\hat{r}_{ji}}{\|\hat{r}_{ji}\|} + c\sum_{j\in\mathcal{N}_i}\mathcal{F}(a_{ij}(\|\hat{r}_{ji}\|))(\hat{v}_j - \hat{v}_i). \tag{6.25}$$

By Lemma 6.10, one obtains

$$\frac{d}{dt}U(\hat{r},\hat{v}) \in \bigcap_{\xi\in\partial U(\hat{r},\hat{v})} \xi^T(\hat{v}_1^T, \dots, \hat{v}_N^T, \mathcal{F}^T(\dot{\hat{v}}_1(t)), \dots, \mathcal{F}^T(\dot{\hat{v}}_N(t)))^T, \; a.e.t.$$

Let $\Xi(\hat{r},\hat{v},\xi) = \xi^T(\hat{v}_1^T, \dots, \hat{v}_N^T, \mathcal{F}^T(\dot{\hat{v}}_1(t)), \dots, \mathcal{F}^T(\dot{\hat{v}}_N(t)))^T$.
If $\|\hat{r}_{ij}\| < r_s$, then by (6.20)–(6.25), one has

$$\begin{aligned}
&\Xi(\hat{r},\hat{v},\xi) \\
&= \sum_{i=1}^N \left(\frac{\partial U(\hat{r},\hat{v})}{\partial\hat{r}_i}\right)^T \hat{v}_i + \sum_{i=1}^N \mathcal{F}^T(\dot{\hat{v}}_i(t))\hat{v}_i \\
&= \sum_{i=1}^N \left(\sum_{j\in\mathcal{N}_i}\frac{\partial\psi(\|\hat{r}_{ij}\|)}{\partial\hat{r}_i}\right)^T \hat{v}_i + \sum_{i=1}^N\sum_{j\in\mathcal{N}_i}\phi(\|\hat{r}_{ji}\|)\frac{\hat{r}_{ji}^T}{\|\hat{r}_{ji}\|}\hat{v}_i
\end{aligned}$$

$$
\begin{aligned}
&+c\sum_{i=1}^{N}\sum_{j\in\mathcal{N}_i} a_{ij}(\hat{v}_j-\hat{v}_i)^T\hat{v}_i \\
&+\sum_{i=1}^{N}(f(r_i,v_i)-f(r_o,v_o)-\hat{D}_iH_i\hat{r}_i-\hat{\tilde{D}}_i\tilde{H}_i\hat{v}_i)^T\hat{v}_i \\
&=-c\hat{v}^T(L\otimes I_n)\hat{v} \\
&+\sum_{i=1}^{N}(f(r_i,v_i)-f(r_o,v_o)-\hat{D}_iH_i\hat{r}_i-\hat{\tilde{D}}_i\tilde{H}_i\hat{v}_i)^T\hat{v}_i.
\end{aligned}
\tag{6.26}
$$

If $\|\hat{r}_{ij}\|=r_s$, then

$$
\begin{aligned}
&\frac{d}{dt}U(\hat{r},\hat{v}) \\
&\in\bigcap_{\xi_i\in\partial_{\hat{r}_i}U(\hat{r},\hat{v})}\left(\sum_{i=1}^{N}\xi_i^T\hat{v}_i+\sum_{i=1}^{N}\mathscr{F}^{\,T}(\dot{\hat{v}}_i(t))\hat{v}_i\right) \\
&=\bigcap_{\xi_{ij}\in[0,\phi(\|r_s^-\|)]}\left(\sum_{i=1}^{N}\sum_{j\in\mathcal{N}_i}\xi_{ij}\frac{\hat{r}_{ij}^T}{\|\hat{r}_{ij}\|}\hat{v}_i+\sum_{i=1}^{N}\sum_{j\in\mathcal{N}_i}[0,\phi(\|r_s^-\|)]\frac{\hat{r}_{ji}^T}{\|\hat{r}_{ji}\|}\hat{v}_i\right. \\
&\left.+c\sum_{i=1}^{N}\sum_{j\in\mathcal{N}_i}\mathscr{F}(a_{ij})(\hat{v}_j-\hat{v}_i)^T\hat{v}_i\right) \\
&+\sum_{i=1}^{N}(f(r_i,v_i)-f(r_o,v_o)-\hat{D}_iH_i\hat{r}_i-\hat{\tilde{D}}_i\tilde{H}_i\hat{v}_i)^T\hat{v}_i \\
&=\bigcap_{\xi_{ij}\in[0,\phi(\|r_s^-\|)]}\left(\sum_{i=1}^{N}\sum_{j\in\mathcal{N}_i}[-\phi(\|r_s^-\|)+\xi_{ij},\xi_{ij}]\frac{\tilde{r}_{ij}^T}{\|\hat{r}_{ij}\|}\hat{v}_i\right. \\
&\left.+c\sum_{i=1}^{N}\sum_{j\in\mathcal{N}_i}\mathscr{F}(a_{ij})(\hat{v}_j-\hat{v}_i)^T\hat{v}_i\right) \\
&+\sum_{i=1}^{N}(f(r_i,v_i)-f(r_o,v_o)-\hat{D}_iH_i\hat{r}_i-\hat{\tilde{D}}_i\tilde{H}_i\hat{v}_i)^T\hat{v}_i.
\end{aligned}
\tag{6.27}
$$

Since

$$
\bigcap_{\xi_{ij}\in[0,\phi(\|r_s^-\|)]}[-\phi(\|r_s^-\|)+\xi_{ij},\xi_{ij}]=\{0\},
\tag{6.28}
$$

it follows that

$$\begin{aligned}
&\frac{d}{dt}U(\widehat{r},\widehat{v}) \\
&= c\sum_{i=1}^{N}\sum_{j\in\mathcal{N}_i}\widehat{a}_{ij}(\widehat{v}_j-\widehat{v}_i)^T\widehat{v}_i \\
&\quad+\sum_{i=1}^{N}(f(r_i,v_i)-f(r_o,v_o)-\widehat{D}_iH_i\widehat{r}_i-\widehat{\widetilde{D}}_i\widetilde{H}_i\widehat{v}_i)^T\widehat{v}_i \\
&= -c\widehat{v}^T(\widehat{L}\otimes I_n)\widehat{v}+\sum_{i=1}^{N}(f(r_i,v_i)-f(r_o,v_o)-\widehat{D}_iH_i\widehat{r}_i-\widehat{\widetilde{D}}_i\widetilde{H}_i\widehat{v}_i)^T\widehat{v}_i, \\
&\quad a.e.t.
\end{aligned} \tag{6.29}$$

where $\widehat{L}_{ij}=-\widehat{a}_{ij}$ for $i\neq j$, and $\widehat{L}_{ii}=\sum_{j\in\mathcal{N}_i}\widehat{a}_{ij}$.

If $\|\widehat{r}_{ij}\|> r_s$, similarly, by (6.26) one has

$$\begin{aligned}
\Xi(\widehat{r},\widehat{v},\xi) &= -c\widehat{v}^T(L\otimes I_n)\widehat{v} \\
&\quad+\sum_{i=1}^{N}(f(r_i,v_i)-f(r_o,v_o)-\widehat{D}_iH_i\widehat{r}_i-\widehat{\widetilde{D}}_i\widetilde{H}_i\widehat{v}_i)^T\widehat{v}_i.
\end{aligned} \tag{6.30}$$

Therefore, from (6.26), (6.29), and (6.30), one obtains

$$\begin{aligned}
&\frac{d}{dt}U(\widehat{r},\widehat{v}) \\
&= -c\widehat{v}^T(\widehat{L}\otimes I_n)\widehat{v}+\sum_{i=1}^{N}(f(r_i,v_i)-f(r_o,v_o)-\widehat{D}_iH_i\widehat{r}_i-\widehat{\widetilde{D}}_i\widetilde{H}_i\widehat{v}_i)^T\widehat{v}_i \\
&\in -c\widehat{v}^T(\mathscr{F}(L)\otimes I_n)\widehat{v}+\sum_{i=1}^{N}(f(r_i,v_i)-f(r_o,v_o)-\widehat{D}_iH_i\widehat{r}_i-\widehat{\widetilde{D}}_i\widetilde{H}_i\widehat{v}_i)^T\widehat{v}_i, \\
&\quad a.e.t.
\end{aligned} \tag{6.31}$$

By Assumptions 6.2 and 6.17, one obtains

$$\begin{aligned}
&\frac{d}{dt}\overline{U}(\widehat{r},\widehat{v}) \\
&= \sum_{i=1}^{N}\widehat{v}_i^T[f(r_i,v_i)-f(r_o,v_o)]+\sum_{i=1}^{l}\widehat{v}_i^T[-(D_iH_i)^\eta\widehat{r}_i-(\widetilde{D}_i\widetilde{H}_i)^\eta\widehat{v}_i] \\
&\quad -c\widehat{v}^T(\widehat{L}\otimes I_n)\widehat{v}+\sum_{m=1}^{n}\sum_{k\in\mathscr{M}_m}\epsilon_{mk}\widehat{r}_{i_km}\widehat{v}_{i_km}
\end{aligned}$$

$$\leq \hat{v}^T[(\theta_2 I_N - c\hat{L}) \otimes I_n]\hat{v} - \sum_{k=1}^{p} \hat{r}_{i_k}^T(t)(D_{i_k}H_{i_k})^\eta \hat{v}_{i_k} - \sum_{\tilde{k}=1}^{q} \hat{v}_{j_{\tilde{k}}}^T(t)(\tilde{D}_{j_{\tilde{k}}}\tilde{H}_{j_{\tilde{k}}})^\eta \hat{v}_{j_{\tilde{k}}}$$

$$+ \sum_{m=1}^{n} \sum_{k\in\mathcal{M}_m} c_{mk}\hat{r}_{i_k m}\hat{v}_{i_k m}$$

$$\leq \hat{v}^T[(\theta_2 I_N - c\hat{L}) \otimes I_n]\hat{v} - \sum_{\tilde{m}=1}^{n} \sum_{\tilde{k}\in\widetilde{\mathcal{M}}_{\tilde{m}}} \epsilon_{\tilde{m}\tilde{k}}\hat{v}_{j_{\tilde{k}}\tilde{m}}^2, a.e.t. \quad (6.32)$$

Let $\acute{v}_{\tilde{m}} = (\hat{v}_{1\tilde{m}}, \hat{v}_{2\tilde{m}}, \dots, \hat{v}_{N\tilde{m}})$. Then, one has

$$\frac{d}{dt}\overline{U}(\hat{r},\hat{v}) \leq \sum_{\tilde{m}=1}^{n} \acute{v}_{\tilde{m}}^T(\theta_2 I_N - c\hat{L})\acute{v}_{\tilde{m}} - \sum_{\tilde{m}=1}^{n} \acute{v}_{\tilde{m}}^T \Xi_{\tilde{m}}\acute{v}_{\tilde{m}}$$

$$\in \sum_{\tilde{m}=1}^{n} \acute{v}_{\tilde{m}}^T(\theta_2 I_N - \Xi_{\tilde{m}} - c\mathcal{F}(\hat{L}))\acute{v}_{\tilde{m}}, a.e.t. \quad (6.33)$$

From (6.17) and by Assumption 6.16, one knows that $\frac{d}{dt}\overline{U}(\hat{r},\hat{v}) \leq 0, a.e.t$, which indicates that $\hat{r}$ and $\hat{v}$ are bounded. Then, by applying Lemma 2.32, one has $\Xi_{\tilde{m}} + c\hat{L} > 0$, and it follows from LaSalle invariance principle in [26] that all the solutions converge to the largest invariant set in $S = \{(\hat{r},\hat{v}) |\ \hat{v} = 0\}$. The proof of (ii) is completed. Note that every solution of the system converges to an equilibrium point $(r^*,0)$, where r^* is a local minimum of $\overline{U}(\hat{r},0)$. This completes the proof of (ii).

Next, suppose that at least $k+1$ agents collide at time t_1. Then, one has

$$V(\hat{r}(t_1)) \geq (k+1)\psi(0).$$

Since $\overline{U}$ is absolutely continuous and nonincreasing almost everywhere, one obtains

$$V(\hat{r}(t_1)) \leq \overline{U}(\hat{r}(0),\hat{v}(0)) < (k+1)\psi(0).$$

This is a contradiction.

Finally, on the largest invariant set $S = \{(\hat{r},\hat{v}) |\ \hat{v} = 0\}$, one has

$$\dot{\hat{r}}_i(t) = 0,$$

$$\sum_{i=1}^{N} \hat{D}_i H_i \hat{r}_i = \sum_{k=1}^{p} D_{i_k}H_{i_k}\hat{r}_{i_k} = 0. \quad (6.34)$$

It follows that (6.19) is satisfied. □

Remark 11. Even if the states of the leader cannot be observed by one agent in the group, many agents can share their information with the neighbors and the position of the leader can still be followed by the other agents in a distributed way. In addition, as long as the velocity vector of the leader can be observed, all the agents in the group can move with the same velocity as the leader. Here, it is noted that the average position of the informed agents, with only partial information about the leader's position, can still converge to that of the leader as shown by (iv) of Theorem 6.18.

If the network structure is changing very slowly, Assumption 6.17 can also be relaxed but it is omitted here. In addition, even if Assumption 6.16 is not satisfied and the network is not connected, the above theoretical results can still be used to study the flocking problem for various connected components.

6.3 Simulation Examples

Suppose that the leader moves in the following periodic manner:

$$\dot{r}_o(t) = v_o,$$
$$\dot{v}_o(t) = Av_o, \tag{6.35}$$

where $r_o(t), v_o(t) \in R^2$ and $A = 0.1 \begin{pmatrix} 0 & -1 \\ 1 & 0 \end{pmatrix}$.

Consider the same multi-agent dynamical system as follows:

$$\begin{aligned}
\dot{r}_i(t) &= v_i, \\
\dot{v}_i(t) &= Av_i - D_i H_i(r_i - r_o) - \widetilde{D}_i \widetilde{H}_i(v_i - v_o) \\
&\quad + \sum_{j\in\mathcal{N}_i} a_{ij}(\| r_j - r_i \|) \left[a - b\, e^{-\frac{\|r_{ij}\|^2}{\bar{c}}} \right] (r_j - r_i) \\
&\quad + c \sum_{j\in\mathcal{N}_i} a_{ij}(\| r_j - r_i \|)(v_j - v_i), i = 1, 2, \ldots, l, \\
\dot{v}_i(t) &= f(r_i, v_i) + \sum_{j\in\mathcal{N}_i} a_{ij}(\| r_j - r_i \|) \left[a - b\, e^{-\frac{\|r_{ij}\|^2}{\bar{c}}} \right] (r_j - r_i) \\
&\quad + c \sum_{j\in\mathcal{N}_i} a_{ij}(\| r_j - r_i \|)(v_j - v_i), i = l + 1, l + 2, \ldots, N,
\end{aligned} \tag{6.36}$$

where $N = 20$, $l = 6$, $a = 1$, $b = 20$, $\bar{c} = 0.2$, $r_s = 1$, $c = 5$, and $d = \sqrt{c \ln (b/a)} = 0.7740$, $H_i = \widetilde{H}_i = \begin{pmatrix} 1 & 0 \end{pmatrix}$ for $i = 1, 2, 3$, and $H_i = \widetilde{H}_i = \begin{pmatrix} 0 & 1 \end{pmatrix}$ for $i = 4, 5, 6$. Choose $D_i = \widetilde{D}_i = \begin{pmatrix} 1 \\ 0 \end{pmatrix}$ for $i = 1, 2, 3$, and $D_i = \widetilde{D}_i = \begin{pmatrix} 0 \\ 1 \end{pmatrix}$ for $i = 4, 5, 6$. It is easy to see that three agents can observe the first position (v_{o1}) and velocity (r_{o1}) state component of the leader and the other three agents can observe the second position (v_{o2}) and velocity (r_{o2}) state component of the leader.

In the simulation, the initial positions and velocities of the 20 agents are chosen randomly from $[-1, 1] \times [-1, 1]$ and $[0, 2] \times [0, 2]$, respectively. From Theorem 6.18, it

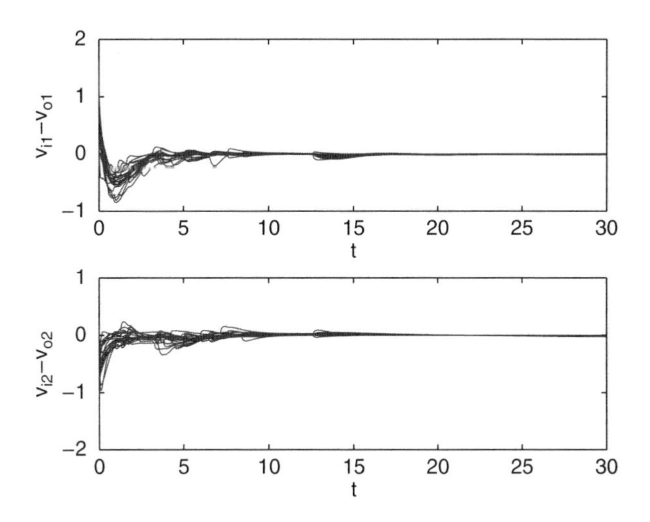

Figure 6.1 Relative velocities of all agents to the leader, $i = 1, 2, \ldots, 20$

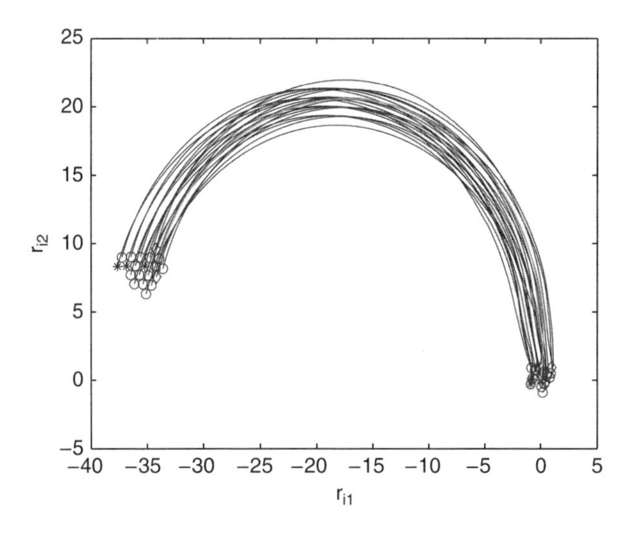

Figure 6.2 Positions of all agents, $i = 1, 2, \ldots, 20$

follows that all agents move with the same velocity, as shown in Fig 6.1. The positions of agents converge to a local minimum of $\overline{U}(r^*, 0)$ in (6.17), as illustrated in Fig 6.2. The figure in greater detail at time 30s is shown in Fig 6.3, where the circle represents the agent without navigation feedback, the diamond denotes the agent ($i = 1, 2, 3$) with pinning navigation feedback of the first state component of the leader, the star ($i = 4, 5, 6$) denotes the agent with pinning navigation feedback of the second state component of the leader, and the square is the leader. By condition (iv) in Theorem 6.18, it follows that

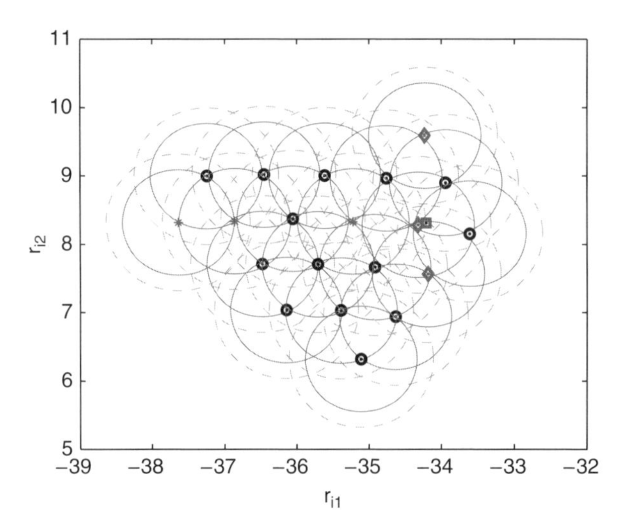

Figure 6.3 Positions of all agents at time 30s, $i = 1, 2, \ldots, 20$

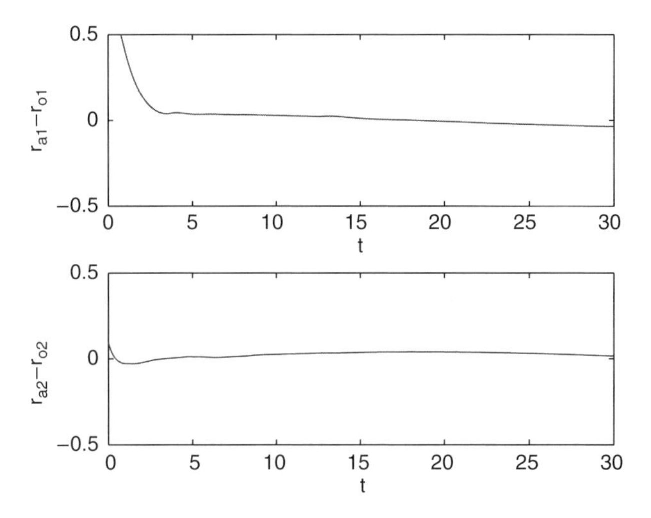

Figure 6.4 Error state between the average position of agents, with pinning observer-based navigation feedback, and the leader

$$\frac{1}{3}\sum_{i=1}^{3} r_{l1}(t) \to r_{o1}(t), \frac{1}{3}\sum_{i=4}^{6} r_{l2}(t) \to r_{o2}(t), t \to \infty,$$

with $\frac{1}{3}\sum_{i=1}^{3} r_{l1}(t) = r_{a1}(t)$ and $\frac{1}{3}\sum_{i=4}^{6} r_{l2}(t) = r_{a2}(t)$, the error state $r_a(t) - r_o(t)$ is shown in Fig (6.4).

In this example, the agents 1–3 and 4–6 have the information of the first and second state component of the leader, respectively. All agents share this information and

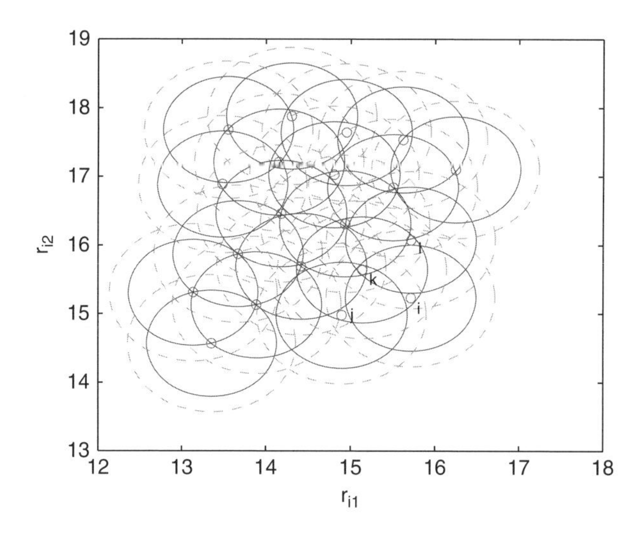

Figure 6.5 Positions of all the agents at time 15s, $i = 1, 2, \ldots, 20$

communicate with their neighbors, so that the whole group can follow the state of the leader. Note that only some agents in the group have partial information about the leader, but all agents move with the same velocity as the leader and the average position of the informed agents follows the position of the leader.

Let $f_i = 0$ and $\widetilde{f}_i = 0$. In the simulation, the initial positions and velocities of the 20 agents are chosen randomly from $[-1, 1] \times [-1, 1]$ and $[0, 2] \times [0, 2]$, respectively. From similar analysis as in Theorem 6.18, all agents move with the same velocity asymptotically.

Of particular interest is the final position state, where $t = 15s$. The position state at $t = 15s$ is shown in higher resolution in Fig 6.5. The solid circles are the positions where the attraction and repulsion are balanced ($d = 0.7740$) and the dashed circles represent the sensing ranges ($r_s = 1$) of the agents. If an agent i is in the solid circle of agent j, then the repulsion plays a key role; if it is between the solid and dashed curves, then the attraction dominates; otherwise, they do not have any influence on each other. The agents i, j, k, and l in Fig 6.5 are amplified in Fig 6.6. It is easy to see that agent i is attracted to agents j and l, and repulsed from agent k. These attractions and repulsion are balanced as illustrated by Fig 6.6, which means that the position state may not form an α-lattice but reach the local minimum of the potential function.

It should also be pointed out that the configuration of the α-lattice is a local minimum of the potential function, but it does not imply that the local minimum of the potential function necessarily forms an α-lattice. Therefore, Lemma 3 in [82] is incorrect. Mathematically, as it stands, one can only have

$$f_i^{\alpha} = \sum_{j \in \mathcal{N}_i} \phi(\| r_j - r_i \|) \frac{r_j - r_i}{\| r_j - r_i \|} = 0,$$

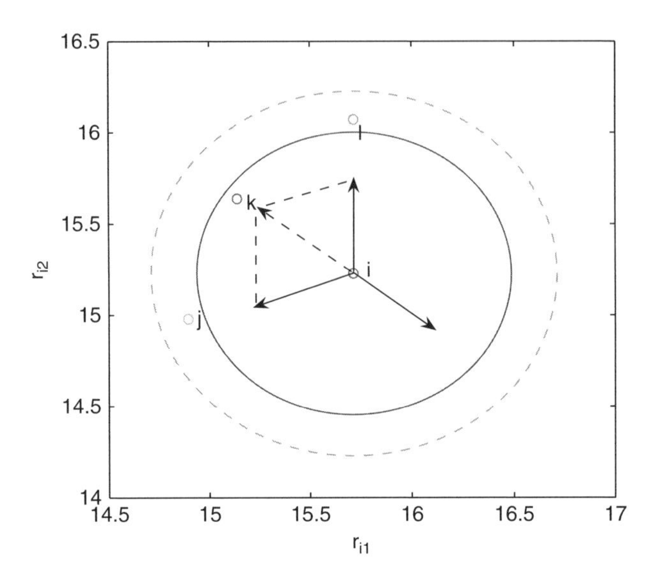

Figure 6.6 External forces around agent i

which does not imply that $\phi(\| r_j - r_i \|) = 0$ if there is a connection between agents i and j. The correct conclusion is that if all the external forces on each agent are balanced, then the formed configuration attains a local minimum of the potential function.

6.4 Notes

In this chapter, a distributed leader-follower algorithm for multi-agent dynamical systems has been developed and analyzed, which is applicable to the situation where the group has one virtual leader and the asymptotic velocity is time-varying. In addition, observer-based pinning navigation feedback has been derived, where each informed agent only senses partial states of the leader. It has been proved that although each informed agent can obtain only partial information about the leader, the velocity of the whole group converges to that of the leader and the centroid of those informed agents. Namely, some informed agents, having the leader's position information, can force the whole group to follow the trajectory of the leader asymptotically.

It has been the goal of this chapter to analyze different flocking algorithms by using tools from nonsmooth analysis in combination with ideas from the study of synchronization in complex networks. The leader-follower algorithm considered in this chapter falls into the category of free-flocking where no obstacles are considered. One may further investigate constrained-flocking with obstacle avoidance because it may lead to a more challenging scenario where the group sometimes splits.

7

Consensus of Multi-agent Systems with Sampled Data Information

This chapter studies second-order consensus in multi-agent dynamical systems using sampled data [127, 158, 159].

First, based on sampled full information, a distributed linear consensus protocol with second-order dynamics is designed, where both sampled position and velocity data are utilized. A necessary and sufficient condition based on the sampling period, the coupling gains, and the spectrum of the Laplacian matrix, is established for reaching consensus of the system. It is found that second-order consensus in such a multi-agent system can be achieved by appropriately choosing the sampling period determined by a third-order polynomial function of time. In particular, it is shown that second-order consensus cannot be reached for a sufficiently large sampling period but it can be reached for a sufficiently small one under some conditions. Then, the coupling gains are designed under the given network structure and the sampling period. Furthermore, the consensus regions are characterized for the spectrum of the Laplacian matrix. Next, second-order consensus in delayed undirected networks with sampled position and velocity data is discussed. A necessary and sufficient condition is given, by which an appropriate sampling period can be chosen to achieve consensus in the multi-agent system.

Second, a distributed linear consensus protocol with second-order dynamics is designed, where both the current and some sampled past position data are utilized. It is found that second-order consensus in such a multi-agent system cannot be reached without any sampled position data using the given protocol, while it can be achieved by appropriately choosing the sampling period. A necessary and sufficient condition for reaching consensus of the system in this setting is established, based on which, consensus regions are then characterized. It is shown that if all the eigenvalues of the Laplacian matrix are real, then second-order consensus in the multi-agent system can be reached for any sampling period, except at some critical points depending on the

Distributed Cooperative Control of Multi-agent Systems, First Edition.
Wenwu Yu, Guanghui Wen, Guanrong Chen, and Jinde Cao.

spectrum of the Laplacian matrix. However, if there exists at least one eigenvalue of the Laplacian matrix with a nonzero imaginary part, second-order consensus cannot be reached for both sufficiently small or sufficiently large sampling periods. In such cases, it might be possible to find some disconnected stable consensus regions determined by choosing appropriate sampling periods.

The problem of consensus in directed networks of multiple agents with intrinsic nonlinear dynamics and sampled-data information is discussed next. A new protocol is deduced from a class of continuous-time linear consensus protocols by implementing a data-sampling technique and a zero-order hold circuit. From a delayed-input approach, the sampled-data multi-agent system is converted to an equivalent nonlinear system with a time-varying delay. Theoretical analysis on this time-delayed system shows that consensus with asymptotic time-varying velocities in a strongly connected network can be achieved over some suitable sampled-data intervals. A multi-step procedure is further presented to estimate the upper bound of the maximal allowable sampling intervals. The results are then extended to a network topology with a directed spanning tree. For the case of a topology without a directed spanning tree, it is shown that the new protocol can still guarantee the system to achieve consensus by appropriately informing a fraction of agents.

7.1 Problem Statement

Recently, there has been growing interest in consensus algorithms, where all the agents are governed by second-order dynamics. In [146], some basic theoretical analysis was carried out for this case, where for each agent the second-order dynamics are governed by the position and velocity terms of the agents, and the asymptotic velocity is constant. A necessary and sufficient condition is given to ensure second-order consensus and it was found that both the real and imaginary parts of the eigenvalues of the Laplacian matrix of the corresponding network play key roles in reaching consensus.

However, in most of the works for sampled-data control in multi-agent systems [18, 41, 65, 163], it is difficult to see how the network structure, sampling period, and control gains affect the network dynamics since the authors mainly showed that consensus can be reached if the network parameters were in some derived regions, which are hard to apply.

In order to understand how the sampling period can affect the consensus problem, in [158] a distributed linear consensus protocol with second-order dynamics was designed, where both the current and some sampled past position data were utilized. It was found that second-order consensus in such a multi-agent system cannot be reached without any sampled position data using the given protocol, while it can be achieved by appropriately choosing the sampling period. However, current information is usually unavailable. Therefore, in this chapter, only sampled data without current data will be introduced into the protocol, that is, both sampled position and velocity data will be used instead. This is a memoryless scheme since only information at some particular past time instants is needed. By using the sampled data instead of the current information, one can utilize less information and save energy.

Usually, the velocity information of neighboring agents is unavailable to be measured in some real situations. By using only sampled position data in this chapter, without requiring the velocity information of agents in second-order dynamics as in [50, 51, 95], it is also found that second-order consensus in multi-agent system can be reached by appropriately choosing the sampling period. Note that this approach, relying on sampled position data allows one to study a general directed network topology, while that in [50, 51, 95] relying on observers design typically deals with only undirected network topologies.

7.2 Second-Order Consensus of Multi-agent Systems with Sampled Full Information

The second-order consensus protocol in multi-agent dynamical systems is described by [93, 95, 146]

$$\begin{aligned} \dot{x}_i(t) &= v_i, \\ \dot{v}_i(t) &= \alpha \sum_{j=1,\, j\neq i}^{N} G_{ij}(x_j(t) - x_i(t)) + \beta \sum_{j=1,\, j\neq i}^{N} G_{ij}(v_j(t) - v_i(t)), \\ i &= 1, 2, \ldots, N, \end{aligned} \tag{7.1}$$

where $x_i \in R^n$ and $v_i \in R^n$ are the position and velocity states of the ith node, respectively, $\alpha > 0$ and $\beta > 0$ are the coupling gains, and $G = (G_{ij})_{N\times N}$ is the coupling configuration matrix representing the topological structure of the network and thus is the weighted adjacency matrix of the network. The Laplacian matrix $L = (L_{ij})_{N\times N}$ is defined by

$$L_{ii} = -\sum_{j=1,\, j\neq i}^{N} L_{ij}, \qquad L_{ij} = -G_{ij}, \quad i \neq j, \tag{7.2}$$

which satisfies the diffusion property that $\sum_{j=1}^{N} L_{ij} = 0$. For notational simplicity, $n = 1$ is considered throughout this chapter, but all the obtained results can be easily generalized to the case with $n > 1$ by using the Kronecker product operations [53].

In the above case, all the current position and velocity states have to be utilized. But in a real situation, agents in the system usually communicate with each other only on some particular time intervals. Therefore, in order to utilize less information and save energy, it is desirable to use sampled data instead of the current data [158]:

$$\dot{x}_i(t) = v_i,$$
$$\dot{v}_i(t) = \alpha \sum_{j=1,\ j\neq i}^{N} G_{ij}(x_j(t_k) - x_i(t_k)) + \beta \sum_{j=1,\ j\neq i}^{N} G_{ij}(v_j(t_k) - v_i(t_k)),$$
$$t \in [t_k, t_{k+1}), \quad i = 1, 2, \ldots, N, \tag{7.3}$$

where t_k are the sampling instants satisfying $0 = t_0 < t_1 < \cdots < t_k < \cdots$, and $\alpha > 0$ and $\beta > 0$ are the coupling gains. For simplicity, assume that $t_{k+1} - t_k = T$, where $T > 0$ is the sampling period.

The concept of consensus is the same as before, which is re-stated in the following.

Definition 7.1 Second-order consensus in multi-agent system (7.3) is said to be achieved if, for any initial conditions,

$$\lim_{t\to\infty} \| x_i(t) - x_j(t) \| = 0, \ \lim_{t\to\infty} \| v_i(t) - v_j(t) \| = 0, \forall i, j = 1, 2, \ldots, N.$$

Because of (7.2), system (7.3) can be equivalently rewritten as

$$\dot{x}_i(t) = v_i,$$
$$\dot{v}_i(t) = -\alpha \sum_{j=1}^{N} L_{ij} x_j(t_k) - \beta \sum_{j=1}^{N} L_{ij} v_j(t_k), t \in [t_k, t_{k+1}), i = 1, 2, \ldots, N. \tag{7.4}$$

The following notations will be used throughout. Let $\mathscr{R}(u)$ and $\mathscr{L}(u)$ be the real and imaginary parts of the complex number u, $0 = \mu_1 \leq \mathscr{R}(\mu_2) \leq \cdots \leq \mathscr{R}(\mu_N)$ be the N eigenvalues of the Laplacian matrix L, $I_m \in R^{m\times m}(0_m \in R^{m\times m})$ be the m-dimensional identity (zero) matrix, $1_m \in R^m$ ($0_m \in R^m$) be the vector with all entries being 1 (0).

Lemma 7.2 *([88]) Given a complex coefficient polynomial of order two as follows:*

$$g(s) = s^2 + (\xi_1 + \boldsymbol{i}\gamma_1)s + \xi_0 + \boldsymbol{i}\gamma_0,$$

where ξ_1, γ_1, ξ_0, and γ_0 are real constants, $g(s)$ is stable if and only if $\xi_1 > 0$ and $\xi_1\gamma_1\gamma_0 + \xi_1^2\xi_0 - \gamma_0^2 > 0$.

Lemma 7.3 *([88]) Given a real coefficient polynomial of order three as follows:*

$$f(s) = a_3 s^3 + a_2 s^2 + a_1 s + a_0,$$

Then, $f(s)$ is stable if and only if a_3, a_2, a_1, a_0 are positive and $a_1 a_2 > a_0 a_3$.

7.2.1 *Second-Order Consensus of Multi-agent Systems with Sampled Full Information*

Let $\eta_i = (x_i, v_i)^T, A = \begin{pmatrix} 0 & 1 \\ 0 & 0 \end{pmatrix}, B = \begin{pmatrix} 0 & 0 \\ 1 & 0 \end{pmatrix}$, and $C = \begin{pmatrix} 0 & 0 \\ 0 & 1 \end{pmatrix}$. Then, system (4) can be rewritten as

$$\dot{\eta}_i(t) = A\eta_i(t) - \alpha \sum_{j=1}^{N} L_{ij}B\eta_j(t_k) - \beta \sum_{j=1}^{N} L_{ij}C\eta_j(t_k),$$
$$t \in [t_k, t_{k+1}), i = 1, 2, \dots, N. \tag{7.5}$$

Note that a solution of an isolated node of system (7.5) satisfies

$$\dot{s}(t) = As(t), \qquad t \in [t_k, t_{k+1}), \tag{7.6}$$

where $s(t) = (s_1, s_2)^T$ is the state vector. Let $\eta = (\eta_1^T, \dots, \eta_N^T)^T$ and rewrite system (7.5) in a matrix form:

$$\dot{\eta}(t) = (I_N \otimes A)\eta(t) - [L \otimes (\alpha B + \beta C)]\eta(t_k), t \in [t_k, t_{k+1}), \tag{7.7}$$

where $\otimes$ is the Kronecker product [53].

Let J be the Jordan form associated with the Laplacian matrix L, i.e., $L = PJP^{-1}$, where P is a nonsingular matrix. By Lemma 2.8,

$$\begin{aligned} \dot{y}(t) &= (P^{-1} \otimes I_2)(I_N \otimes A)\eta(t) - (P^{-1} \otimes I_2)[L \otimes (\alpha B + \beta C)]\eta(t_k), \\ &= (I_N \otimes A)y(t) - [J \otimes (\alpha B + \beta C)]y(t_k), t \in [t_k, t_{k+1}), \end{aligned} \tag{7.8}$$

where $y(t) = (P^{-1} \otimes I_2)\eta(t)$. If the graph $\mathcal{G}$ is undirected, then L is symmetric and J is a diagonal matrix with real eigenvalues. However, when $\mathcal{G}$ is directed, some eigenvalues of L may be complex and $J = \text{diag}(J_1, J_2, \dots, J_r)$, where

$$J_l = \begin{pmatrix} \mu_l & 0 & 0 & 0 \\ 1 & \ddots & 0 & 0 \\ 0 & \ddots & \ddots & 0 \\ 0 & 0 & 1 & \mu_l \end{pmatrix}_{N_l \times N_l} \tag{7.9}$$

in which μ_l are the eigenvalues of the Laplacian matrix L, with multiplicity N_l, $l = 1, 2, \dots, r$ and $N_1 + N_2 + \cdots + N_r = N$.

Let $P = (p_1, \dots, p_N)$, $P^{-1} = (q_1, \dots, q_N)^T$, $y(t) = (P^{-1} \otimes I_2)\eta(t) = (y_1^T, \dots, y_N^T)^T$, and $y_i = (y_{i1}, y_{i2})^T$. Note that if the network $\mathcal{G}$ contains a directed spanning tree, then by Lemma 2.7, 0 is a simple eigenvalue of the Laplacian matrix L, so

$$\dot{y}_1(t) = Ay_1(t), \quad t \in [t_k, t_{k+1}). \tag{7.10}$$

Theorem 7.4 *([158]) Suppose that the network $\mathcal{G}$ contains a directed spanning tree. Then, seocnd-order consensus in system (7.3) can be reached if and only if, in (7.8),*

$$\lim_{t\to\infty} \| y_i \| \to 0, \quad i = 2, \ldots, N. \tag{7.11}$$

Corollary 7.5 *Suppose that the network $\mathcal{G}$ contains a directed spanning tree. Then, second-order consensus in system (7.3) can be reached if and only if the following $N-1$ systems are asymptotically stable:*

$$\dot{z}_i(t) = Az_i(t) - \mu_i(\alpha B + \beta C)z_i(t_k), t \in [t_k, t_{k+1}), i = 2, \ldots, N. \tag{7.12}$$

Proof. This can be proved by using the similar analysis in [158]. □

It is generally hard to check the conditions (7.11) and (7.12) in Theorem 7.4 and Corollary 7.5, which do not reveal how network structure affects the consensus behavior.

Next, a theorem is derived to ensure consensus, depending on the control gains, spectra of the Laplacian matrix, and the sampling period.

Theorem 7.6 *Suppose that the network $\mathcal{G}$ contains a directed spanning tree. Then, second-order consensus in system (7.3) can be reached if and only if*

$$\frac{2\beta}{\alpha} > T, \tag{7.13}$$

and

$$\begin{aligned} & f(\alpha, \beta, \mu_i, T) \\ & = \left(\frac{2\beta}{\alpha T} - 1\right)^2 \left(\frac{4\mathcal{R}(\mu_i)}{\alpha \| \mu_i \|^2 T^2} - \frac{2\beta}{\alpha T}\right) - \frac{16\mathcal{L}^2(\mu_i)}{\alpha^2 \| \mu_i \|^4 T^4} \\ & > 0, \quad i = 2, \ldots, N. \end{aligned} \tag{7.14}$$

Proof. It suffices to prove that system (7.12) is asymptotically stable if and only if the conditions (7.13) and (7.14) are satisfied.

From (7.12), it follows that

$$(e^{-At} z_i(t))' = -e^{-At} \mu_i(\alpha B + \beta C)z_i(t_k), \quad t \in [t_k, t_{k+1}). \tag{7.15}$$

Simple calculation gives $e^{At} = \begin{pmatrix} 1 & t \\ 0 & 1 \end{pmatrix}$. Integrating both sides of (7.15) from t_k to t gives

$$\begin{aligned} & z_i(t) \\ & = e^{A(t-t_k)} z_i(t_k) - e^{At} \int_{t_k}^{t} e^{-As} \mu_i(\alpha B + \beta C)z_i(t_k)ds \end{aligned}$$

$$= \begin{pmatrix} 1 & t-t_k \\ 0 & 1 \end{pmatrix} z_i(t_k) - \begin{pmatrix} 1 & t \\ 0 & 1 \end{pmatrix} \begin{pmatrix} t-t_k & -\frac{t^2}{2}+\frac{t_k^2}{2} \\ 0 & t-t_k \end{pmatrix} \mu_i(\alpha B+\beta C) z_i(t_k)$$

$$= \begin{pmatrix} 1 & \delta_k \\ 0 & 1 \end{pmatrix} z_i(t_k) - \begin{pmatrix} \frac{\alpha}{2}\mu_i\delta_k^2 & \frac{\beta}{2}\mu_i\delta_k^2 \\ \alpha\mu_i\delta_k & \beta\mu_i\delta_k \end{pmatrix} z_i(t_k)$$

$$= \begin{pmatrix} 1-\frac{\alpha}{2}\mu_i\delta_k^2 & \delta_k-\frac{\beta}{2}\mu_i\delta_k^2 \\ -\alpha\mu_i\delta_k & 1-\beta\mu_i\delta_k \end{pmatrix} z_i(t_k), t \in [t_k, t_{k+1}), \tag{7.16}$$

where $\delta_k = t - t_k$.

Let $D(t) = \begin{pmatrix} 1-\frac{\alpha}{2}\mu_i t^2 & t-\frac{\beta}{2}\mu_i t^2 \\ -\alpha\mu_i t & 1-\beta\mu_i t \end{pmatrix}$. It is easy to see that $D(t)$ is bounded on $[0, T]$. So, for $0 = t_0 < t_1 < \cdots < t_k < \cdots$ and $t_{k+1} - t_k = T$, one has

$$z_i(t) = D(t-t_k) D^k(T) z_i(t_0), \quad t \in [t_k, t_{k+1}). \tag{7.17}$$

Since $D(t - t_k)$ is bounded when $t \in [t_k, t_{k+1})$, $z_i(t) \to 0$ if and only if all eigenvalues λ of $D(T)$ satisfy $\| \lambda \| < 1$.

Let $|\lambda I_2 - D(T)| = 0$. Then

$$\lambda^2 - \left(2 - \frac{\alpha}{2}\mu_i T^2 - \beta\mu_i T\right)\lambda + 1 - \beta\mu_i T + \frac{\alpha}{2}\mu_i T^2 = 0. \tag{7.18}$$

Let $\lambda = \frac{s+1}{s-1}$. Then, (7.18) can be transformed to

$$s^2 + \left(\frac{2\beta}{\alpha T} - 1\right)s + \frac{4}{\alpha\mu_i T^2} - \frac{2\beta}{\alpha T} = 0. \tag{7.19}$$

It is well known that $\| \lambda \| < 1$ in (7.18) if and only if $\mathscr{R}(s) < 0$ in (7.19). Therefore, $z_i(t) \to 0$ if and only if all the roots in (7.19) have negative real parts.

By Lemma 7.2, (7.19) is stable if and only if

$$\frac{2\beta}{\alpha T} - 1 > 0, \tag{7.20}$$

and

$$\left(\frac{2\beta}{\alpha T} - 1\right)^2 \left(\frac{4\mathscr{R}(\mu_i)}{\alpha \| \mu_i \|^2 T^2} - \frac{2\beta}{\alpha T}\right) - \frac{16\mathscr{L}^2(\mu_i)}{\alpha^2 \| \mu_i \|^4 T^4} > 0. \tag{7.21}$$

Therefore, $z_i(t) \to 0$ if and only if (7.13) and (7.14) are satisfied. By Corollary 7.5, second-order consensus in system (7.3) is reached if and only if (7.13) and (7.14) are satisfied. □

Although a necessary and sufficient condition for second-order consensus in multi-agent system (7.3) is established in Theorem 7.6, it is still difficult to see how to design the network parameters in system (7.3) to achieve this goal. In particular, for a given network, one can design appropriate parameters α, β, and T such that the conditions (7.13) and (7.14) in Theorem 7.6 are satisfied. However, since the condition (7.14) holds for all $i = 2, \cdots, N$, one can apply the consensus regions [31, 153] to further simplify the condition (7.14) so as to get a more applicable result.

Next, the design for choosing the sampling period T, the coupling gains α and β, and the spectrum of the Laplacian matrix μ_i, are discussed.

7.2.2 Selection of Sampling Periods

The sampling period T plays a key role in reaching consensus in multi-agent system (7.13). It is still unclear from Theorem 7.6 about how to choose an appropriate sampling period T. This subsection aims to solve this problem.

Different from the results in [158], where the hyperbolic functions and trigonometric functions on the sampling period T were derived, the function on the sampling period T may be simpler in the condition (7.14) of Theorem 7.6. Note that the inequality (14) can be equivalently written as

$$\left(T - \frac{2\beta}{\alpha}\right)^2 \left(T - \frac{2\mathscr{R}(\mu_i)}{\beta \parallel \mu_i \parallel^2}\right) + \frac{8\mathscr{L}^2(\mu_i)}{\alpha\beta \parallel \mu_i \parallel^4} < 0, i = 2, \ldots, N. \tag{7.22}$$

Corollary 7.7 *Suppose that the network $\mathscr{G}$ contains a directed spanning tree. Then, second-order consensus in system (7.3) can be reached if and only if (7.13) and (7.22) hold.*

Actually, from (7.22), it is easy to choose the appropriate sampling period T since the left-hand side of inequality (7.22) is a polynomial of T. Let

$$g_i(T) = \left(T - \frac{2\beta}{\alpha}\right)^2 \left(T - \frac{2\mathscr{R}(\mu_i)}{\beta \parallel \mu_i \parallel^2}\right) + \frac{8\mathscr{L}^2(\mu_i)}{\alpha\beta \parallel \mu_i \parallel^4},$$
$$i = 2, \ldots, N, \tag{7.23}$$

with $\mathbb{S}_i = \{T | g_i(T) < 0\}$ and $\mathbb{S} = \bigcap_{i=1}^{N} \mathbb{S}_i$. Note that $g_i(T) = 0$ has at most three roots, which indicates that $\mathbb{S}_i$ can be easily solved. The following corollary shows how to choose the sampling period.

Theorem 7.8 *Suppose that the network $\mathscr{G}$ contains a directed spanning tree. Then, second-order consensus in system (7.3) can be reached if and only if the sampling period $T \in \mathbb{S} \bigcap \left(0, \frac{2\beta}{\alpha}\right)$.*

For a general sampling period T, one can apply Theorem 7.8 to check if it is useful for reaching consensus in multi-agent system (7.3). But, does a small or a large sampling period work?

Corollary 7.9 *Suppose that the network $\mathcal{G}$ contains a directed spanning tree. Then, second-order consensus in system (7.3) cannot be reached for a sufficiently large sampling period T while it can be reached for a sufficiently small sampling period if (7.13) and*

$$\frac{\mathcal{L}^2(\mu_i)}{\mathcal{R}(\mu_i) \parallel \mu_i\|^2} < \frac{\beta^2}{\alpha} \tag{7.24}$$

hold for all $i = 2, \dots, N$.

Proof. If the sampling period T is sufficiently large, larger than $\frac{2\beta}{\alpha}$, the condition (7.22) or (7.13) is not satisfied, which indicates that consensus cannot be reached in system (7.3).

While the sampling period T is sufficiently small, condition (7.22) holds if (7.24) is satisfied. □

If the network is undirected, a simplified result can be obtained.

Corollary 7.10 *Suppose that the connected network $\mathcal{G}$ is undirected. Then, second-order consensus in system (7.3) can be reached if and only if*

$$T < \min\left(\frac{2\beta}{\alpha}, \frac{2}{\beta\mu_N}\right). \tag{7.25}$$

Proof. Since the network $\mathcal{G}$ is undirected, $\mathcal{L}(\mu_i) = 0$ and $\mathcal{R}(\mu_i) = \mu_i > 0$ for all $i = 2, 3, \dots, N$, due to the symmetry of the Laplacian matrix L. Thus, (7.22) is equivalent to

$$\left(T - \frac{2\beta}{\alpha}\right)^2 \left(T - \frac{2}{\beta\mu_i}\right) < 0, i = 2, 3, \dots, N. \tag{7.26}$$

Combining it with the condition (7.13), one concludes that second-order consensus in system (7.3) can be reached if and only if (7.25) holds. □

7.2.3 Design of Coupling Gains

Though the sampling period can be appropriately chosen from Theorem 7.8, it is still unknown as to how to design the coupling gains α and β under the given sampling period T. Multiplying by α^3 on both sides of (7.14), inequality (14) can be equivalently written as

$$\left(\alpha - \frac{2\beta}{T}\right)^2 \left(\frac{4\mathcal{R}(\mu_i)}{\parallel \mu_i\|^2 T^2} - \frac{2\beta}{T}\right) - \alpha\frac{16\mathcal{L}^2(\mu_i)}{\parallel \mu_i\|^4 T^4} > 0, i = 2, \dots, N. \tag{7.27}$$

Theorem 7.11 *Suppose that the network $\mathcal{G}$ contains a directed spanning tree. Then, second-order consensus in system (7.3) can be reached if and only if*

$$\alpha < \min_{i=2,\dots,N} \left(\frac{2\beta}{T} + \gamma_i - \sqrt{\gamma_i^2 + \frac{4\beta\gamma_i}{T}}\right) \tag{7.28}$$

and

$$\beta < \min_{i=2,\ldots,N} \left(\frac{2\mathscr{R}(\mu_i)}{\| \mu_i \|^2 T} \right), \tag{7.29}$$

where $\gamma_i = \dfrac{4\mathscr{L}^2(\mu_i)}{\| \mu_i\|^2 T^2 (2\mathscr{R}(\mu_i) - \beta \| \mu_i\|^2 T)}$, $i = 2, \ldots, N$.

Proof. From (7.14) and (7.22), it is easy to verify that $\frac{2\mathscr{R}(\mu_i)}{\|\mu_i\|^2 T} > \beta$. Then, (7.27) can be rewritten as

$$\left(\alpha - \frac{2\beta}{T} \right)^2 - 2\alpha\gamma_i > 0. \tag{7.30}$$

The left-hand side of (7.30) has two roots, that is,

$$\alpha_{i1,i2} = \frac{2\beta}{T} + \gamma_i \pm \sqrt{\gamma_i^2 + \frac{4\beta\gamma_i}{T}}. \tag{7.31}$$

By solving (7.30), one obtains that $\alpha < \alpha_{i1}$ or $\alpha > \alpha_{i2}$. Since $\alpha_{i2} > \frac{2\beta}{T}$ contradicts condition $\alpha < \frac{2\beta}{T}$ in (7.13), one finally has that $\alpha < \alpha_{i1}$ for all $i = 2, \ldots, N$. □

Given the network structure, one can first choose the coupling gain β satisfying the condition (7.29) and then find the appropriate parameter α such that (7.28) holds under the chosen β for reaching consensus in multi-agent system (7.3). If the network is undirected, a similar simplified result can be obtained since $\mathscr{L}(\mu_i) = 0$ in Theorem 7.11.

Corollary 7.12 *Suppose that the connected network $\mathscr{G}$ is undirected. Then, second-order consensus in system (7.3) can be reached if and only if*

$$\alpha < \frac{2\beta}{T} \quad \text{and} \quad \beta < \frac{2}{\mu_N T}. \tag{7.32}$$

Note that the condition (7.25) in Corollary 7.10 and the condition (7.32) in Corollary 7.12 are the same. Since $T < \frac{2\beta}{\alpha}$ and $T < \frac{2}{\beta\mu_N}$, one can choose the parameter β such that T attains its maximum value, leading to consensus in system (7.3).

Corollary 7.13 *Suppose that the connected network $\mathscr{G}$ is undirected. By designing $\beta = \sqrt{\alpha/\mu_N}$, second-order consensus in system (7.3) can be reached if and only if the sampling period T satisfies*

$$T < \frac{2}{\sqrt{\alpha\mu_N}}. \tag{7.33}$$

7.2.4 Consensus Region for the Network Spectrum

Since the condition (7.14) holds for all $i = 2, \cdots, N$, one can find a stable consensus region [31, 153] as follows:

$$S = \{\mu | f(\alpha, \beta, \mu, T) > 0\}. \tag{7.34}$$

Then, the problem is transformed to finding if all the nonzero eigenvalues of the Laplacian matrix lie in the stable consensus region S, i.e., $\mu_i \in S$ for all $i = 2, \cdots, N$.

Theorem 7.14 *Suppose that the network $\mathcal{G}$ contains a directed spanning tree. Then, second-order consensus in system (7.3) can be reached if and only if (7.13) holds and $\mu_i \in S$ for all $i = 2, \cdots, N$.*

7.2.5 Second-Order Consensus in Delayed Undirected Networks with Sampled Position and Velocity Data

In some real situations, the input time delays always exist, which cannot be ignored. When time delays are introduced into the protocol, one can consider the system as follows:

$$\begin{aligned}
\dot{x}_i(t) &= v_i, \\
\dot{v}_i(t) &= -\alpha \sum_{j=1}^{N} L_{ij} x_j(t_k - \tau) - \beta \sum_{j=1}^{N} L_{ij} v_j(t_k - \tau), t \in [t_k, t_{k+1}), \\
& i = 1, 2, \ldots, N,
\end{aligned} \tag{7.35}$$

where $\tau > 0$ is the time delay.

Let $\eta_i = (x_i, v_i)^T$, $A = \begin{pmatrix} 0 & 1 \\ 0 & 0 \end{pmatrix}$, $B = \begin{pmatrix} 0 & 0 \\ 1 & 0 \end{pmatrix}$, and $C = \begin{pmatrix} 0 & 0 \\ 0 & 1 \end{pmatrix}$. Then, system (7.35) can be rewritten as

$$\begin{aligned}
\dot{\eta}_i(t) &= A\eta_i(t) - \alpha \sum_{j=1}^{N} L_{ij} B\eta_j(t_k - \tau) - \beta \sum_{j=1}^{N} L_{ij} C\eta_j(t_k - \tau), \\
& t \in [t_k, t_{k+1}), i = 1, 2, \ldots, N.
\end{aligned} \tag{7.36}$$

Let $\eta = (\eta_1^T, \ldots, \eta_N^T)^T$ and rewrite system (7.36) in a matrix form:

$$\dot{\eta}(t) = (I_N \otimes A)\eta(t) - [L \otimes (\alpha B + \beta C)]\eta(t_k - \tau), t \in [t_k, t_{k+1}), \tag{7.37}$$

where $\otimes$ is the Kronecker product [53]. Let J be the Jordan form associated with the Laplacian matrix L, i.e., $L = PJP^{-1}$, where P is a nonsingular matrix. By Lemma 2.8, one has

$$\begin{aligned}
\dot{y}(t) &= (P^{-1} \otimes I_2)(I_N \otimes A)\eta(t) - (P^{-1} \otimes I_2)[L \otimes (\alpha B + \beta C)]\eta(t_k - \tau), \\
&= (I_N \otimes A)y(t) - [J \otimes (\alpha B + \beta C)]y(t_k - \tau), t \in [t_k, t_{k+1}),
\end{aligned} \tag{7.38}$$

where $y(t) = (P^{-1} \otimes I_2)\eta(t)$. Here, for simplicity, only undirected networks are discussed, i.e., L is symmetric and J is a diagonal matrix with real eigenvalues.

By the same calculation, one can also show that the asymptotical behavior of the system (7.35) is dominated by the stability of the system, as follows:

$$\dot{z}_i(t) = Az_i(t) - \mu_i(\alpha B + \beta C)z_i(t_k - \tau),$$
$$t \in [t_k, t_{k+1}), i = 2, \ldots, N. \tag{7.39}$$

Theorem 7.15 *Suppose that the connected network $\mathcal{G}$ is undirected and $\tau < T$. Then, second-order consensus in system (7.35) can be reached if and only if*

$$3 - a_1 - a_2 + 3a_3 > 0,$$
$$1 - a_1 + a_2 - a_3 > 0,$$
$$1 - a_2 + a_1 a_3 - a_3^2 > 0, \tag{7.40}$$

and

$$\tau < \frac{\beta}{\alpha}, \tag{7.41}$$

where $a_1 = \frac{\alpha}{2}\mu_i(T-\tau)^2 + \beta\mu_i(T-\tau) - 2$, $a_2 = 1 + \beta\mu_i T + \alpha\mu_i T(T-\tau) + \frac{\alpha}{2}\mu_i T^2 - \alpha\mu_i(T-\tau)^2 - 2\beta\mu_i(T-\tau)$, *and* $a_3 = \frac{\alpha}{2}\mu_i T^2 + \frac{\alpha}{2}\mu_i(T-\tau)^2 + \beta\mu_i(T-\tau) - \beta\mu_i T - \alpha\mu_i T(T-\tau)$.

Proof. From (7.39), it follows that

$$(e^{-At}z_i(t))' = -e^{-At}\mu_i(\alpha B + \beta C)z_i(t_k - \tau), t \in [t_k, t_{k+1}). \tag{7.42}$$

Integrating both sides of (7.42) from t_k to t gives

$$z_i(t) = \begin{pmatrix} 1 & t - t_k \\ 0 & 1 \end{pmatrix} z_i(t_k) - \begin{pmatrix} \frac{\alpha}{2}\mu_i(t-t_k)^2 & \frac{\beta}{2}\mu_i(t-t_k)^2 \\ \alpha\mu_i(t-t_k) & \beta\mu_i(t-t_k) \end{pmatrix} z_i(t_k - \tau),$$
$$t \in [t_k, t_{k+1}). \tag{7.43}$$

Let $E(t) = \begin{pmatrix} 1 & t \\ 0 & 1 \end{pmatrix}$ and $F(t) = \begin{pmatrix} \frac{\alpha}{2}\mu_i t^2 & \frac{\beta}{2}\mu_i t^2 \\ \alpha\mu_i t & \beta\mu_i t \end{pmatrix}$. It is easy to see that $E(t)$ and $F(t)$ are both bounded on $[0, T]$. So, for $0 = t_0 < t_1 < \cdots < t_k < \cdots$, $t_{k+1} - t_k = T$ and $\tau < T$,

$$z_i(t_k) = E(T)z_i(t_{k-1}) - F(T)z_i(t_{k-1} - \tau),$$

and

$$z_i(t_k - \tau) = E(T - \tau)z_i(t_{k-1}) - F(T - \tau)z_i(t_{k-1} - \tau).$$

Let $v_i(t) = (z_i(t), z_i(t-\tau))$ and $G(t) = \begin{pmatrix} E(t) & -F(t) \\ E(t-\tau) & -F(t-\tau) \end{pmatrix}$. Then

$$v_i(t_k) = G(T)v(t_{k-1}) = G^k(T)v_i(t_0),$$
$$v_i(t) = G(t-t_k)G^k(T)v_i(t_0).$$

Thus, $z_i(t) \to 0$ if and only if $v_i(t_k) \to 0$, i.e., all eigenvalues λ of $G(T)$ satisfy $\| \lambda \| < 1$.

Let $|\lambda I_4 - G(T)| = 0$. This gives

$$\begin{aligned}
&\lambda^4 + \left[\frac{\alpha}{2}\mu_i(T-\tau)^2 + \beta\mu_i(T-\tau) - 2\right]\lambda^3 \\
&+ \left[1 + \beta\mu_i T + \alpha\mu_i T(T-\tau) + \frac{\alpha}{2}\mu_i T^2 \right. \\
&\left. -\alpha\mu_i(T-\tau)^2 - 2\beta\mu_i(t-\tau)\right]\lambda^2 \\
&- \left[\beta\mu_i T + \alpha\mu_i T(T-\tau) - \frac{\alpha}{2}\mu_i T^2 - \frac{\alpha}{2}\mu_i(T-\tau)^2 \right. \\
&\left. -\beta\mu_i(T-\tau)\right]\lambda = 0.
\end{aligned} \tag{7.44}$$

It is easy to see that $G(t)$ has an eigenvalue $\lambda = 0$. Let $\lambda = \frac{s+1}{s-1}$, $a_1 = \frac{\alpha}{2}\mu_i(T-\tau)^2 + \beta\mu_i(T-\tau) - 2$, $a_2 = 1 + \beta\mu_i T + \alpha\mu_i T(T-\tau) + \frac{\alpha}{2}\mu_i T^2 - \alpha\mu_i(T-\tau)^2 - 2\beta\mu_i(t-\tau)$, and $a_3 = \frac{\alpha}{2}\mu_i T^2 + \frac{\alpha}{2}\mu_i(T-\tau)^2 + \beta\mu_i(T-\tau) - \beta\mu_i T - \alpha\mu_i T(T-\tau)$. This gives

$$\begin{aligned}
&(1 + a_1 + a_2 + a_3)s^3 + (3 + a_1 - a_2 + 3a_3)s^2 \\
&+(3 - a_1 - a_2 + 3a_3)s + 1 - a_1 + a_2 - a_3 = 0.
\end{aligned} \tag{7.45}$$

It is well known that $\| \lambda \| < 1$ in (7.44) if and only if $\mathcal{R}(s) < 0$. Therefore, $z_i(t) \to 0$ if and only if all the roots in (7.45) have negative real parts.

By Lemma 7.3, (7.45) is stable if and only if $1 + a_1 + a_2 + a_3 > 0, 3 + a_1 - a_2 + 3a_3 > 0, 3 - a_1 - a_2 + 3a_3 > 0, 1 - a_1 + a_2 - a_3 > 0$, and $(3 + a_1 - a_2 + 3a_3)(3 - a_1 - a_2 + 3a_3) - (1 + a_1 + a_2 + a_3)(1 - a_1 + a_2 - a_3) > 0$.

Solving the first two polynomial inequalities, gives the condition (7.41). Therefore, $z_i(t) \to 0$ if and only if (7.40) and (7.41) are satisfied. By Corollary 7.5, second-order consensus in system (7.35) can be reached if and only if (7.40) and (7.41) are satisfied. □

Remark 7.16 In Theorem 7.15, a necessary and sufficient condition for second-order consensus in the multi-agent system (7.35) with time delay is established. For a given network, one can design appropriate parameters α, β, T, and τ, such that the conditions (7.40) and (7.41) in Theorem 7.15 are satisfied.

7.2.6 Simulation Examples

In this subsection, some examples are given to verify the theoretical analysis.

Example 1. Second-order consensus in a multi-agent system with an undirected topology

Consider the multi-agent system (7.3) with an undirected topology, where $L = \begin{pmatrix} 3 & -1 & -1 & -1 \\ -1 & 2 & -1 & 0 \\ -1 & -1 & 3 & -1 \\ -1 & 0 & -1 & 2 \end{pmatrix}$, $\alpha = 1$, and $\beta = 0.8$. By simple calculation, we have $\mu_1 = 0, \mu_2 = 2, \mu_3 = 4$, and $\mu_4 = 4$. From Corollary 7.10, multi-agent system (7.3) can reach second-order consensus if and only if $T < 0.625$. It is easy to see that second-order consensus in system (7.3) can be reached if $T = 0.05$ for a sufficiently small T as in Corollary 7.9 and $T = 0.62$, while it cannot be reached for $T = 0.63$. The position and velocity states of all the agents are shown in Fig. 7.1.

According to Corollary 7.13, one can choose $\beta = 0.5$ so as to get a maximum value for T that can tolerate the sampling period $T = 1$. The position and velocity states of all the agents are shown in Fig. 7.2, where second-order consensus can be reached if $T = 0.95$ while it cannot be reached when $T = 1.05$.

Example 2. Second-order consensus in a multi-agent system with a directed topology

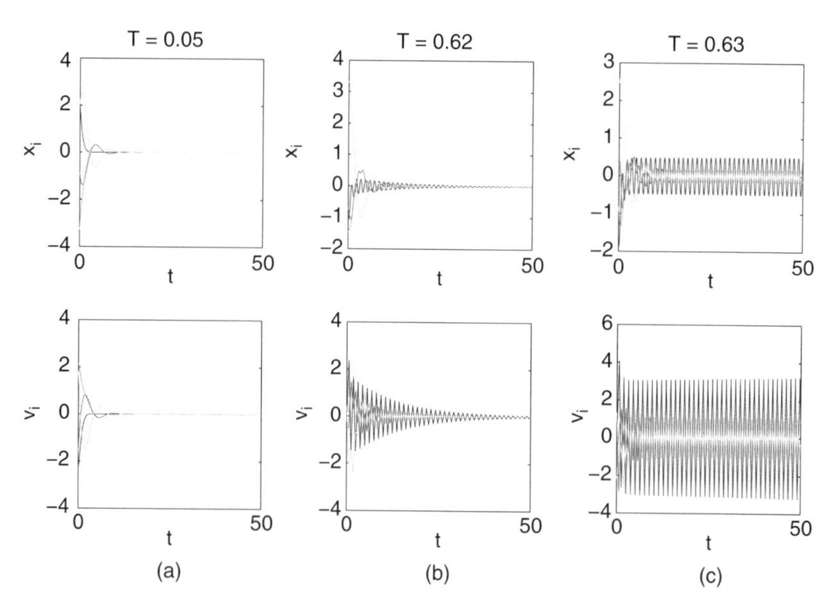

Figure 7.1 Position and velocity states of agents, where $T = 0.05(a), T = 0.62(b)$, and $T = 0.63(c)$

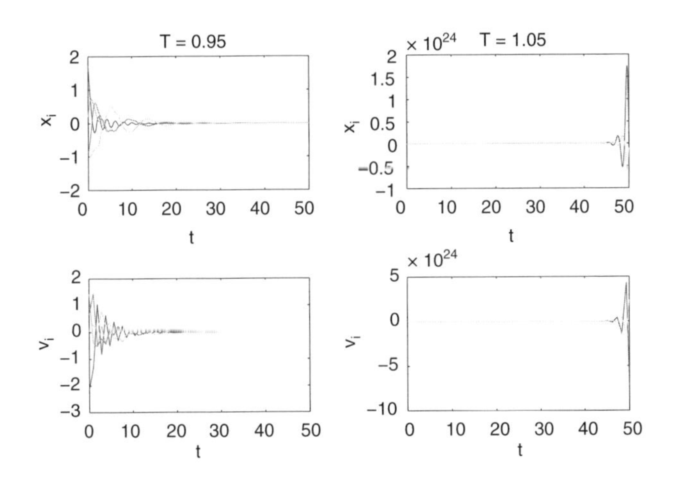

Figure 7.2 Position and velocity states of agents, where $T = 0.95(a)$ and $T = 1.05(b)$

1. *Section of the sampling period*

Consider the multi-agent system (7.3) with a directed topology, where $L = \begin{pmatrix} 2 & 0 & -2 & 0 \\ -1 & 1 & 0 & 0 \\ 0 & -3 & 3 & 0 \\ 0 & -2 & 0 & 2 \end{pmatrix}$, $\alpha = 1$, and $\beta = 0.8$. By simple calculation. One has $\mu_1 = 0, \mu_2 = 2, \mu_3 = 3 + 1.4142i$, and $\mu_4 = 3 - 1.4142i$. From Theorem 7.8, multi-agent system (7.3) can reach second-order consensus if and only if $T \in \mathbb{S} \bigcap (0, \frac{2\beta}{\alpha})$. By (7.23), one has $\mathbb{S}_2 = (0, 0.5359)$ and $\mathbb{S}_3 = \mathbb{S}_4 = (0, 1.6) \bigcup (1.6, 2.5)$. Then, second-order consensus can be reached in multi-agent system (7.3) if and only if $T < 0.5359$.

Consider the sampling period T as a variable of $g_i(T)$. In Fig. 7.3, it is easy to see that second-order consensus in the system can be reached if $T = 0.5$, while it

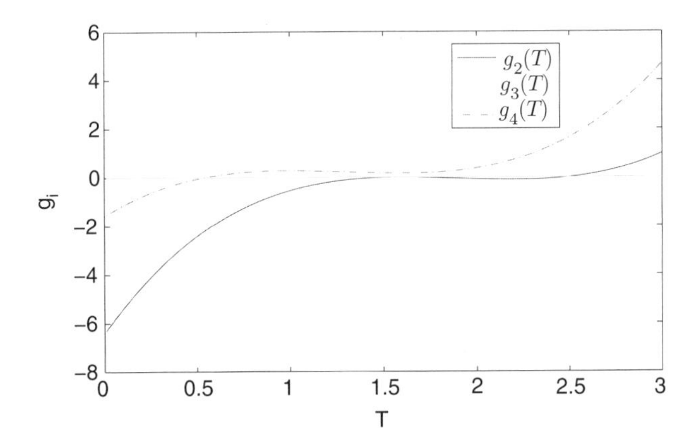

Figure 7.3 States of $g_i(T)$ vs. the sampling period $T, i = 2, 3, 4$

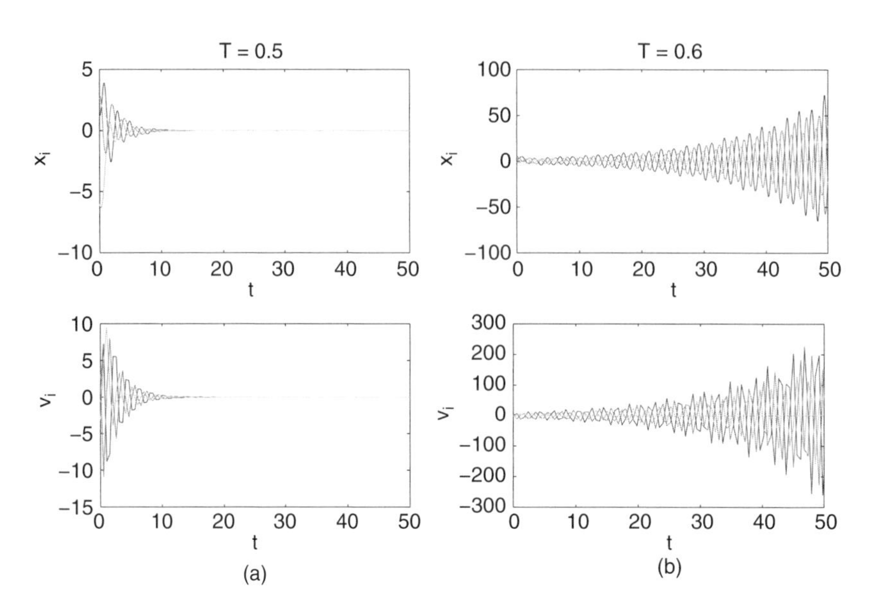

Figure 7.4 Position and velocity states of agents, where $T = 0.5(a)$ and $T = 0.6(b)$

cannot be reached for $T = 0.6$, where $g_2(T) > 0$. The position and velocity states of all the agents are shown in Fig. 7.4.

2. *Design of the coupling gains*

   Consider the same example in this subsection, except that the sampling period $T = 0.5$ is fixed. The aim is to choose two appropriate coupling gains α and β such that second-order consensus in multi-agent system (7.3) can be reached. By Theorem 7.11, consensus can be achieved if $\alpha < 4.3636$ and $\beta < 1.0909$. Then, one can first fix $\beta = 0.8$ and second-order consensus can be reached if and only if $\alpha < 1.1527$. The position and velocity states of all the agents are shown in Fig. 7.5, where it is easy to see that second-order consensus in system (7.3) can be reached if $\alpha = 1.05$ while it cannot be reached if $\alpha = 1.25$.

Example 3. Second-order consensus in delayed undirected networks

Consider the multi-agent system (7.35) with an undirected topology, where $L = \begin{pmatrix} 3 & -1 & -1 & -1 \\ -1 & 2 & -1 & \\ -1 & -1 & 3 & -1 \\ -1 & 0 & -1 & 2 \end{pmatrix}$, $T = 0.5$, $\alpha = 1$, and $\beta = 0.8$. By simple calculation, one has $\mu_1 = 0, \mu_2 = 2, \mu = 4, \mu = 4$. From Theorem 7.15, second-order consensus in multi-agent system (7.35) can be reached if $\tau = 0.1$, while it cannot be reached if $\tau = 0.4$. The position and velocity states of all the agents are shown in Fig. 7.6.

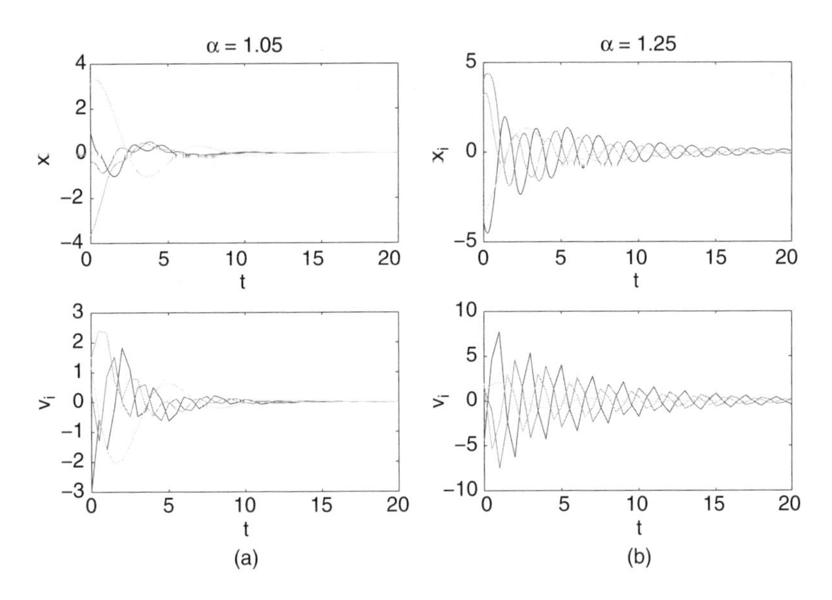

Figure 7.5 Position and velocity states of agents, where $T = 1.05$ (a) and $T = 1.25$ (b)

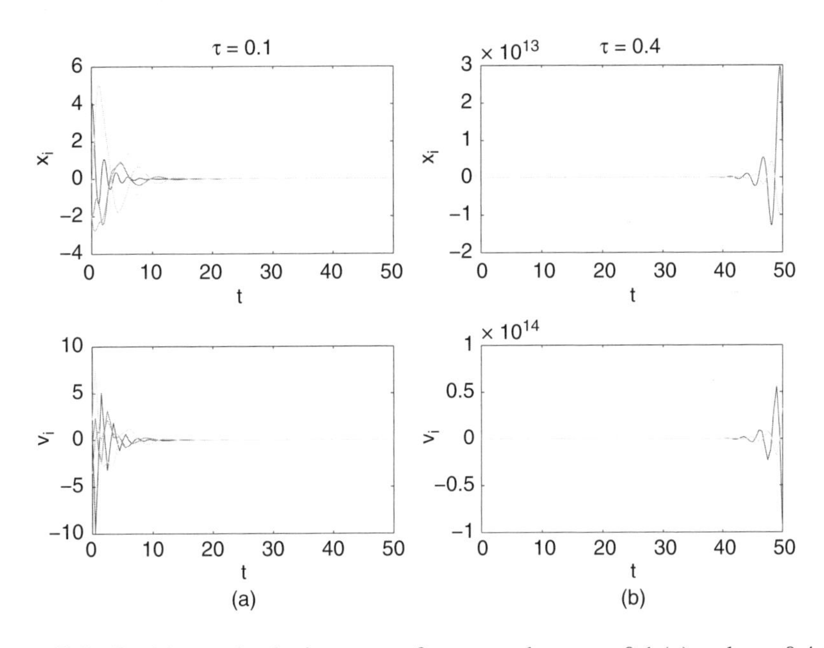

Figure 7.6 Position and velocity states of agents, where $\tau = 0.1$ (a) and $\tau = 0.4$ (b)

7.3 Second-Order Consensus of Multi-agent Systems with Sampled Position Information

In many cases, it is difficult to measure the relative velocity difference between two neighboring agents. In [50, 51, 95], distributed observers were designed for second-order multi-agent systems, where the velocity states were assumed to be unavailable, and some slack variables were introduced and a higher-order controller was designed. In this section, only sampled position data is used. It will be shown that second-order consensus can be reached in multi-agent dynamical systems under some conditions even if the velocity states are unavailable. To do so, the following consensus protocol with both current and sampled position data is considered:

$$\begin{aligned} \dot{x}_i(t) &= v_i, \\ \dot{v}_i(t) &= \alpha \sum_{j=1,\ j\neq i}^{N} G_{ij}(x_j(t) - x_i(t)) - \beta \sum_{j=1,\ j\neq i}^{N} G_{ij}(x_j(t_k) - x_i(t_k)), \\ & t \in [t_k, t_{k+1}), i = 1, \cdots, N, \end{aligned} \tag{7.46}$$

where t_k are the sampling instants satisfying $0 = t_0 < t_1 < \ldots < t_k < \ldots$, and α and β are the coupling strengths. For simplicity, assume that $t_{k+1} - t_k = T$, where $T > 0$ is the sampling period.

In [19, 152], consensus in multi-agent systems was studied, where information of both the current and delayed position states data was used. In the existing delay-involved consensus algorithms [19, 152], all the position states in the time interval $[t - \tau, t]$ have to be kept in memory, where τ is the time delay constant. In order to utilize less information and save energy, it is desirable to use only sampled data instead of delayed information in a spectrum of a time interval.

System (7.46) can be equivalently rewritten as follows:

$$\begin{aligned} \dot{x}_i(t) &= v_i, \\ \dot{v}_i(t) &= -\alpha \sum_{j=1}^{N} L_{ij} x_j(t) + \beta \sum_{j=1}^{N} L_{ij} x_j(t_k), t \in [t_k, t_{k+1}), \\ & i = 1, 2, \cdots, N. \end{aligned} \tag{7.47}$$

7.3.1 Second-Order Consensus in Multi-agent Dynamical Systems with Sampled Position Data

Let $\eta_i = (x_i, v_i)^T$, $A = \begin{pmatrix} 0 & 1 \\ 0 & 0 \end{pmatrix}$, and $B = \begin{pmatrix} 0 & 0 \\ 1 & 0 \end{pmatrix}$. Then, system (7.47) can be rewritten as

$$\begin{aligned} \dot{\eta}_i(t) &= A\eta_i(t) - \alpha \sum_{j=1}^{N} L_{ij} B\eta_j(t) + \beta \sum_{j=1}^{N} L_{ij} B\eta_j(t_k), t \in [t_k, t_{k+1}), \\ & i = 1, 2, \ldots, N. \end{aligned} \tag{7.48}$$

Note that a solution of an isolated node of system (7.48) satisfies

$$\dot{s}(t) = As(t), t \in [t_k, t_{k+1}), \tag{7.49}$$

where $s(t) = (s_1, s_2)^T$ is the state vector. Let $\eta = (\eta_1^T, \dots, \eta_N^T)^T$ and rewrite system (7.48) into a matrix form:

$$\dot{\eta}(t) = [(I_N \otimes A) - \alpha(L \otimes B)]\eta(t) + \beta(L \otimes B)\eta(t_k), t \in [t_k, t_{k+1}), \tag{7.50}$$

where $\otimes$ is the Kronecker product [53]. Let J be the Jordan form associated with the Laplacian matrix L, i.e., $L = PJP^{-1}$, where P is a nonsingular matrix. By Lemma 2.8, we have

$$\begin{aligned}\dot{y}(t) &= (P^{-1} \otimes I_2)[(I_N \otimes A) - \alpha(L \otimes B)]\eta(t) + \beta(P^{-1} \otimes I_2)(L \otimes B)\eta(t_k)\\ &= [(P^{-1} \otimes A) - \alpha(JP^{-1} \otimes B)]\eta(t) + \beta(JP^{-1} \otimes B)\eta(t_k)\\ &= [(I_N \otimes A) - \alpha(J \otimes B)]y(t) + \beta(J \otimes B)y(t_k), t \in [t_k, t_{k+1}),\end{aligned} \tag{7.51}$$

where $y(t) = (P^{-1} \otimes I_2)\eta(t)$. If the graph $\mathscr{G}$ is undirected, then L is symmetric and J is a diagonal matrix with real eigenvalues. However, when $\mathscr{G}$ is directed, some eigenvalues of L may be complex, and $J = \text{diag}(J_1, J_2, \dots, J_r)$, where

$$J_l = \begin{pmatrix} \mu_l & 0 & 0 & 0 \\ 1 & \ddots & 0 & 0 \\ 0 & \ddots & \ddots & 0 \\ 0 & 0 & 1 & \mu_l \end{pmatrix}_{N_l \times N_l}, \tag{7.52}$$

in which μ_l are the eigenvalues of the Laplacian matrix L, with multiplicity N_l, $l = 1, 2, \dots, r$, and $N_1 + N_2 + \dots + N_r = N$.

Let $P = (p_1, \dots, p_N)$, $P^{-1} = (q_1, \dots, q_N)^T$, $y(t) = (P^{-1} \otimes I_2)\eta(t) = (y_1^T, \dots, y_N^T)^T$, and $y_i = (y_{i1}, y_{i2})^T$. Note that if the network $\mathscr{G}$ contains a directed spanning tree, then by Lemma 2.7, 0 is a simple eigenvalue of the Laplacian matrix L, so

$$\dot{y}_1(t) = Ay_1(t), t \in [t_k, t_{k+1}). \tag{7.53}$$

Theorem 7.17 *Suppose that the network $\mathscr{G}$ contains a directed spanning tree. Then, second-order consensus in system (7.46) can be reached if and only if, in (7.51),*

$$\lim_{t\to\infty} \parallel y_i \parallel \to 0, i = 2, \dots, N. \tag{7.54}$$

Proof. (Sufficiency). Since the network $\mathscr{G}$ contains a directed spanning tree, $p_1 = 1_N/\sqrt{N}$ is the unit right eigenvector of the Laplacian matrix L associated with the simple zero eigenvalue $\mu_1 = 0$, where $LP = PJ$ and $P = (p_1, \dots, p_N)$. From $\lim_{t\to\infty} \parallel y_i \parallel \to 0$ for $i = 2, \dots, N$, one has

$$\lim_{t\to\infty}\left\|\eta(t)-\frac{1}{\sqrt{N}}(y_1(t)^T,\ldots,y_1(t)^T)^T\right\|$$
$$=\lim_{t\to\infty}\left\|(P\otimes I_2)y(t)-\frac{1}{\sqrt{N}}(y_1(t)^T,\ldots,y_1(t)^T)^T\right\|=0,$$

where $y_1(t)$ satisfies (7.53). Therefore, second-order consensus in system (7.46) is reached.

(Necessity). If second-order consensus in system (7.46) can be reached, then there exists a vector $\eta^*(t)\in R^2$ such that $\lim_{t\to\infty}\parallel\eta(t)-1_N\otimes\eta^*(t)\parallel=0$. Then, one has $0_N=P^{-1}L1_N=JP^{-1}1_N=J(q_1^T1_N,\ldots,q_N^T1_N)^T$. From the Jordan form (7.52) and $0=\mu_1<\mathscr{R}(\mu_2)\leq\ldots\leq\mathscr{R}(\mu_N)$, one gets $q_i^T1_N=0$ for $i=2,\ldots,N$. Therefore, $\parallel y_i(t)\parallel=\parallel(q_i^T\otimes I_2)\eta(t)\parallel\rightarrow\parallel(q_i^T1_N)\otimes\eta^*(t)\parallel=0$, as $t\to\infty$, for all $i=2,\ldots,N$. □

Corollary 7.18 *Suppose that the network $\mathscr{G}$ contains a directed spanning tree. Then, second-order consensus in system (7.46) can be reached if and only if the following $N-1$ systems are asymptotically stable:*

$$\dot{z}_i(t)=(A-\alpha\mu_iB)z_i(t)+\beta\mu_iBz_i(t_k),t\in[t_k,t_{k+1}),i=2,\ldots,N. \tag{7.55}$$

Proof. (Necessity). If $\lim_{t\to\infty}\parallel y_i\parallel\rightarrow 0$ for $i=2,\ldots,N$, then the $N-1$ systems (7.55) are asymptotically stable since the variables in (7.55) are the first term of each Jordan block in system (7.51).

(Sufficiency). It suffices to prove that if the $N-1$ systems (7.55) are asymptotically stable, then $\lim_{t\to\infty}\parallel y_i\parallel\rightarrow 0$ for $i=2,\ldots,N$. From the properties of the Jordan form (7.52), the asymptotical behavior in system (7.51) is dominated by the diagonal terms, therefore the conclusion follows. □

It is still very hard to check the conditions (7.54) and (7.55) in Theorem 7.17 and Corollary 7.18, which do not reveal how network structure affects the consensus behavior. Next, a theorem is derived to ensure consensus, depending on the control gains, the spectra of the Laplacian matrix, and the sampling period.

Theorem 7.19 *Suppose that the network $\mathscr{G}$ contains a directed spanning tree. Then, second-order consensus in system (7.46) can be reached if and only if*

$$0<\frac{\beta}{\alpha}<1, \tag{7.56}$$

and

$$\begin{aligned}&f(\alpha,\beta,\mu_i,T)\\&=\frac{(\beta/\alpha)^2}{1-(\beta/\alpha)}(\sin^2(d_iT)-\sinh^2(c_iT))(\cosh(c_iT)-\cos(d_iT))^2\\&-4\sin^2(d_iT)\sinh^2(c_iT)>0,i=2,\ldots,N,\end{aligned} \tag{7.57}$$

where $c_i=\sqrt{\frac{|\alpha|(\|\mu_i\|-sign(\alpha)\mathscr{R}(\mu_i))}{2}}$ *and* $d_i=\sqrt{\frac{|\alpha|(\|\mu_i\|+sign(\alpha)\mathscr{R}(\mu_i))}{2}}$.

Proof. It suffices to prove that system (7.55) is asymptotically stable if and only if the conditions (7.56) and (7.57) are satisfied.

From (7.55), it follows that

$$(e^{-(A-\alpha\mu_iB)t}z_i(t))' = e^{-(A-\alpha\mu_iB)t}\beta\mu_iBz_i(t_k), t \in [t_k, t_{k+1}). \tag{7.58}$$

Let $z_i = (z_{i1} \quad z_{i2})^T$ in (7.55). If $\alpha = 0$, then system (7.55) is $\dot{z}_{i1} = \beta\mu_i z_{i1}(t_k)$ for $t \in [t_k, t_{k+1})$, which is unstable. Thus, $\alpha \neq 0$.

Since $\mu_i \neq 0$ for $i = 2, \ldots, N$, matrix $(A - \alpha\mu_iB) = \begin{pmatrix} 0 & 1 \\ -\alpha\mu_i & 0 \end{pmatrix}$ is nonsingular with $(A - \alpha\mu_iB)^{-1} = \begin{pmatrix} 0 & -1/\alpha\mu_i \\ 1 & 0 \end{pmatrix}$. Integrating both sides of (7.58) from t_k to t gives

$$\begin{aligned} z_i(t) &= e^{(A-\alpha\mu_iB)(t-t_k)}(I_2 + (A - \alpha\mu_iB)^{-1}\beta\mu_iB)z_i(t_k) \\ &\quad -(A - \alpha\mu_iB)^{-1}\beta\mu_iBz_i(t_k) \\ &= e^{(A-\alpha\mu_iB)(t-t_k)}\begin{pmatrix} 1-\beta/\alpha & 0 \\ 0 & 1 \end{pmatrix}z_i(t_k) \\ &\quad + \begin{pmatrix} \beta/\alpha & 0 \\ 0 & 0 \end{pmatrix}z_i(t_k), t \in [t_k, t_{k+1}). \end{aligned} \tag{7.59}$$

If the sampled data is missing, i.e., $\beta = 0$, it is easy to see that $A - \alpha\mu_iB$ has at least one eigenvalue with a nonnegative real part and it thus follows from (7.59) that $z_i(t) = e^{(A-\alpha\mu_iB)(t-t_k)}z_i(t_k)$, which is unstable.

Next, it is to show that, by introducing sampled position data, the state in (7.59) is asymptotically stable.

Let $a_j + \mathbf{i}b_j$ be the eigenvalues of $(A - \alpha\mu_iB)$, $j = 1, 2$. Then,

$$\begin{aligned} a_j^2 - b_j^2 &= -\alpha\mathscr{R}(\mu_i), \\ 2a_jb_j &= -\alpha\mathscr{I}(\mu_i). \end{aligned}$$

The solutions of $a_j + \mathbf{i}b_j$ can be classified according to two cases, i.e., $\alpha > 0$ and $\alpha < 0$.

Case I ($\alpha > 0$). Simple calculation gives

$$\begin{aligned} a_1 + \mathbf{i}b_1 &= c_i - \mathbf{i}d_i, a_2 + \mathbf{i}b_2 = -c_i + \mathbf{i}d_i, \text{ if } \mathscr{I}(\mu_i) \geq 0, \\ a_1 + \mathbf{i}b_1 &= c_i + \mathbf{i}d_i, a_2 + \mathbf{i}b_2 = -c_i - \mathbf{i}d_i, \text{ if } \mathscr{I}(\mu_i) < 0, \end{aligned} \tag{7.60}$$

where $c_i = \sqrt{\frac{\alpha(\|\mu_i\|-\mathscr{R}(\mu_i))}{2}}$ and $d_i = \sqrt{\frac{\alpha(\|\mu_i\|+\mathscr{R}(\mu_i))}{2}}$. Without loss of generality, assume that $\mathscr{I}(\mu_i) \geq 0$. For the case of $\mathscr{I}(\mu_i) < 0$, the derived conditions in (7.56) and (7.57) of the asymptotical stability of system (7.59) for both $\mathscr{I}(\mu_i) \geq 0$ and $\mathscr{I}(\mu_i) < 0$ are the same since $f(\alpha, \beta, \mu_i, T)$ is an even function on c_i and d_i.

Let $\rho_1(t) = \cosh((c_i - \mathbf{i}d_i)(t))$ and $\rho_2(t) = \sinh((c_i - \mathbf{i}d_i)(t))$. By calculation,

$$e^{(A-\alpha\mu_i B)(t-t_k)} = \begin{pmatrix} \rho_1(t-t_k) & \rho_2(t-t_k)/(c_i - \mathbf{i}d_i)) \\ (c_i - \mathbf{i}d_i)\rho_2(t-t_k) & \rho_1(t-t_k) \end{pmatrix}. \tag{7.61}$$

Substituting (7.61) into (7.59), one obtains

$$z_i(t) = \begin{pmatrix} \rho_1(t-t_k)(1-\beta/\alpha)+\beta/\alpha & \rho_2(t-t_k)/(c_i - \mathbf{i}d_i) \\ (c_i - \mathbf{i}d_i)\rho_2(t-t_k)(1-\beta/\alpha) & \rho_1(t-t_k) \end{pmatrix} z_i(t_k),$$

$$t \in [t_k, t_{k+1}). \tag{7.62}$$

Let $C(t) = \begin{pmatrix} \rho_1(t)(1-\beta/\alpha)+\beta/\alpha & \rho_2(t)/(c_i - \mathbf{i}d_i) \\ (c_i - \mathbf{i}d_i)\rho_2(t)(1-\beta/\alpha) & \rho_1(t) \end{pmatrix}$. It is easy to see that $C(t)$ is bounded on $[0, t]$. So, for $0 = t_0 < t_1 < \ldots < t_k < \ldots$ and $t_{k+1} - t_k = T$, one has

$$z_i(t) = C(t-t_k)C^k(T)z_i(t_0), t \in [t_k, t_{k+1}). \tag{7.63}$$

Since $C(t-t_k)$ is bounded when $t \in [t_k, t_{k+1})$, $z_i(t) \to 0$ if and only if all eigenvalues λ of $C(T)$ satisfy $\| \lambda \| < 1$.

Let $|\lambda I_2 - C(T)| = 0$, so that

$$\lambda^2 - \left(2\cosh((c_i - \mathbf{i}d_i)T) + \frac{\beta}{\alpha}(1 - \cosh((c_i - \mathbf{i}d)_i T))\right)\lambda$$

$$+\left(1 - \frac{\beta}{\alpha}\right) + \frac{\beta}{\alpha}\cosh((c_i - \mathbf{i}d_i)T) = 0. \tag{7.64}$$

Let $\lambda = \frac{s+1}{s-1}$. Then, (7.64) can be transformed to

$$0 = (1 - \cosh((c_i - \mathbf{i}d_i)T))\left(1 - \frac{\beta}{\alpha}\right)s^2 + \frac{\beta}{\alpha}(1$$

$$- \cosh((c_i - \mathbf{i}d_i)T))s + (1 + \cosh((c_i - \mathbf{i}d_i)T)). \tag{7.65}$$

It is well known that $\| \lambda \| < 1$ in (7.64) if and only if $\mathscr{R}(s) < 0$ in (7.65). Therefore, $z_i(t) \to 0$ if and only if all the roots in (7.65) have negative real parts.

If $\cosh((c_i - \mathbf{i}d_i)T) = 1$ or $\frac{\beta}{\alpha} = 1$, then $C(T)$ has an eigenvalue $\lambda = 1$. Therefore, $(1 - \cosh((c_i - \mathbf{i}d_i)T))\left(1 - \frac{\beta}{\alpha}\right) \neq 0$ and (7.65) can be simplified to

$$s^2 + \frac{\beta/\alpha}{1-\beta/\alpha}s + \frac{(1 + \cosh((c_i - \mathbf{i}d_i)T))}{(1-\beta/\alpha)(1 - \cosh((c_i - \mathbf{i}d_i)T))} = 0. \tag{7.66}$$

Let $D_i = \frac{1+\cosh((c_i - \mathbf{i}d_i)T)}{1-\cosh((c_i - \mathbf{i}d_i)T)}$. By Lemma 7.2, (7.66) is stable if and only if

$$\frac{\beta}{\alpha}\left(\frac{\beta}{\alpha} - 1\right) < 0 \tag{7.67}$$

and

$$\frac{(\beta/\alpha)^2}{1-(\beta/\alpha)}\mathscr{R}(D_i) > \mathscr{I}^2(D_i). \tag{7.68}$$

By solving (7.68), one obtains the condition (7.57). Therefore, $z_i(t) \to 0$ if and only if both (7.56) and (7.57) are satisfied. By Corollary 7.18, second-order consensus in system (7.46) is reached if and only if both (7.56) and (7.57) are satisfied.

Case II ($\alpha < 0$). Simple calculation gives

$$\begin{aligned} a_1 + \mathbf{i}b_1 &= c_i - \mathbf{i}d_i, a_2 + \mathbf{i}b_2 = -c_i + \mathbf{i}d_i, \text{ if } \mathscr{I}(\mu_i) \geq 0, \\ a_1 + \mathbf{i}b_1 &= c_i + \mathbf{i}d_i, a_2 + \mathbf{i}b_2 = -c_i - \mathbf{i}d_i, \text{ if } \mathscr{I}(\mu_i) < 0, \end{aligned} \tag{7.69}$$

where $c_i = \sqrt{\frac{-\alpha(\|\mu_i\| + \mathscr{R}(\mu_i))}{2}}$ and $d_i = \sqrt{\frac{-\alpha(\|\mu_i\| - \mathscr{R}(\mu_i))}{2}}$. The rest of the proof is similar to that of Case I. □

Remark 7.20 In Theorem 7.19, a necessary and sufficient condition for second-order consensus in multi-agent dynamical system (7.46) is established. For a given network, we can design appropriate α, β, and T, such that the conditions (7.56) and (7.57) in Theorem 7.19 are satisfied. It is interesting to see that f increases as the parameter β/α increases. Thus, one can choose a large value of β/α such that (7.57) holds. Since the condition (7.57) holds for all $i = 2, \ldots, N$, one can find a stable consensus region as follows: $\mathscr{S} = \{c + \mathbf{i}d \mid f(\alpha, \beta, c + \mathbf{i}d, T) > 0\}$, where c and d are real. Then, the problem is transformed to finding if all the nonzero eigenvalues of the Laplacian matrix lie in the stable consensus region $\mathscr{S}$, i.e., $\mu_i \in \mathscr{S}$ for all $i = 2, \ldots, N$. In [31, 65], disconnected synchronization regions of complex networks were discussed. It was shown that there indeed exist some disconnected synchronization regions for several particular complex networks when the synchronous state is an equilibrium point. In this subsection, by introducing sampled position data in the consensus algorithm, it will be shown by simulation that there exist some disconnected regions for choosing appropriate sampling periods.

Corollary 7.21 *Suppose that the network $\mathscr{G}$ contains a directed spanning tree and all the eigenvalues of its Laplacian matrix are real. Then, second-order consensus in system (7.46) can be reached if and only if*

$$0 < \beta < \alpha \tag{7.70}$$

and

$$\sqrt{\alpha\mu_i}T \neq k\pi, i = 2, \ldots, N, k = 0, 1, \ldots. \tag{7.71}$$

Proof. If $\alpha > 0$, then $c_i = 0$ and $d_i = \sqrt{\alpha\mu_i}$. Therefore, $\sinh(c_iT) = 0$ and $\cosh(c_iT) = 1$. The condition in (7.57) can be simplified to

$$\frac{(\beta/\alpha)^2}{1-(\beta/\alpha)}\sin^2(d_iT)(1-\cos(d_iT))^2 > 0, i = 2, \dots, N, \tag{7.72}$$

which is equivalent to the conditions (7.70) and (7.71).

If $\alpha < 0$, then $c_i = \sqrt{-\alpha\mu_i}$ and $d_i = 0$. The condition in (7.57) is

$$\frac{(\beta/\alpha)^2}{1-(\beta/\alpha)}(-\sinh^2(c_iT))(\cosh(c_iT)-1)^2 > 0, i = 2, \dots, N, \tag{7.73}$$

which cannot be satisfied since $\beta/\alpha < 1$ according to the condition (7.56). □

Remark 7.22 If all the eigenvalues of the Laplacian matrix are real, which includes undirected networks as a special case, then the condition (7.71), i.e., $T \neq \frac{k\pi}{\sqrt{\alpha\mu_i}}$, is very easy to verify and apply. It is quite interesting to see that second-order consensus in the multi-agent system (7.46) can be reached if and only if $0 < \beta < \alpha$ and the sampling period T is not of some particular values.

Usually, the convergence rate around the critical points $T = \frac{k\pi}{\sqrt{\alpha\mu_i}}$ is very low. Therefore, it is hard to achieve good performance for a large T in a very large-scale network. A corollary is given below to simplify the above theoretical analysis.

Corollary 7.23 *Suppose that the network $\mathcal{G}$ contains a directed spanning tree and all the eigenvalues of its Laplacian matrix are real. Then, second-order consensus in system (7.46) can be reached if (7.70) is satisfied and*

$$0 < T < \frac{\pi}{\sqrt{\alpha\mu_N}}. \tag{7.74}$$

Corollary 7.23 implies that if the network $\mathcal{G}$ contains a directed spanning tree and all the eigenvalues of the Laplacian matrix are real, then second-order consensus in system (7.46) can be reached provided that the sampling period is less than the critical value $\frac{\pi}{\sqrt{\alpha\mu_N}}$ depending on the largest eigenvalue of the Laplacian matrix. However, surprisingly, second-order consensus in system (7.46) cannot be reached in a general directed network with complex Laplacian eigenvalues for a sufficiently small or a sufficiently large sampling period T, as shown blow.

Corollary 7.24 *Suppose that the network $\mathcal{G}$ contains a directed spanning tree and there is at least one eigenvalue of its Laplacian matrix with a nonzero imaginary part. Then, second-order consensus in the system (7.46) cannot be reached for a sufficiently small or a sufficiently large sampling period T.*

Proof. Without loss of generality, suppose that $\mathscr{I}(\mu_k) \neq 0$, $2 \leq k \leq N$. Then, from the condition (7.57), $c_k \neq 0$ and $d_k \neq 0$. Consider the following Taylor series for a sufficiently small T:

$$\begin{aligned}
\sinh(c_k T) &= c_k T + \frac{(c_k T)^3}{3!} + o(T^3) \\
\cosh(c_k T) &= 1 + \frac{(c_k T)^2}{2!} + o(T^3) \\
\sin(d_k T) &= d_k T - \frac{(d_k T)^3}{3!} + o(T^3) \\
\cos(d_k T) &= 1 - \frac{(d_k T)^2}{2!} + o(T^3).
\end{aligned} \tag{7.75}$$

Substituting (7.75) into (7.57) gives

$$\begin{aligned}
& f(\alpha, \beta, \mu_k, T) \\
&= \frac{(\beta/\alpha)^2}{1-(\beta/\alpha)}(\sin^2(d_k T) - \sinh^2(c_k T)) \\
&\times(\cosh(c_k T) - \cos(d_k T))^2 - 4\sin^2(d_k T)\sinh^2(c_k T) \\
&= \frac{(\beta/\alpha)^2}{4(1-(\beta/\alpha))}(c_k^2 - d_k^2)^3 T^6 + o(T^6) - 4c_k^2 d_k^2 T^2 - o(T^2) > 0.
\end{aligned} \tag{7.76}$$

which cannot be satisfied for a sufficiently small sampling period T since $c_k \neq 0$ and $d_k \neq 0$.

If the sampling period T is sufficiently large, then $\sin^2(d_k T) - \sinh^2(c_k T) < 0$ for $c_k \neq 0$ and $d_k \neq 0$, and thus the condition (7.57) is not satisfied. □

Remark 7.25 If the network $\mathscr{G}$ contains a directed spanning tree and all the eigenvalues of the Laplacian matrix are real, then second-order consensus in system (7.46) can be reached for a sufficiently small sampling period T as stated in Corollary 7.21. However, if there is at least one eigenvalue of the Laplacian matrix having a nonzero imaginary part, then second-order consensus cannot be reached for a sufficiently small sampling period T as shown in Corollary 7.24, which is inconsistent with the common intuition that the consensus protocol (7.46) should be better if the sampled information is more accurate for a small sampling period. Interestingly, the nonzero imaginary part of the eigenvalue of the Laplacian matrix leads to possible instability of consensus.

7.3.2 Simulation Examples

Example 1. Second-order consensus in a multi-agent system with an undirected topology Consider the multi-agent system (7.46) with an undirected topology, where

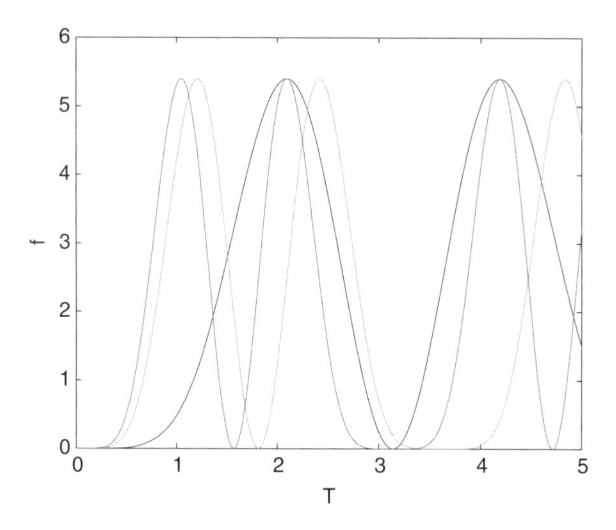

Figure 7.7 States of $f(\alpha, \beta, \mu_i, \cdot)$ vs. the sampling period T, $i = 2, 3, 4$

$L = \begin{pmatrix} 3 & -1 & -1 & -1 \\ -1 & 2 & -1 & 0 \\ -1 & -1 & 2 & 0 \\ -1 & 0 & 0 & 1 \end{pmatrix}$, $\alpha = 1$, and $\beta = 0.8$. By simple calculation, $\mu_1 = 0$, $\mu_2 = 1$, $\mu_3 = 3$, and $\mu_4 = 4$. From Corollary 7.21, the system can reach second-order consensus if and only $T \neq \frac{k\pi}{\sqrt{\alpha\mu_i}}$, for $i = 2, 3, 4$ and $k = 0, 1, \ldots$. It is easy to obtain $\frac{k\pi}{\sqrt{\alpha\mu_4}} \approx 1.5708$, $\frac{k\pi}{\sqrt{\alpha\mu_3}} \approx 1.8138$, and $\frac{k\pi}{\sqrt{\alpha\mu_2}} \approx 3.1416$. Consider the sampling period T as a variable of $f(\alpha, \beta, \mu_i, \cdot)$. The states of T versus f are shown in Fig. 7.7. It is easy to see that second-order consensus in the system can be reached if $T = 1.0$, $T = 1.57$, or $T = 2.5$, while the convergence is not good when $T = 1.57$ around the critical point $\frac{k\pi}{\sqrt{\alpha\mu_4}}$. The position and velocity states of all the agents are shown in Fig. 7.8.

Example 2. Second-order consensus in a multi-agent system with a directed topology Consider the multi-agent system (7.46) with a directed topology, where $L = \begin{pmatrix} 2 & 0 & -2 & 0 \\ -1 & 1 & 0 & 0 \\ 0 & -5 & 5 & 0 \\ 0 & -1 & 0 & 1 \end{pmatrix}$, $\alpha = 1$, and $\beta = 0.8$. By simple calculation, $\mu_1 = 0$, $\mu_2 = 1$, $\mu_3 = 4 + \mathbf{i}$, and $\mu_4 = 4 - \mathbf{i}$. From Theorem 7.19, the multi-agent system can reach second-order consensus if and only $f(\alpha, \beta, \mu_i, T) > 0$, for $i = 2, 3, 4$. Consider the sampling period T as a variable of $f(\alpha, \beta, \mu_i, \cdot)$. The states of T versus f are shown in Fig. 7.9. It is easy to see that second-order consensus in the system can be reached if $T = 1.0$, or $T = 2.0$, while it cannot be reached for a sufficiently small $T = 0.1$ according to Corollary 7.24 or $T = 1.5$ where $f(\alpha, \beta, \mu_3, T) < 0$. Moreover, when Fig. 7.9 is amplified around the origin, as shown in the inner Fig (a), where there are two same lines with complex eigenvalues, i.e., $f(\alpha, \beta, \mu_3, \cdot)$ and $f(\alpha, \beta, \mu_4, \cdot)$ under

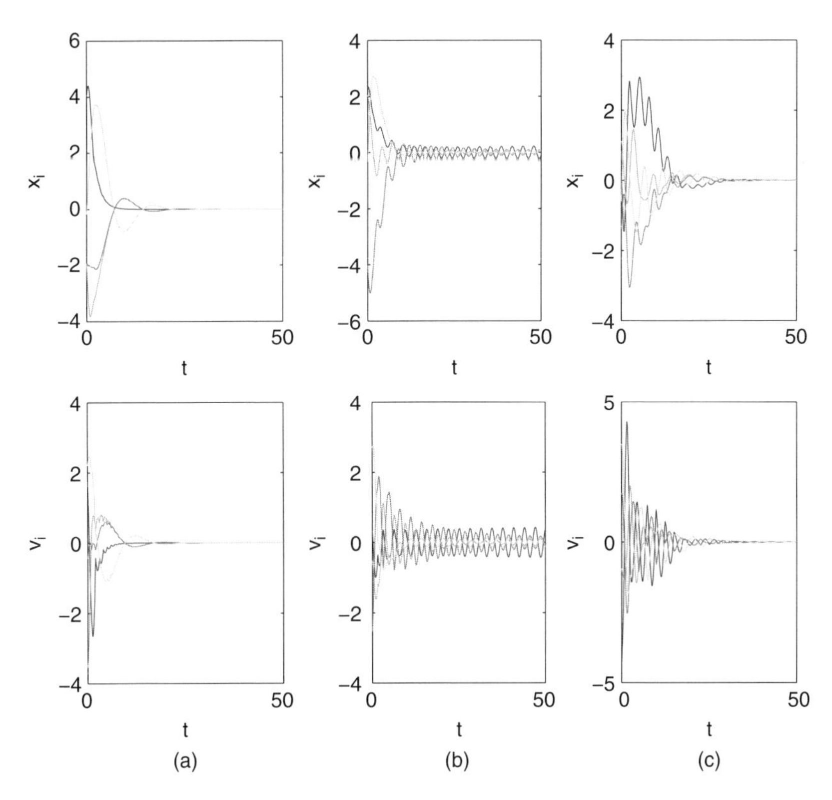

Figure 7.8 Position and velocity states of agents, where $T = 1.0$ (a), $T = 1.57$ (b), and $T = 2.5$ (c)

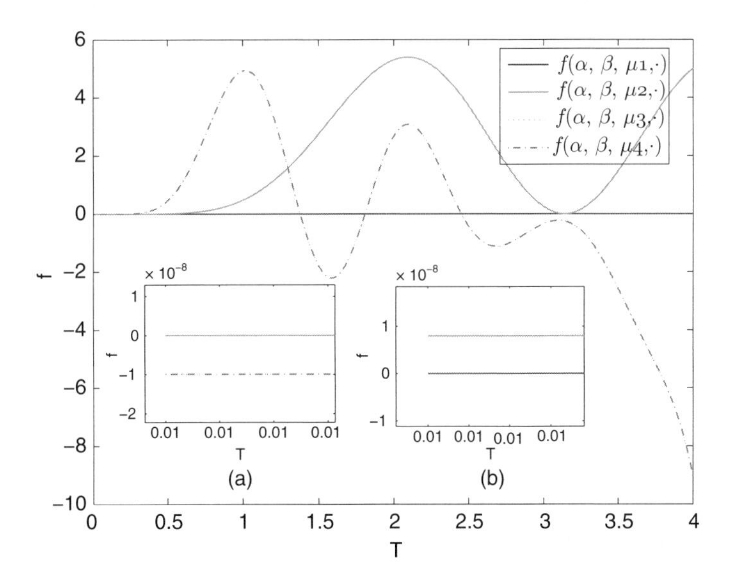

Figure 7.9 States of $f(\alpha, \beta, \mu_i, \cdot)$ vs. the sampling period T, $i = 1, 2, 3, 4$

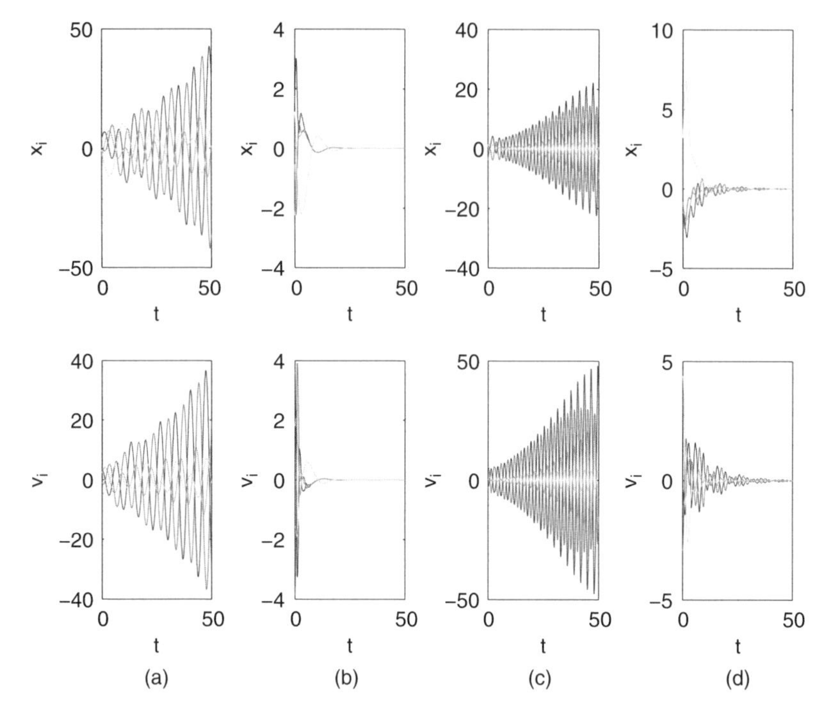

Figure 7.10 Position and velocity states of agents, where $T = 0.1$ (a), $T = 1.0$ (b), $T = 1.5$ (c), and $T = 2.0$ (d)

the zero. If Fig. 7.9 is further amplified around the origin, as shown in the inner Fig (b), where there is one line with real eigenvalue, i.e., $f(\alpha, \beta, \mu_2, \cdot)$, above zero. The position and velocity states of all the agents are shown in Fig. 7.10.

From Figs. 7.7 and 7.9, one can see that if the eigenvalue μ_i of the Laplacian matrix is real, then $f(\alpha, \beta, \mu_i, T) > 0$, except at some critical points $T = \frac{k\pi}{\sqrt{\alpha\mu_i}}$, where $f = 0$. However, if the eigenvalue μ_i of the Laplacian matrix is complex with a nonzero imaginary part, then $f(\alpha, \beta, \mu_i, T) \leq 0$ for a sufficiently small T or a sufficiently large T. Even in this case, there may be some disconnected stable consensus regions by choosing an appropriate sampling period T, as shown in Fig. 7.10. Therefore, the design of an appropriate sampling period T plays a key role in reaching consensus. In addition, one can design appropriate coupling strengths α and β such that second-order consensus in the multi-agent system (7.46) can be reached, as guaranteed by theory. Details are omitted for brevity.

7.4 Consensus of Multi-agent Systems with Nonlinear Dynamics and Sampled Information

In the previous subsection, consensus for second-order multi-agent systems with sampled-data information was studied, but it has been observed that only static consensus can be achieved. There is little work reported in the literature on consensus with time-varying final velocities for multi-agent systems. Very recently, a new class of consensus algorithms for coupled harmonic oscillators with a directed topology was introduced and discussed in [96], where time-varying asymptotic oscillatory velocities were generated. For multi-agent systems with intrinsic nonlinear dynamics, it is the agents' intrinsic dynamics that determine the final consensus states as the coupling term gradually vanishes when consensus is achieved asymptotically. Obviously, a consensus trajectory is a solution of the isolated system. And, the consensus trajectory may be an isolated equilibrium point, a periodic orbit, or even a chaotic orbit in some applications. Pertinent works along this line include [133–147], where consensus problems of continuous-time coupled systems are investigated.

Prior to the protocols proposed in this chapter, the coupling laws employed in the above-mentioned works are based on a common assumption that each agent can communicate with its neighbors all the time, which is clearly impractical in many cases. In this section, the focus is on solving the sampled-data consensus problem for multi-agent systems with intrinsic nonlinear dynamics, where each agent's dynamics are driven by both a nonlinear term depending on its own state and a navigational feedback term based on the sampled relative states between its own and the neighbors' ones. Based on a novel delayed-input approach, the sampled-data multi-agent system is converted to a nonlinear system with delay in the feedback.

The commonly studied first-order continuous-time multi-agent system model is described by [81, 98]

$$\dot{x}_i(t) = -\alpha \sum_{j=1}^{N} l_{ij} x_j(t), \quad i = 1, 2, \cdots, N, \tag{7.77}$$

where $x_i \in \mathbb{R}^n$ is the position state of the ith agent, α represents the coupling strength, and $L = [l_{ij}]_{N\times N}$ is the Laplacian matrix of the communication topology $\mathcal{G}(\mathcal{A})$. Suppose that $\mathcal{G}(\mathcal{A})$ is strongly connected. When the states of multiple agents reach consensus, all the position states of agents converge to $\sum_{j=1}^{N} \xi_j x_j(0)$, which depends only on the initial positions of the agents with $\xi = (\xi_1, \cdots, \xi_N)^T$ being the nonnegative left eigenvector of L associated with the eigenvalue 0, satisfying $\xi^T \mathbf{1}_N = 1$, where $\mathbf{1}_N$ is the N-dimensional column vector with all entries being 1.

However, in practical applications of multi-agent formations, the velocity of each agent is generally not a constant but a time-varying variable. To cope with this, consider the following multi-agent dynamical system:

$$\dot{x}_i(t) = f(x_i(t), t) + u_i(t), \quad i = 1, 2, \cdots, N, \tag{7.78}$$

where $f(x_i(t), t) \in \mathbb{R}^n$ describes the intrinsic nonlinear dynamics of the ith agent, and $u_i(t) \in \mathbb{R}^n$ is a control input to be designed. It is assumed that the relative states between each pair of neighboring agents are measured at discrete sampling times t_k, $k \in \mathbb{N}$, and the control inputs are generated based on zero-order hold. It is also assumed that there exists a positive constant h such that $t_{k+1} - t_k \leq h$, $k \in \mathbb{N}$. The objective here is to design a distributed control protocol $u_i(t)$ based on the relative sampling-data information such that $\lim_{t\to\infty} \| x_i(t) - x_j(t) \| = 0$, for all $i, j = 1, 2, \cdots, N$. An admissible protocol can be deduced from (7.77) by employing sampling technique and a zero-order hold circuit, that is,

$$u_i(t) = -\alpha \sum_{j=1}^{N} l_{ij} x_j(t_k), \quad t_k \le t < t_{k+1}, \quad k \in \mathbb{N}, \quad i = 1, 2, \cdots, N. \tag{7.79}$$

Substituting (7.79) into (7.78) gives

$$\dot{x}_i(t) = f(x_i(t), t) - \alpha \sum_{j=1}^{N} l_{ij} x_j(t_k), \quad t_k \le t < t_{k+1}, \quad k \in \mathbb{N}, \quad i = 1, 2, \cdots, N. \tag{7.80}$$

Clearly, if consensus can be achieved, it is natural to require a consensus state $s(t) \in \mathbb{R}^n$ of the system (7.80) to be a trajectory of an isolated node satisfying

$$\dot{s}(t) = f(s(t), t) \tag{7.81}$$

Here, $s(t)$ may be an isolated equilibrium point, a periodic orbit, or even a chaotic orbit in some applications.

Assumption 7.26 *There exists a nonnegative constant ρ such that*

$$\| f(x_1, t) - f(x_2, t) \| \le \rho \| x_1 - x_2 \|, \; \forall\, x_1, x_2 \in \mathbb{R}^n; \; \forall\, t \ge 0. \tag{7.82}$$

Note that Assumption 7.26 is a Lipschitz condition, satisfied by many well-known systems.

Let $d_k(t) = t - t_k$, for $t \in [t_k, t_{k+1})$, $k \in \mathbb{N}$. Then, $t_k = t - d_k(t)$ with $0 \le d_k(t) < h$, so system (7.80) can be written as

$$\begin{aligned} \dot{x}_i(t) = & f(x_i(t), t) - \alpha \sum_{j=1}^{N} l_{ij} x_j(t - d_k(t)), \quad t_k \le t < t_{k+1}, \\ & i = 1, 2, \cdots, N, \quad k \in \mathbb{N}. \end{aligned} \tag{7.83}$$

Before moving forward, the following lemma is provided.

Lemma 7.27 *[39] Suppose that $x \in \mathbb{R}^n$, $P = P^T \in \mathbb{R}^{n\times n}$, and $H \in \mathbb{R}^{m\times n}$ such that* $\mathrm{Rank}(H) = l < n$. *Then, the following statements are equivalent:*

1. $x^T P x < 0, \forall x \in \{x : Hx = 0, x \neq 0\}$,
2. $P - \sigma H^T H < 0$, *for some scalar* $\sigma \in \mathbb{R}$,
3. $\exists X \in \mathbb{R}^{n\times m}$ *such that* $P + XH + H^T X^T < 0$,
4. ${H^{\perp}}^T P H^{\perp} < 0$, *where* $H^{\perp}$ *is the kernel of* H, *i.e.,* $HH^{\perp} = 0$.

Remark 7.28 From a delayed-input approach, the sampled-data feedback system (7.80) is equivalently transformed into a continuous-time system with a time-varying delay in the feedback as shown in (7.83). In the following, theoretical analysis is performed based on the time-varying delay system (7.83).

7.4.1 The Case with a Fixed and Strongly Connected Topology

In this subsection, consensus in strongly connected networks of multiple agents with sampled-data information is studied.

Let $e_i(t) = x_i(t) - x_0(t)$ represent the position vector of the ith agent relative to the weighted average position of all the agents in system (7.80), where $x_0(t) = \sum_{j=1}^{N} \xi_j x_j(t)$, and $\xi = (\xi_1, \xi_2, \cdots, \xi_N)^T$ is the positive left eigenvector of Laplacian matrix L associated with its zero eigenvalue, satisfying $\xi^T \mathbf{1}_N = 1$. Then, for $t \in [t_k, t_{k+1})$ and arbitrarily given $k \in \mathbb{N}$, one has the following error dynamical system:

$$\begin{aligned}\dot{e}_i(t) =& f(x_i(t), t) - f(x_0(t), t) - \sum_{j=1}^{N} \xi_j (f(x_j(t), t) - f(x_0(t), t)) \\ & -\alpha \sum_{j=1}^{N} l_{ij} e_j(t - d_k(t)), \quad i = 1, 2, \cdots, N. \end{aligned} \tag{7.84}$$

Let $e(t) = (e_1^T(t), \cdots, e_N{}^T(t))^T$ and $e(t - d_k(t)) = (e_1^T(t - d_k(t)), \cdots, e_N^T(t - d_k(t)))^T$. Then, system (7.84) can be written as

$$\dot{e}(t) = [(I - \mathbf{1}_N \xi^T) \otimes I_n] F(x(t), t) - \alpha (L \otimes I_n) e(t - d_k(t)), \quad t \in [t_k, t_{k+1}), \tag{7.85}$$

where $F(x(t), t) = f(x(t), t) - \mathbf{1}_N \otimes f(x_0(t), t), f(x(t), t) = (f^T(x_1(t), t), \cdots, f^T(x_N(t), t))^T$.

Theorem 7.29 *Suppose that the network is strongly connected and Assumption 7.26 holds. Then, consensus in system (7.80) is achieved if there exist symmetric matrices $P, Q \in \mathbb{R}^{N\times N}$ such that $E^T P E > 0$, $E^T Q E > 0$, and the following LMI holds:*

$$\begin{pmatrix} 2h^2 E^T Q E - E^T E & * & * & * \\ E^T P E & \rho E^T E - E^T Q E & * & * \\ O & -E^T Q E & -2E^T Q E & * \\ -\alpha h^2 E^T L^T Q E & -\alpha E^T L^T P E & -E^T Q E & \alpha^2 h^2 E^T L^T Q L E - E^T Q E \end{pmatrix} < 0, \tag{7.86}$$

where

$$E = \begin{pmatrix} I_{N-1} \\ -\frac{\bar{\xi}^T}{\xi_N} \end{pmatrix} \in \mathbb{R}^{N\times(N-1)}, \quad \bar{\xi} = (\xi_1, \cdots, \xi_{N-1})^T \in \mathbb{R}^{N-1}, \tag{7.87}$$

and $\xi = (\xi_1, \xi_2, \cdots, \xi_N)^T$ is the positive left eigenvector of Laplacian matrix L associated with its zero eigenvalue, satisfying $\xi^T \mathbf{1}_N = 1$.

Proof. Construct the following Lyapunov–Krasovskii functional:

$$V(t) = e^T(t)(P \otimes I_n)e(t) + 2h \int_{-h}^{0} \int_{t+\theta}^{t} \dot{e}^T(s)(Q \otimes I_n)\dot{e}(s)dsd\theta, \tag{7.88}$$

where symmetric matrices P and $Q \in \mathbb{R}^{N\times N}$ satisfy $E^T PE > 0$ and $E^T QE > 0$, with

$$E = \begin{pmatrix} I_{N-1} \\ -\frac{\bar{\xi}^T}{\xi_N} \end{pmatrix} \in \mathbb{R}^{N\times(N-1)}, \tag{7.89}$$

in which $\bar{\xi} = (\xi_1, \cdots, \xi_{N-1})^T \in \mathbb{R}^{N-1}$, $\xi = (\xi_1, \xi_2, \cdots, \xi_N)^T \in \mathbb{R}^N$ is the positive left eigenvector of Laplacian matrix L associated with its zero eigenvalue, satisfying $\xi^T \mathbf{1}_N = 1$. Since $(\xi^T \otimes I_n)e(t) = 0$, $V(t)$ defined by (7.88) is a valid Lyapunov–Krasovskii functional for system (7.85).

For $t \in [t_k, t_{k+1})$ and arbitrarily given k, taking the time derivative of $V(t)$ along the trajectories of (7.85) gives

$$\begin{aligned}
\dot{V}(t) =& 2e^T(t)(P \otimes I_n)\dot{e}(t) + 2h^2\dot{e}^T(t)(Q \otimes I_n)\dot{e}(t) - 2h \int_{t-h}^{t} \dot{e}^T(s)(Q \otimes I_n)\dot{e}(s)ds \\
=& 2e^T(t)(P \otimes I_n)\{[(I - \mathbf{1}_N\xi^T) \otimes I_n]F(x(t), t) - \alpha(L \otimes I_n)e(t - d_k(t))\} \\
&+2h^2\dot{e}^T(t)(Q \otimes I_n)\dot{e}(t) - h \int_{t-h}^{t} \dot{e}^T(s)(Q \otimes I_n)\dot{e}(s)ds \\
&-h \int_{t-h}^{t} \dot{e}^T(s)(Q \otimes I_n)\dot{e}(s)ds \\
\leq& 2e^T(t)(P \otimes I_n)\{[(I - \mathbf{1}_N\xi^T) \otimes I_n]F(x(t), t) - \alpha(L \otimes I_n)e(t - d_k(t))\} \\
&+2h^2\dot{e}^T(t)(Q \otimes I_n)\dot{e}(t) - h \int_{t-h}^{t} \dot{e}^T(s)(Q \otimes I_n)\dot{e}(s)ds \\
&-(h - d_k(t)) \int_{t-h}^{t-d_k(t)} \dot{e}^T(s)(Q \otimes I_n)\dot{e}(s)ds \\
&-d_k(t) \int_{t-d_k(t)}^{t} \dot{e}^T(s)(Q \otimes I_n)\dot{e}(s)ds.
\end{aligned}$$

It follows from Jensen's inequality [48] that

$$\begin{aligned}
&-h \int_{t-h}^{t} \dot{e}^T(s)(Q \otimes I_n)\dot{e}(s)ds \\
&\leq -[e(t) - e(t-h)]^T(Q \otimes I_n)[e(t) - e(t-h)],
\end{aligned} \tag{7.90}$$

$$-(h-d_k(t))\int_{t-h}^{t-d_k(t)} \dot{e}^T(s)(Q\otimes I_n)\dot{e}(s)ds$$
$$\leq -[e(t-d_k(t))-e(t-h)]^T(Q\otimes I_n)[e(t-d_k(t))-e(t-h)],$$
$$\text{and } -d_k(t)\int_{t-d_k(t)}^{t} \dot{e}^T(s)(Q\otimes I_n)\dot{e}(s)ds \tag{7.91}$$
$$\leq -[e(t)-e(t-d_k(t))]^T(Q\otimes I_n)[e(t)-e(t-d_k(t))]. \tag{7.92}$$

Let $e(t)-e(t-h)=\mu(t)$, $e(t-d_k(t))-e(t-h)=\nu(t)$ and $e(t)-e(t-d_k(t))=\omega(t)$. Then, according to (7.90)–(7.92) and Assumption 7.26, one has

$$\begin{aligned}\dot{V}(t) \leq\ & 2e^T(t)(P\otimes I_n)\{[(I-\mathbf{1}_N\xi^T)\otimes I_n]F(x(t),t)-\alpha(L\otimes I_n)e(t-d_k(t))\}\\ &+h^2\dot{e}^T(t)(Q\otimes I_n)\dot{e}(t)-\mu^T(t)(Q\otimes I_n)\mu(t)-\nu^T(t)(Q\otimes I_n)\nu(t)\\ &-\omega^T(t)(Q\otimes I_n)\omega(t)\\ \leq\ & 2e^T(t)(P\otimes I_n)\{[(I-\mathbf{1}_N\xi^T)\otimes I_n]F(x(t),t)-\alpha(L\otimes I_n)e(t-d_k(t))\}\\ &-F^T(x(t),t)F(x(t),t)+\rho e^T(t)e(t)+h^2\dot{e}^T(t)(Q\otimes I_n)\dot{e}(t)\\ &-\mu^T(t)(Q\otimes I_n)\mu(t)-\nu^T(t)(Q\otimes I_n)\nu(t)-\omega^T(t)(Q\otimes I_n)\omega(t).\end{aligned}$$

Let $y(t)=[\dot{e}^T(t),e^T(t),e^T(t-d_k(t)),F^T(x(t),t),\mu^T(t),\nu^T(t),\omega^T(t)]^T$. Then,

$$\dot{V}(t)\leq y^T(t)(\Omega\otimes I_n)y(t), \tag{7.93}$$

where

$$\Omega=\begin{pmatrix} 2h^2Q & O & O & O & O & O & O\\ O & \rho I_N & -\alpha PL & P(I-\mathbf{1}_N\xi^T) & O & O & O\\ O & -\alpha L^TP^T & O & O & O & O & O\\ O & [P(I-\mathbf{1}_N\xi^T)]^T & O & -I & O & O & O\\ O & O & O & O & -Q & O & O\\ O & O & O & O & O & -Q & O\\ O & O & O & O & O & O & -Q\end{pmatrix}.$$

Furthermore, it follows from (7.80) and the fact $[(\mathbf{1}_7^T\otimes\xi^T)\otimes I_n]y(t)=0$ that

$$Ay(t)=0_{7Nn}, \tag{7.94}$$

where $A=\begin{pmatrix}S\\T\end{pmatrix}\otimes I_n$, $S=\begin{pmatrix} I_N & O_N & \alpha L & -(I-\mathbf{1}_N\xi^T) & O_N & O_N & O_N\\ O_N & I_N & -I_N & O_N & -I_N & I_N & O_N\end{pmatrix}\in\mathbb{R}^{2N\times 7N}$, and $T=(I_7^T\otimes\xi^T)\in\mathbb{R}^{7\times 7N}$. System (7.80) is asymptotically stable if, for all $y(t)$ satisfying $Ay(t)=0$,

$$y^T(t)(\Omega\otimes I_n)y(t)<0. \tag{7.95}$$

According to Lemma 7.27, $y^T(t)(\Omega \otimes I_n)y(t) < 0$ is equivalent to

$$A^{\perp^T}(\Omega \otimes I_n)A^{\perp} < 0, \tag{7.96}$$

where

$$A^{\perp} = \begin{pmatrix} E & O & O & -\alpha LE & O \\ O & E & O & O & O \\ O & O & O & E & O \\ E & O & O & O & O \\ O & E & E & O & O \\ O & O & E & E & O \\ O & O & O & O & E \end{pmatrix} \otimes I_n. \tag{7.97}$$

Thus, inequality (7.96) can be rewritten as

$$\begin{pmatrix} \Phi_{11} & * & * & * & * \\ E^TPE & \Phi_{22} & * & * & * \\ O & -E^TQE & \Phi_{33} & * & * \\ -\alpha h^2E^TL^TQE & -\alpha E^TL^TPE & -E^TQE & \Phi_{44} & * \\ O & O & O & O & -E^TQE \end{pmatrix} < 0, \tag{7.98}$$

where

$$\begin{aligned} \Phi_{11} &= 2h^2E^TQE - E^TE, \\ \Phi_{22} &= \rho E^TE - E^TQE, \\ \Phi_{33} &= -2E^TQE, \\ \Phi_{44} &= \alpha^2h^2E^TL^TQLE - E^TQE, \end{aligned}$$

which altogether are equivalent to (7.86) because $E^TQE > 0$. Therefore, $\dot{V}(t) < -\epsilon \parallel e(t)\parallel^2$ for some sufficiently small $\epsilon > 0$, which ensures the achievement of consensus in multi-agent system (7.80), see e.g. [49]. This completes the proof. □

Remark 7.30 Note that the symmetric matrices P, Q in (7.88) may not be positive definite, each of which could have a simple non-positive eigenvalue since $E^TPE > 0$, $E^TQE > 0$, according to the Sylvester's law of inertia [52]. The inequality constraints $E^TPE > 0$ and $E^TQE > 0$ make the quadratic functional $V(t)$ defined by (7.88) be a valid Lyapunov–Krasovskii functional for system (7.85) and also make it possible to derive consensus conditions in terms of strict linear matrix inequalities. Compared with constructing Lyapunov–Krasovskii functional $V(t)$ with positive definite matrices P and Q, the construction of $V(t)$ here has the advantage of reducing the conservativeness of the consensus conditions obtained by solving some linear matrix inequalities.

Remark 7.31 The maximal allowable $h_{\max}$ guaranteeing consensus in Theorem 7.29 can be obtained by following the two-step procedure below:

1. Set $h_{\max} = h_0$ and step size $\tau = \tau_0$, where h_0 and τ_0 are specified positive constants.
2. Search symmetric matrices P and Q such that $E^T PE > 0$, $E^T QE > 0$, and (7.86) holds. If the conditions are satisfied, set $h_{\max} = h_{\max} + \tau_0$ and return to Step 2). Otherwise, stop and h is the maximal allowable sampling interval.

7.4.2 *The Case with Topology Containing a Directed Spanning Tree*

In this subsection, consensus in multi-agent systems (7.80) whose topology contains a directed spanning tree is studied.

According to the Frobenius normal form [13] in (2.29), without loss of generality, it is assumed that the Laplacian matrix L is in its Frobenius normal form. Furthermore, let $\overline{L}_{ii} = \overline{L}_i + A_i$, where $\overline{L}_i$ is a zero-row-sum matrix and $A_i \geq 0$ is a diagonal matrix, $i = 1, \cdots, m$. By Lemma 2.24, there exists a positive vector $\overline{\xi}_1 = (\overline{\xi}_{1_1}, \cdots, \overline{\xi}_{1_{q_1}})^T$ such that $\overline{\xi}_1^T \overline{L}_1 = 0$ and $\xi_1^T \mathbf{1}_{q_1} = 1$.

Theorem 7.32 *Suppose that the communication topology $\mathscr{G}(\mathscr{A})$ contains a directed spanning tree and Assumption 7.26 holds. Then, consensus in system (7.80) is achieved if there exist symmetric matrices P_1, $Q_1 \in \mathbb{R}^{q_1}$, and positive-definite matrices P_i, $Q_i \in \mathbb{R}^{q_i}$, $i = 2, \cdots, m$, such that $\overline{E}_1^T P_1 \overline{E}_1 > 0$, $\overline{E}_1^T Q_1 \overline{E}_1 > 0$, and the following LMIs hold:*

$$\begin{pmatrix} 2h^2\overline{E}_1^T Q_1 \overline{E}_1 - \overline{E}_1^T \overline{E}_1 & * & * & * \\ \overline{E}_1^T P_1 \overline{E}_1 & \rho \overline{E}_1^T \overline{E}_1 - \overline{E}_1^T Q_1 \overline{E}_1 & * & * \\ O & -\overline{E}_1^T Q_1 \overline{E}_1 & -2\overline{E}_1^T Q_1 \overline{E}_1 & * \\ -\alpha h^2 \overline{E}_1^T \overline{L}_{11}^T Q_1 \overline{E}_1 & -\alpha \overline{E}_1^T \overline{L}_{11}^T P_1 \overline{E}_1 & -\overline{E}_1^T Q_1 \overline{E}_1 & \Theta \end{pmatrix} < 0, \tag{7.99}$$

$$\begin{pmatrix} 2h^2 Q_i - I & * & * & * \\ P_i & \rho I - Q_i & * & * \\ O & -Q_i & -2Q_i & * \\ -\alpha h^2 \overline{L}_{ii}^T Q_i & -\alpha \overline{L}_{ii}^T P_i & -Q_i & \alpha^2 h^2 \overline{L}_{ii}^T Q_i \overline{L}_{ii} - Q_i \end{pmatrix} < 0, \tag{7.100}$$

where

$$\overline{E}_1 = \begin{pmatrix} I_{q_1 - 1} \\ -\frac{\overline{\eta}^T}{\overline{\xi}_{1_{q_1}}} \end{pmatrix} \in \mathbb{R}^{q_1 \times (q_1 - 1)}, \quad \overline{\eta} = (\overline{\xi}_{1_1}, \cdots, \overline{\xi}_{1_{q_1 - 1}})^T \in \mathbb{R}^{q_1 - 1},$$

with $\Theta = \alpha^2 h^2 \overline{E}_1^T \overline{L}_{11}^T Q_1 \overline{L}_{11} \overline{E}_1 - \overline{E}_1^T Q_1 \overline{E}_1$, and $\overline{\xi}_1 = (\overline{\xi}_{1_1}, \cdots, \overline{\xi}_{1_{q_1}})^T$ is the positive left eigenvector of Laplacian matrix $\overline{L}_{11}$ associated with its zero eigenvalue, satisfying $\overline{\xi}_1^T \mathbf{1}_{q_1} = 1$.

Proof. It is noted that the condensation network of $\mathscr{G}(\mathscr{A})$, denoted by $\mathscr{G}^*(\mathscr{A}^*)$, is itself a directed spanning tree. The dynamics of the agents belonging to the node set of the root of $\mathscr{G}^*(\mathscr{A}^*)$ will not be affected by the others and the local topology among them is strongly connected. According to (7.99), and by Theorem 7.29, the states of these agents will reach consensus with an asymptotic decay rate, i.e., there exists $\epsilon_1 > 0$, such that $x_{1_i}(t) = \overline{x}(t) + \mathcal{O}(t^{-\epsilon_1})$, $i = 1, 2, \cdots, q_1$, where $\overline{x}(t)$ represents the asymptotic consensus state satisfying $\dot{\overline{x}}(t) = f(\overline{x}(t), t) + \mathcal{O}(t^{-\epsilon_1})$.

Based on the above analysis, consider the dynamics of the agents, denoted by $v_{i_1}, v_{i_2}, \cdots, v_{i_{q_i}}$, $2 \le i \le m$, belonging to the ith node in $\mathscr{G}^*(\mathscr{A}^*)$. It is only affected by these nodes, such that there exist directed paths from them to v_{i_s}, $s = 1, 2, \cdots, q_i$. Suppose that such nodes excluding v_{i_s}, $s = 1, 2, \cdots, q_i$, are $v_{j_1}, v_{j_2}, \cdots, v_{j_{k_i}}$. Furthermore, assume that the states of nodes $v_{j_1}, v_{j_2}, \cdots, v_{j_{k_i}}$ have already achieved consensus, and the consensus position state is $s(t) = \overline{x}(t)$. For $t \in [t_k, t_{k+1})$ and arbitrarily given k, simple calculations give the following error dynamical system:

$$
\begin{aligned}
\dot{x}_{i_r}(t) =& f(x_{i_r}(t), t) - f(\overline{x}(t), t) - \alpha \sum_{j=1}^{q_i} l_{i_r i_j}(x_{i_j}(t - d_k(t)) - x_{i_r}(t - d_k(t))) \\
& - \alpha \sum_{p=1}^{k_i} l_{i_r j_p}(\overline{x}(t - d_k(t)) - x_{i_r}(t - d_k(t))) + \mathcal{O}(e^{-\epsilon t}), \\
& r = 1, 2, \cdots, q_i,
\end{aligned} \tag{7.101}
$$

for some $\epsilon > 0$. Let $\hat{x}_{i_r}(t) = x_{i_r}(t) - \overline{x}(t)$, $r = 1, 2, \cdots, q_i$. It then follows from (7.101) that

$$
\begin{aligned}
\dot{\hat{x}}_{i_r}(t) =& f(x_{i_r}(t), t) - f(\overline{x}(t), t) - \alpha \sum_{j=1}^{q_i} l_{i_r i_j}(\hat{x}_{i_j}(t - d_k(t)) - \hat{x}_{i_r}(t - d_k(t))) \\
& - \alpha \sum_{p=1}^{k_i} l_{i_r j_p} \hat{x}_{i_r}(t - d_k(t)) + \mathcal{O}(e^{-\epsilon t}), \quad t \in [t_k, t_{k+1}),
\end{aligned} \tag{7.102}
$$

where $r = 1, 2, \cdots, q_i$.

Let $\hat{x}(t) = (\hat{x}_{i_1}(t), \cdots, \hat{x}_{i_{q_i}}(t))^T$, $\hat{x}(t - d_k(t)) = (\hat{x}_{i_1}^T(t - d_k(t)), \cdots, \hat{x}_{i_{q_i}}^T(t - d_k(t)))^T$. Then, system (7.102) can be written as

$$
\dot{\hat{x}}(t) = F(x(t), t) - \alpha(\overline{L}_{ii} \otimes I_n) e(t - d_k(t)) + \mathcal{O}(e^{-\epsilon t}), \quad t \in [t_k, t_{k+1}), \tag{7.103}
$$

where $F(x(t), t) = f(x(t), t) - \mathbf{1}_{q_i} \otimes f(\overline{x}(t), t)$, $f(x(t), t) = (f^T(x_{i_1}(t), t), \cdots, f^T(x_{i_{q_i}}(t), t))^T$. According to (7.107) and by following the proof of Theorem 1, one can show that the states of agents $v_{i_1}, v_{i_2}, \cdots, v_{i_{q_i}}$, $2 \leq i \leq m$, will reach consensus asymptotically, and converge to $\overline{x}(t)$. This completes the proof. □

Remark 7.33 In the proof of Theorem 7.32, the multiple agents in the first strongly connected component of the network in the Frobenius normal form of the Laplacian matrix can be regarded as a leader while the rest as followers. If the consensus can be achieved in system (7.80), the position states of the followers approach that of the leader asymptotically.

Example 7.1 *Consider multi-agent system (7.80) with the topology $\mathscr{G}(\mathscr{A}_1)$ as shown in Figure 7.11, where the weights are indicated on the edges. Figure 7.11 shows that the topology $\mathscr{G}(\mathscr{A}_1)$ contains a directed spanning tree, where agents 1–3 and 4–7 belong to the first and the second strongly connected components, respectively. In simulations, the following two cases are considered:*

(I) Let $f(x_i(t), t) = (0.15\ \sin(x_{i1}(t)),\ \ 0.15\ \cos(x_{i2}(t)))^T \in \mathbb{R}^2$, *where* $x_i(t) = (x_{i1}(t), x_{i2}(t))^T \in \mathbb{R}^2$, $i = 1, \cdots, N$.

(II) Let $f(x_i(t), t) = (\kappa(-x_{i1}(t) + x_{i2}(t) - \eta(x_{i1}(t))), x_{i1}(t) - x_{i2}(t) + x_{i3}(t), -\rho x_{i2}(t))^T \in \mathbb{R}^3$, *where* $\eta(x_{i1}(t)) = bx_{i1}(t) + 0.5(a-b)(|x_{i1}(t)+1| - |x_{i1}(t)-1|)$, $i =$

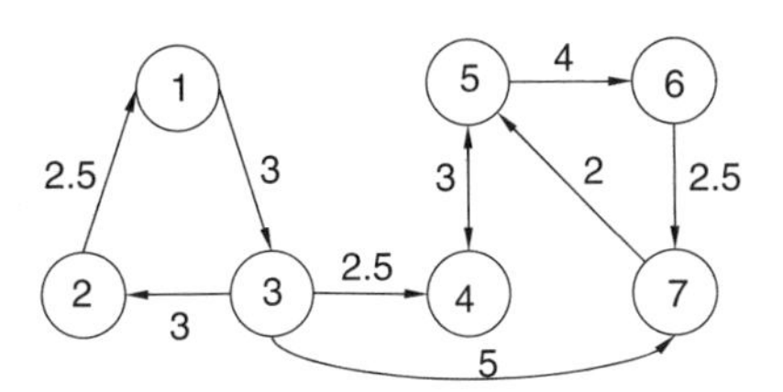

Figure 7.11 Communication topology $\mathscr{G}(\mathscr{A}_1)$

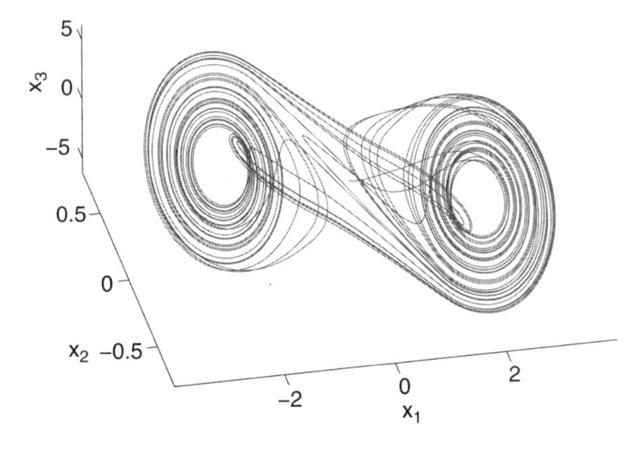

Figure 7.12 The state trajectories of a single agent, for case (II)

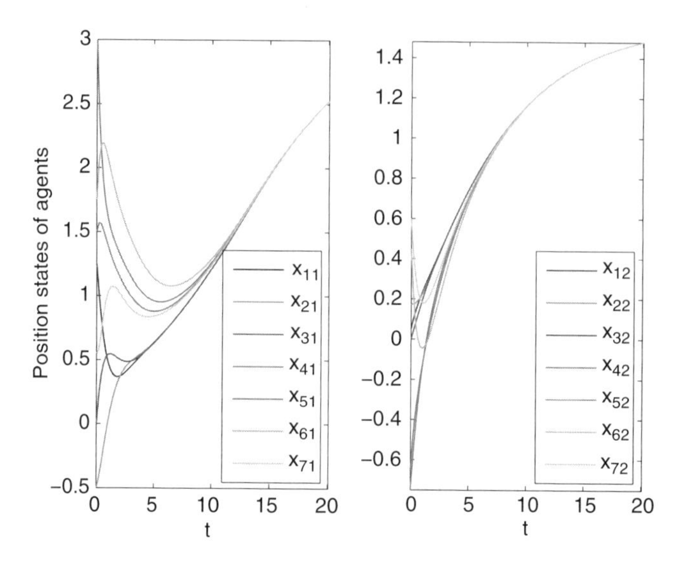

Figure 7.13 Consensus is achieved with sampling interval $h = 0.2000$, for case (I)

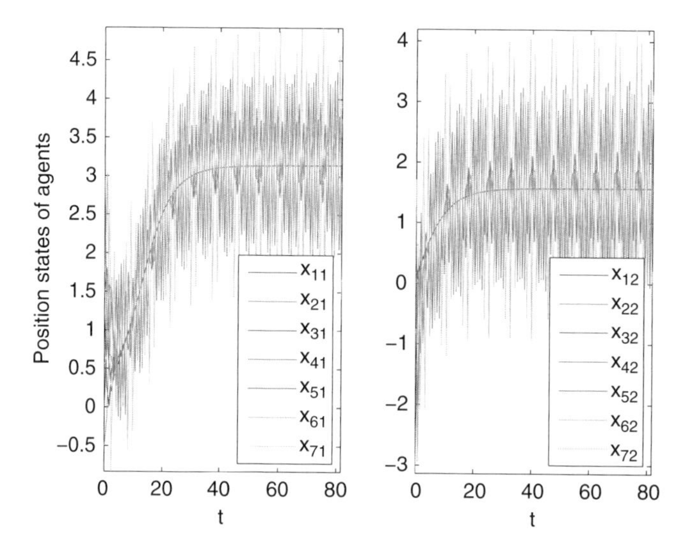

Figure 7.14 Consensus cannot be achieved with sampling interval $h = 0.7500$, for case (I)

$1, \cdots, N$. *In this case, the isolated system is chaotic when* $\kappa = 10$, $\rho = 18$, $a = -4/3$ *and* $b = -3/4$, *as shown in Fig. 2.*

For case (I), let the coupling strength $\alpha = 0.30$. *Some calculations give the maximal allowable sampling intervals* $h_{\max} = 0.2065$. *Taking* $h = 0.2$ *and* $h = 0.75$, *the position states of all agents are obtained as shown in Figs. 7.13 and 7.14, respectively, with the same initial conditions* $x_1(0) = (1.25, 0.05)^T$, $x_2(0) = (-0.5, 0.175)^T$,

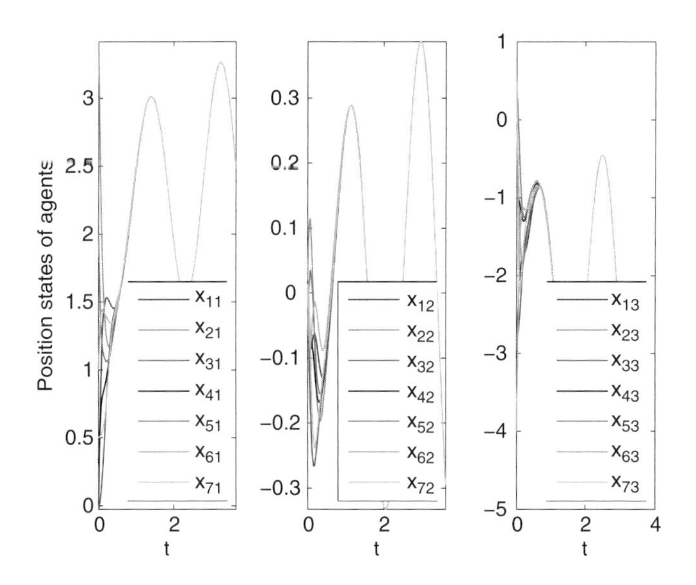

Figure 7.15 Consensus is achieved with sampling interval $h = 0.0500$, for case (II)

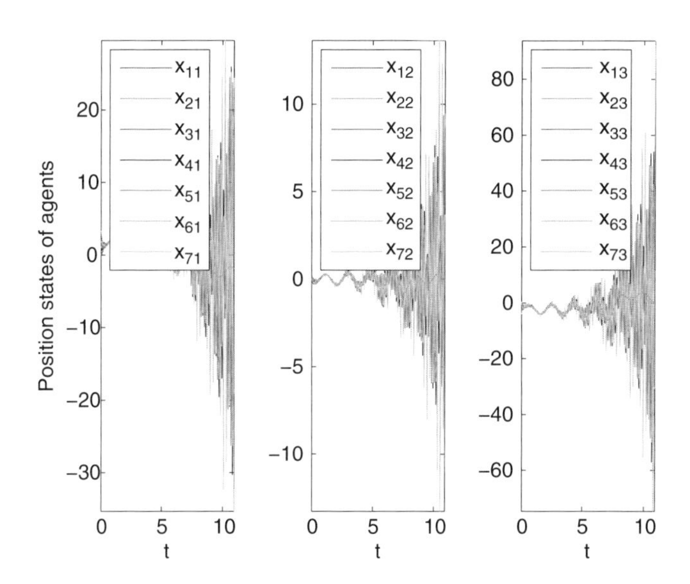

Figure 7.16 Consensus cannot be achieved with sampling interval $h = 0.1550$, for case (II)

$x_3(0) = (0, 0)^T$, $x_4(0) = (1.5, -0.75)^T$, $x_5(0) = (3.0, -0.65)^T$, $x_6(0) = (1.75, 0.45)^T$, *and* $x_7(0) = (0.55, 0.6)^T$. *The results verify the theoretical analysis very well. For case (II), let the coupling strength* $\alpha = 2.0$. *Some calculations give the maximal allowable sampling intervals* $h_{\max} = 0.0400$. *Taking* $h = 0.04$ *and* $h = 0.12$, *the position states of all agents are shown in Figs. 7.15 and 7.16, respectively, with the same initial*

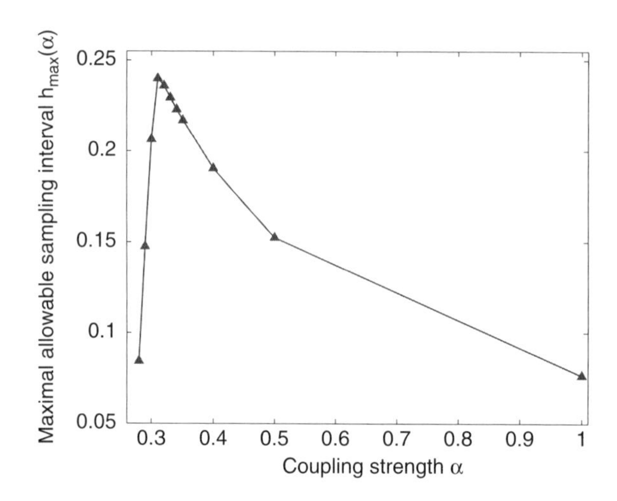

Figure 7.17 Relation between the maximal allowable sampling interval $h_{\max}$ and the coupling strength α, for case (I)

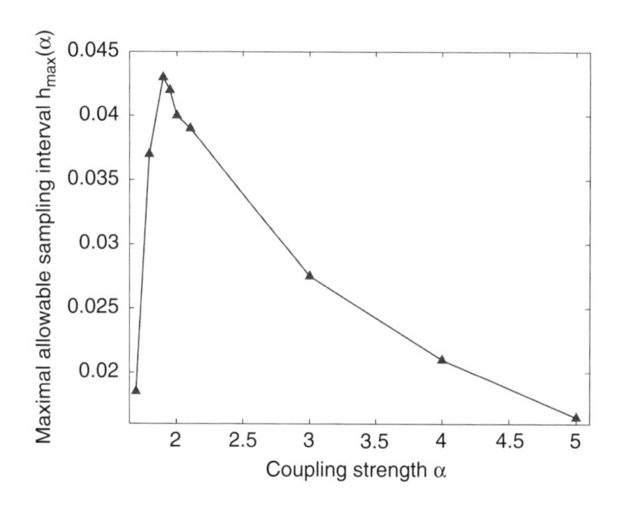

Figure 7.18 Relation between the maximal allowable sampling interval $h_{\max}$ and the coupling strength α, for case (II)

conditions $x_1(0) = (0.016, 0.018, -3.05)^T$, $x_2(0) = (3.415, 0.02, -0.04)^T$, $x_3(0) = (-0.025,$
$0.015, -2.8)^T$, $x_4(0) = (0.315, -0.075, -1.01)^T$, $x_5(0) = (2.008, -0.065, 0.40)^T$, $x_6(0) = (1.005, 0.014, -4.002)^T$, *and* $x_7(0) = (0.508, 0.006, 0.50)^T$. *The simulation results are in good agreement with the theoretical analysis. Furthermore, according to the multi-step procedure given in Remark 7.31, we can get a maximal allowable sampling interval* $h_{\max}(\alpha)$ *for each* α *if the conditions of Theorem 7.32 are satisfied. In the following, the nontrivial relationship between the upper bound of the maximal allowable sampling intervals* $h_{\max}(\alpha)$ *and the coupling strength* α *is numerically*

demonstrated (see Figs. 7.17 and 7.18). Particularly, it is shown that there exists a positive constant $\alpha_0 > 0$, such that $h_{\max}(\alpha)$ attains its maximum when $\alpha = \alpha_0$, and monotonically increases by enlarging α with $\alpha \leq \alpha_0$ while monotonically decreases by enlarging α with $\alpha \geq \alpha_0$, for cases (I) and (II). This is contrary to the common view that the larger the coupling strength, the easier the consensus is achieved.

7.4.3 The Case with Topology Having no Directed Spanning Tree

In this subsection, consider how to guarantee the position states of multi-agent system (7.80) to achieve consensus, where the underlying communication topology does not contain any directed spanning tree. One way to achieve this goal is to introduce a virtual leader to the considered multi-agent system and pin a fraction of agents via designing some pinning navigational feedback terms [94]. For the convenience of analysis, it is assumed that there is a dynamical virtual leader $s(t)$, satisfying

$$\dot{s}(t) = f(s(t), t), \quad s(t) \in \mathbb{R}^n, \tag{7.104}$$

in system (7.80).

For $t \in [t_k, t_{k+1})$, and an arbitrarily given k, one has the following dynamical system:

$$\begin{aligned} \dot{x}_i(t) =& f(x_i(t), t) - \alpha \sum_{j=1}^{N} l_{ij} x_j(t - d_k(t)) - c_i \beta (x_i(t - d_k(t)) - s(t - d_k(t))), \\ & i = 1, \cdots, N, \end{aligned} \tag{7.105}$$

where $c_i = 1$ if agent i is informed and 0 otherwise. Let $\widetilde{x}_i(t) = x_i(t) - s(t)$, $i = 1, 2, \cdots, N$. Then, by (7.105),

$$\dot{\widetilde{x}}(t) = F(x(t), t) - (\widetilde{L} \otimes I_n)\widetilde{x}(t - d_k(t)), \quad t \in [t_k, t_{k+1}), \tag{7.106}$$

where $\widetilde{x}(t) = (\widetilde{x}_1^T(t), \cdots, \widetilde{x}_N^T(t))^T$, $F(x(t), t) = f(x(t), t) - \mathbf{1}_N \otimes f(\widetilde{x}(t), t)$, $f(x(t), t) = (f^T(x_1(t), t), \cdots, f^T(x_N(t), t))^T$, $\widetilde{L} = \alpha L + \beta \operatorname{diag}\{c_1, \cdots, c_N\}$.

Theorem 7.34 *Suppose that there exists a path from the virtual leader to each agent i, $i = 1, \cdots, N$, and Assumption 7.26 holds. Then, consensus in system (7.106) is achieved if there exist positive-definite matrices P and $Q \in \mathbb{R}^N$, such that the following LMI holds:*

$$\begin{pmatrix} h^2 Q - I & * & * & * \\ P & \rho I - Q & * & * \\ O & -Q & -2Q & * \\ -\alpha h^2 \widetilde{L}^T Q & -\alpha \widetilde{L}^T P & -Q & \alpha^2 h^2 \widetilde{L}^T Q \widetilde{L} - Q \end{pmatrix} < 0. \tag{7.107}$$

Proof. This theorem can be proved directly from that of Theorem 7.32 above, so the proof is omitted for brevity. □

Remark 7.35 In Theorem 7.34, the condition that the virtual leader has a path from it to each agent is very mild. Suppose that the communication topology $\mathscr{G}(\mathscr{A})$ contains m separated strongly connected components. Then, the condition can be satisfied by informing m agents which are the roots of the corresponding spanning trees in the union of m strongly connected components.

Example 7.2 *Consider multi-agent system (7.80) with topology $\mathscr{G}(\mathscr{A}_2)$ as in Fig. 7.19, where the weights are indicated on the edges. Fig. 7.19 shows that the topology $\mathscr{G}(\mathscr{A}_2)$ does not contain any spanning tree. Suppose that there is a virtual leader labeled V in system (7.80), where agents 1 and 4 belonging to the first and the second strongly connected components are informed agents, as indicated in Fig. 7.19. Let $x_v(t)$ be the position states of the virtual leader, where $x_v(t) = (x_{v1}(t), x_{v2}(t))^T$ with $x_v(0) = (0.5, 0.5)^T$. In simulations, let $f(x_i(t), t) = (0.15\ \sin(x_{i1}(t)), 0.10x_{i2}(t))^T \in \mathbb{R}^2$, where $x_i(t) = (x_{i1}(t), x_{i2}(t))^T \in \mathbb{R}^2$, $i = 1, \cdots, N$. Let the coupling*

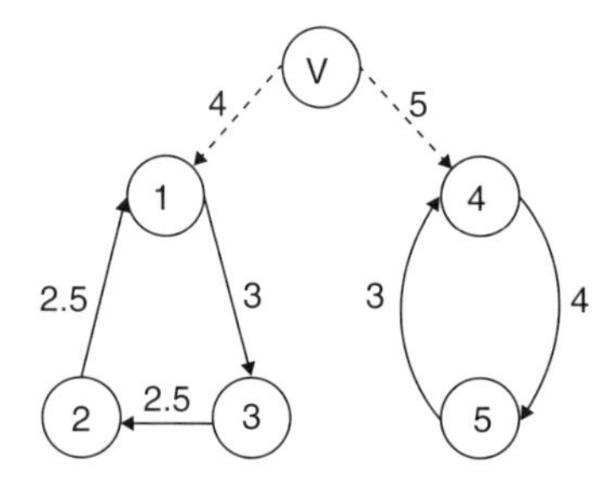

Figure 7.19 Communication topology $\mathscr{G}(\mathscr{A}_2)$

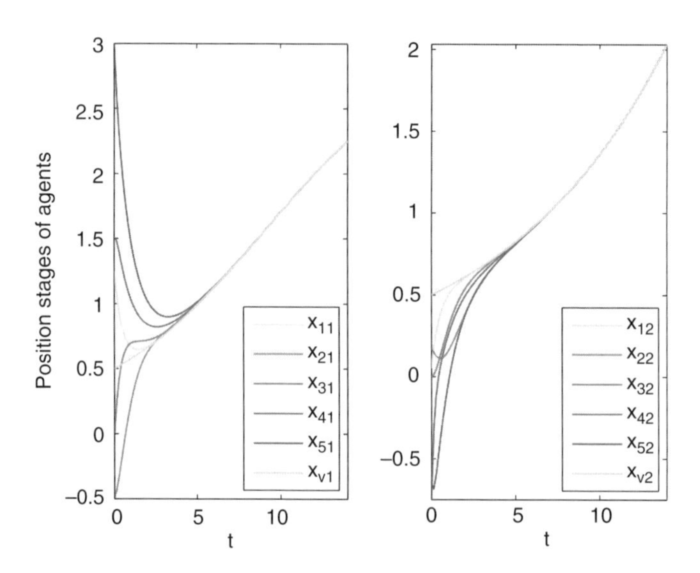

Figure 7.20 Consensus is achieved with sampling interval $h = 0.14$, in Example 7.2

strength $\alpha = 0.48$. *Some calculations give that the maximal allowable sampling intervals* $h_{\max} = 0.1465$. *Taking* $h = 0.14$ *and* $h = 0.425$, *the position state of all agents are shown in Figs. 7.20 and 7.21, respectively, with initial conditions* $x_1(0) = (1.25, 0.05)^T$, $x_2(0) = (-0.5, 0.175)^T$, $x_3(0) = (0, 0)^T$, $x_4(0) = (1.5, -0.75)^T$, *and* $x_5(0) = (3.0, -0.65)^T$. *The results verify the theoretical analysis very well. The dependence of the upper bound of maximal allowable sampling intervals* $h_{\max}$ *on the coupling strength* α *is summarized in Fig. 7.22. It is numerically shown that*

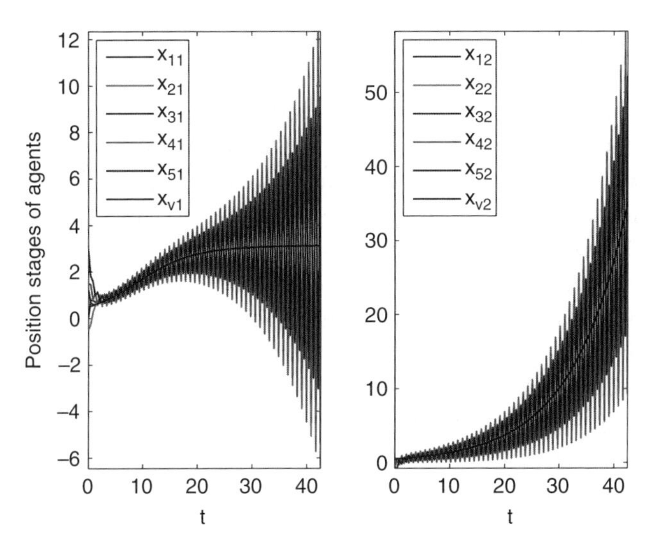

Figure 7.21 Consensus cannot be achieved with sampling interval $h = 0.425$, in Example 7.2

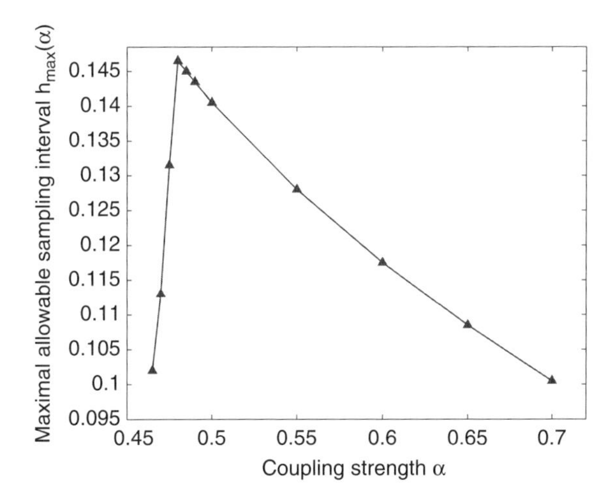

Figure 7.22 Relation between the maximal allowable sampling interval $h_{\max}$ and the coupling strength α, in Example 7.2

there exists a positive constant $\alpha_1 > 0$, such that $h_{\max}(\alpha)$ attains its maximum when $\alpha = \alpha_1$, and monotonically increases by enlarging α with $\alpha \leq \alpha_1$ while monotonically decreases by enlarging α with $\alpha \geq \alpha_1$.

7.5 Notes

In this chapter, second-order consensus in multi-agent systems with sampled position and velocity data has been studied. A distributed linear consensus protocol with second-order dynamics has been designed, where both sampled position and velocity data have been utilized. A necessary and sufficient condition based on the sampling period, the coupling gains, and the spectrum of the Laplacian matrix, has been established for reaching consensus of the system. Second-order consensus in delayed undirected networks with sampled position and velocity data has also been discussed. A necessary and sufficient condition has been given, by which appropriate sampling period can be chosen to achieve consensus.

Second-order consensus in multi-agent dynamical systems with sampled position data has also been investigated. A distributed linear consensus protocol in the second-order dynamics has been designed based on both current and sampled position data. It has been found that second-order consensus in a multi-agent system cannot be reached without sampled position data under the given protocol but it can be achieved by appropriately choosing the sampling period. A necessary and sufficient condition for reaching consensus in multi-agent dynamical systems has been established and demonstrated by simulation examples.

Furthermore, the study for sampled-data control has been extended to the case of multi-agent systems with nonlinear dynamics.

There are still a number of related interesting problems deserving further investigation. For example, it is desirable to study multi-agent systems with nonuniform sampling intervals, nonlinear dynamics with time-varying velocities [149], more general and effective consensus protocols, and so on.

8

Consensus of Second-Order Multi-agent Systems with Intermittent Communication

In some practical multi-agent systems, multiple agents may only communicate with their neighbors intermittently. Motivated by this observation, this chapter introduces and investigates the consensus problem of second-order multi-agent systems with intermittent communication [126]. First, consensus in second-order linear multi-agent systems with intermittent communication under a strongly connected topology or an arbitrarily given topology is studied. Then, consensus in second-order multi-agent systems with nonlinear dynamics and intermittent communication under a strongly connected topology is addressed. It is shown that both the agent dynamics and the connectivity of the communication graph will affect the consensus behaviors of the closed-loop multi-agent systems. Specifically, for second-order multi-agent systems with intermittent communication under a strongly connected topology, it is proved that distributed consensus can be achieved if the general algebraic connectivity of the topology is larger than a critical value and the multiple agents communicate with their neighbors frequently enough as the systems evolve in time.

8.1 Problem Statement

Consensus for second-order multi-agent systems with intermittent communication under a strongly connected topology is studied in this section.

Consider a group of N agents indexed by $1, 2, \cdots, N$. The commonly studied continuous-time second-order protocol of the N networked agents is described by the

Distributed Cooperative Control of Multi-agent Systems, First Edition.
Wenwu Yu, Guanghui Wen, Guanrong Chen, and Jinde Cao.

second-order dynamics [97, 146] as follows:

$$\begin{cases} \dot{x}_i(t) = v_i(t), \\ \dot{v}_i(t) = -\alpha \sum_{j=1}^{N} l_{ij}x_j(t) - \beta \sum_{j=1}^{N} l_{ij}v_j(t), \quad i = 1, 2, \cdots, N, \end{cases} \tag{8.1}$$

where $x_i \in \mathbb{R}^n$ and $v_i \in \mathbb{R}^n$ are the position and velocity states of the ith agent, respectively, α and β represent the coupling strengths, and $L = [l_{ij}]_{N\times N}$ is the Laplacian matrix of the fixed communication topology $\mathscr{G}$. When the agents reach consensus, referred to second-order consensus, the velocities of all agents converge to $\sum_{j=1}^{N} \xi_j v_j(0)$, which depends only on the initial velocities of the agents, where $\xi = (\xi_1, \cdots, \xi_N)$ is the nonnegative left eigenvector of L associated with the eigenvalue 0, satisfying $\xi^T \mathbf{1}_N = 1$ [97, 146].

Note that most of the existing protocols are implemented based on a common assumption that all information is transmitted continuously among agents. However, in some real situations, agents may only communicate with their neighbors over some disconnected time intervals due to the unreliability of communication channels, failure of physical devices, external disturbances and limitations of sensing ranges, etc. Motivated by this observation and based on the above-mentioned works, the following consensus protocol with intermittent measurements is considered here:

$$\begin{cases} \dot{x}_i(t) = v_i(t), \\ \dot{v}_i(t) = -\alpha \sum_{j=1}^{N} l_{ij}x_j(t) - \beta \sum_{j=1}^{N} l_{ij}v_j(t), \quad t \in T, \\ \dot{v}_i(t) = 0, \quad t \in \overline{T}, \quad i = 1, 2, \cdots, N, \end{cases} \tag{8.2}$$

where T represents the union of time intervals over which the agents could communicate with each other and $\overline{T}$ represents the union of time intervals over which the agents could not communicate with each other. Obviously, $T \cup \overline{T} = [0, +\infty)$.

Remark 8.1 If $T = [0, +\infty)$ in system (8.2), that is, each agent can communicate with its neighbors all the time, then system (8.2) becomes the typical second-order system studied in [97, 146], which have been discussed in the previous chapters of this book.

Remark 8.2 For the convenience of theoretical analysis, it is assumed that the multiple agents in system (8.2) could sense the measurements of relative states between their own and the neighbors synchronously, i.e., the intermittent measurements actually are a kind of globally synchronous intermittent measurement. According to [86], this assumption is crucial for possibly constructing a common Lyapunov function for the switching system (8.2).

The following lemma is needed to present the main results of this section.

Lemma 8.3 *([54]) Suppose that $M \in \mathbb{R}^{n \times n}$ is a positive definite matrix and $N \in \mathbb{R}^{n \times n}$ is symmetric. Then, for any vector $x \in \mathbb{R}^n$, the following inequality holds:*

$$\lambda_{\min}(M^{-1}N)x^T Mx \leq x^T Nx \leq \lambda_{\max}(M^{-1}N)x^T Mx. \tag{8.3}$$

8.2 The Case with a Strongly Connected Topology

In this section, the communication topology $\mathscr{G}$ of system (8.2) is assumed to be strongly connected.

Let $\widetilde{x}_i(t) = x_i(t) - \sum_{j=1}^N \xi_j x_j(0) - t\sum_{j=1}^N \xi_j v_j(0)$ and $\widetilde{v}_i(t) = v_i(t) - \sum_{j=1}^N \xi_j v_j(0)$ represent the position and velocity vectors relative to the weighted average position and velocity vectors of the agents in system (8.2), respectively, where $\xi = (\xi_1, \xi_2, \cdots, \xi_N)^T$ is the positive left eigenvector of Laplacian matrix L associated with its zero eigenvalue, satisfying $\xi^T \mathbf{1}_N = 1$. A simple calculation gives the following error dynamical system:

$$\begin{cases} \dot{\widetilde{x}}_i(t) = \widetilde{v}_i(t) \\ \dot{\widetilde{v}}_i(t) = -\alpha \sum\limits_{j=1}^N l_{ij}\widetilde{x}_j(t) - \beta \sum\limits_{j=1}^N l_{ij}\widetilde{v}_j(t), \quad t \in T, \\ \dot{\widetilde{v}}_i(t) = 0, \quad t \in \overline{T}, \quad i = 1, 2, \cdots, N. \end{cases} \tag{8.4}$$

Let $\widetilde{x}(t) = (\widetilde{x}_1^T(t), \cdots, \widetilde{x}_N^T(t))^T$, $\widetilde{v}(t) = (\widetilde{v}_1^T(t), \cdots, \widetilde{v}_N^T(t))^T$ and $\widetilde{y}(t) = (\widetilde{x}^T(t), \widetilde{v}^T(t))^T$. Then, system (8.4) can be written as

$$\begin{cases} \dot{\widetilde{y}}(t) = (B_1 \otimes I_n)\widetilde{y}(t), \quad t \in T, \\ \dot{\widetilde{y}}(t) = (B_2 \otimes I_n)\widetilde{y}(t), \quad t \in \overline{T}, \end{cases} \tag{8.5}$$

where $B_1 = \begin{pmatrix} O_N & I_N \\ -\alpha L & -\beta L \end{pmatrix}$ and $B_2 = \begin{pmatrix} O_N & I_N \\ O_N & O_N \end{pmatrix}$.

Definition 8.4 For a strongly connected network $\mathscr{G}(\mathscr{A})$ with Laplacian matrix L, let

$$a(L) = \min_{x^T\xi = 0, x \neq 0} \frac{x^T \widehat{L} x}{x^T \Xi x},$$

$$b(L) = \max_{x^T\xi = 0, x \neq 0} \frac{x^T \widehat{L} x}{x^T \Xi x},$$

where $\widehat{L} = (1/2)(\Xi L + L^T \Xi)$, $\Xi = \mathrm{diag}(\xi_1, \xi_2, \cdots, \xi_N)$, $\xi = (\xi_1, \xi_2, \cdots, \xi_N)^T > 0$, $\xi^T L = 0$, and $\sum_{i=1}^N \xi_i = 1$. Then, $a(L)$ is called the *general algebraic connectivity* of $\mathscr{G}(\mathscr{A})$.

Theorem 8.5 *Suppose that the communication topology $\mathscr{G}$ is strongly connected. Then, second-order consensus in system (8.2) is achieved if there exists an infinite time sequence of uniformly bounded and nonoverlapping time intervals $[t_k, t_{k+1})$, $k \in \mathbb{N}$,* *with* $\mathbb{N}$ *being the set of positive natural numbers and* $t_1 = 0$*, such that for each time interval* $[t_k, t_{k+1})$*,* $k \in \mathbb{N}$*, the following conditions hold:*

(i) $a(L) > \alpha/\beta^2$,
(ii) $\delta_k > \frac{\gamma_4}{\gamma_3+\gamma_4}\omega_k$,

where δ_k *represents the Lebesgue measure of set* $\{t \mid t \in [t_k, t_{k+1}) \cap T\}$*,* $\omega_k = t_{k+1} - t_k$*,* $\gamma_3 = \frac{4\xi_{\min}\min\{\alpha^2 a(L),(\beta^2 a(L)-\alpha)\}}{\xi_{\max}(2\alpha\beta b(L)+\sqrt{(2\alpha\beta b(L)-\beta)^2+4\alpha^2})}$*,* $\gamma_4 = 2\lambda_{\max}(Q^{-1}P_3)$*,* $Q = \begin{pmatrix} 2\alpha\beta a(L)\Xi & \alpha\Xi \\ \alpha\Xi & \beta\Xi \end{pmatrix}$*, and* $P_3 = \begin{pmatrix} O_N & \frac{1}{2}\alpha\beta(\Xi L + L^T\Xi) \\ \frac{1}{2}\alpha\beta(\Xi L + L^T\Xi) & \alpha\Xi \end{pmatrix}$.

Proof. Construct the following Lyapunov function candidate:

$$V(t) = \frac{1}{2}\widetilde{y}^T(t)(P \otimes I_n)\widetilde{y}(t), \tag{8.6}$$

where $P = \begin{pmatrix} \alpha\beta(\Xi L + L^T\Xi) & \alpha\Xi \\ \alpha\Xi & \beta\Xi \end{pmatrix}$ and $\Xi = \mathrm{diag}\{\xi_1, \xi_2, \cdots, \xi_N\}$. It will be shown that $V(t)$ is a valid Lyapunov function for analyzing the error dynamics described by system (8.5). Then,

$$\begin{aligned} V(t) =& \frac{\alpha\beta}{2}\widetilde{x}^T(t)((\Xi L + L^T\Xi) \otimes I_n)\widetilde{x}(t) + \alpha\widetilde{x}^T(t)(\Xi \otimes I_n)(t)\widetilde{v}(t) \\ &+ \frac{\beta}{2}\widetilde{v}^T(t)(\Xi \otimes I_n)\widetilde{v}(t) \\ \geq& \frac{1}{2}\widetilde{y}^T(t)(Q \otimes I_n)\widetilde{y}(t), \end{aligned} \tag{8.7}$$

where $Q = \begin{pmatrix} 2\alpha\beta a(L)\Xi & \alpha\Xi \\ \alpha\Xi & \beta\Xi \end{pmatrix}$. By the Schur complement argument, $Q > 0$ is equivalent to both $\beta > 0$ and $a(L) > \alpha/(2\beta^2)$. From condition (i), one obtains that $Q > 0$, $V(t) \geq 0$ and $V(t) = 0$ if and only if $\widetilde{y}(t) = 0_{2Nn}$.

For $t \in [t_k, t_{k+1}) \cap T$, for arbitrarily given $k \in \mathbb{N}$, taking the time derivative of $V(t)$ along the trajectories of (8.5) gives

$$\begin{aligned} \dot{V}(t) =& \frac{1}{2}\widetilde{y}^T(t)[(PB_1 + B_1^T P) \otimes I_n]\widetilde{y}(t) \\ =& \frac{1}{2}\widetilde{y}^T(t)\left[\begin{pmatrix} -\alpha^2(\Xi L + L^T\Xi) & O_N \\ O_N & -\beta^2(\Xi L + L^T\Xi) + 2\alpha\Xi \end{pmatrix} \otimes I_n\right]\widetilde{y}(t) \\ =& -\frac{\alpha^2}{2}\widetilde{x}^T(t)[(\Xi L + L^T\Xi) \otimes I_n]\widetilde{x}(t) - \frac{\beta^2}{2}\widetilde{v}^T(t)[(\Xi L + L^T\Xi) \otimes I_n]\widetilde{v}(t) \\ &+ \alpha\widetilde{v}^T(t)(\Xi \otimes I_n)\widetilde{v}(t) \\ \leq& -\alpha^2 a(L)\widetilde{x}^T(t)(\Xi \otimes I_n)\widetilde{x}(t) - \beta^2 a(L)\widetilde{v}^T(t)(\Xi \otimes I_n)\widetilde{v}(t) \end{aligned}$$

$$+\alpha\widetilde{v}^T(t)(\Xi \otimes I_n)\widetilde{v}(t)$$

$$= -\widetilde{y}^T(t)[(P_1 \otimes \Xi) \otimes I_n]\widetilde{y}(t), \tag{8.8}$$

where $P_1 = \begin{pmatrix} \alpha^2 a(L) & O_N \\ O_N & \beta^2 a(L) - \alpha \end{pmatrix}$. On the other hand,

$$V(t) = \frac{1}{2}\widetilde{y}^T(t)(P \otimes I_n)\widetilde{y}(t)$$

$$= \frac{\alpha\beta}{2}\widetilde{x}^T(t)[(\Xi L + L^T\Xi) \otimes I_n]\widetilde{x}(t) + \alpha\widetilde{x}^T(t)(\Xi \otimes I_n)\widetilde{v}(t)$$

$$+\frac{\beta}{2}\widetilde{v}^T(t)(\Xi \otimes I_n)\widetilde{v}(t)$$

$$\leq \alpha\beta b(L)\widetilde{x}^T(t)(\Xi \otimes I_n)\widetilde{x}(t) + \alpha\widetilde{x}^T(t)(\Xi \otimes I_n)\widetilde{v}(t) + \frac{\beta}{2}\widetilde{v}^T(t)(\Xi \otimes I_n)\widetilde{v}(t)$$

$$= \widetilde{y}^T(t)[(P_2 \otimes \Xi) \otimes I_n]\widetilde{y}(t), \tag{8.9}$$

where the positive matrix $P_2 = \begin{pmatrix} \alpha\beta b(L) & \alpha/2 \\ \alpha/2 & \beta/2 \end{pmatrix}$. Thus, according to the following inequality:

$$V(t) \leq \lambda_{\max}(P_2)\xi_{\max}\widetilde{y}^T(t)\widetilde{y}(t) = \gamma_1\widetilde{y}^T(t)\widetilde{y}(t), \tag{8.10}$$

where $\gamma_1 = \frac{2\alpha\beta b(L)+\sqrt{(2\alpha\beta b(L)-\beta)^2+4\alpha^2}}{4}\xi_{\max}$, one has

$$\dot{V}(t) \leq -\lambda_{\min}(P_1)\xi_{\min}\widetilde{y}^T(t)\widetilde{y}(t) = -\gamma_2\widetilde{y}^T(t)\widetilde{y}(t), \tag{8.11}$$

where $\gamma_2 = \min\{\alpha^2 a(L)\xi_{\min}, (\beta^2 a(L) - \alpha)\xi_{\min}\}$. Consequently,

$$\dot{V}(t) \leq -\gamma_3 V(t), \tag{8.12}$$

where $\gamma_3 = \gamma_2/\gamma_1$.

For $t \in \{[t_k, t_{k+1}) \cap \overline{T}\}$, with arbitrarily given $k \in \mathbb{N}$, taking the time derivative of $V(t)$ along the trajectories of (8.5) gives

$$\dot{V}(t) = \widetilde{y}^T(t)[(PB_2) \otimes I_n]\widetilde{y}(t)$$

$$= \frac{1}{2}\widetilde{y}^T(t)[(PB_2 + B_2^T P) \otimes I_n]\widetilde{y}(t)$$

$$= \widetilde{y}^T(t)(P_3 \otimes I_n)\widetilde{y}(t), \tag{8.13}$$

where

$$P_3 = \begin{pmatrix} O_N & \frac{1}{2}\alpha\beta(\Xi L + L^T\Xi) \\ \frac{1}{2}\alpha\beta(\Xi L + L^T\Xi) & \alpha\Xi \end{pmatrix}.$$

According to Lemma 8.3 and inequality (8.7), it follows that

$$\dot{V}(t) \leq \lambda_{\max}((Q \otimes I_n)^{-1}(P_3 \otimes I_n))\widetilde{y}^T(t)(Q \otimes I_n)\widetilde{y}(t)$$

$$= \lambda_{\max}(Q^{-1}P_3)\widetilde{y}^T(t)(Q \otimes I_n)\widetilde{y}(t)$$
$$\leq \gamma_4 V(t), \tag{8.14}$$

where $\gamma_4 = 2\lambda_{\max}(Q^{-1}P_3)$.

Based on the above analysis,

$$V(t_2) \leq V(0)e^{-\Delta_1}, \tag{8.15}$$

where $\Delta_1 = \gamma_3\delta_1 - \gamma_4(\omega_1 - \delta_1)$. Then, according to condition (ii), $\Delta_1 > 0$. By recursion, for any positive integer k,

$$V(t_{k+1}) \leq V(0)e^{-\sum_{j=1}^{k}\Delta_j}, \tag{8.16}$$

where $\Delta_j = r_3\delta_j - r_4(\omega_j - \delta_j) > 0$, $j = 1, 2, \cdots, k$. For arbitrary $t > 0$, there exists a positive integer s such that $t_{s+1} < t \leq t_{s+2}$. Furthermore, since $[t_k, t_{k+1})$, $k \in \mathbb{N}$, is an uniformly bounded and non-overlapping time sequence, one can let $\omega_{\max} = \max_{i\in\mathbb{N}} \omega_i$ and $\kappa = \min_{i\in\mathbb{N}} \Delta_i > 0$. Thus, it follows that

$$V(t) \leq V(t_{s+1})e^{\omega_{\max}r_4}$$
$$\leq e^{\omega_{\max}r_4}V(0)e^{-\sum_{j=1}^{s}\Delta_j}$$
$$\leq e^{\omega_{\max}r_4}V(0)e^{-s\kappa}$$
$$\leq e^{\omega_{\max}r_4}V(0)e^{-(\kappa/\omega_{\max})t}, \tag{8.17}$$

that is,

$$V(t) \leq K_0 e^{-K_1 t}, \quad \text{for all } t > 0, \tag{8.18}$$

where $K_0 = e^{r_4}V(0)$ and $K_1 = \frac{\kappa}{\omega_{\max}}$, which indicates that the states of agents exponentially converge, thereby achieving consensus. Furthermore, the finial consensus value of the position state $x_{\text{con}} = \sum_{j=1}^{N}\xi_j x_j(0) + t\sum_{j=1}^{N}\xi_j v_j(0)$, and the final consensus value of velocity state $v_{\text{con}} = \sum_{j=1}^{N}\xi_j v_j(0)$. This completes the proof. □

Corollary 8.6 *Suppose that the communication topology $\mathcal{G}$ is an undirected connected network. Then, second-order consensus in system (8.2) is achieved if there exists an infinite time sequence of uniformly bounded and non-overlapping time intervals $[t_k, t_{k+1})$, $k \in \mathbb{N}$, with $t_1 = 0$, such that for each time interval $[t_k, t_{k+1})$, $k \in \mathbb{N}$, the following conditions hold:*

(i) $\lambda_2(L) > \alpha/\beta^2$,
(ii) $\delta_k > \frac{\gamma_4}{\gamma_3+\gamma_4}\omega_k$,

where $\lambda_2(L)$ is the second smallest eigenvalue of L, δ_k represents the Lebesgue measure of set $\{t | t \in [t_k, t_{k+1}) \cap T\}$, $\omega_k = t_{k+1} - t_k$, $\gamma_3 = \frac{4\min\{\alpha^2 a(L), (\beta^2 a(L)-\alpha)\}}{2\alpha\beta b(L)+\sqrt{(2\alpha\beta b(L)-\beta)^2+4\alpha^2}}$, $\gamma_4 = 2\lambda_{\max}(Q^{-1}P_3)$, $Q = \begin{pmatrix} 2\alpha\beta\lambda_2(L) & \alpha \\ \alpha & \beta \end{pmatrix} \otimes I_N$, and $P_3 = \begin{pmatrix} O_N & \alpha\beta L \\ \alpha\beta L & \alpha I_N \end{pmatrix}$.

Proof. Construct the same Lyapunov function candidate $V(t)$ as that in the proof of Theorem 1. Under conditions (i) and (ii), the corollary can be proved by following the proof of Theorem 8.5. □

8.3 The Case with a Topology Having a Directed Spanning Tree

Consensus of second-order multi-agent systems with intermittent communication and fixed strongly connected topology was addressed in the previous section. In this section, focus is on the case where the topology only contains a directed spanning tree but may not be strongly connected.

Actually, we can obtain the Frobenius normal form [13] in (2.29) by changing the order of the node indexes. In the following analysis, without loss of generality, it is assumed that the Laplacian matrix L is in its Frobenius normal form:

$$L = \begin{pmatrix} \overline{L}_{11} & O & \cdots & O \\ \overline{L}_{21} & \overline{L}_{22} & \cdots & O \\ \vdots & \vdots & \vdots & O \\ \overline{L}_{m1} & \overline{L}_{m2} & \cdots & \overline{L}_{mm} \end{pmatrix} \tag{8.19}$$

where $\overline{L}_{11} \in \mathbb{R}^{q_1 \times q_1}, \overline{L}_{22} \in \mathbb{R}^{q_2 \times q_2}, \cdots, \overline{L}_{mm} \in \mathbb{R}^{q_m \times q_m}$ are irreducible square matrices, which are uniquely determined to within a simultaneous permutation of their lines but their ordering is not necessarily unique. Furthermore, let $\overline{L}_{ii} = \overline{L}_i + A_i$, where $\overline{L}_i$ is a zero-row-sum matrix and $A_i \geq 0$ is a diagonal matrix. By Lemma 2.25, there exists a positive vector $\overline{\xi}_i = (\overline{\xi}_{i1}, \overline{\xi}_{i2}, \cdots, \overline{\xi}_{iq_i})^T$ of appropriate dimension such that $\overline{\xi}_i^T \overline{L}_i = 0$.

Definition 8.7 For a network $\mathscr{G}$ containing a directed spanning tree and the Laplacian matrix L in the form of (2.29), define

$$c(\overline{L}_{ii}) = \min_{x \neq 0} \frac{x^T \widehat{\overline{L}}_{ii} x}{x^T \overline{\Xi}_i x}, \tag{8.20}$$

$$d(\overline{L}_{ii}) = \max_{x \neq 0} \frac{x^T \widehat{\overline{L}}_{ii} x}{x^T \overline{\Xi}_i x}, \tag{8.21}$$

where $\widehat{\overline{L}}_{ii} = (\overline{\Xi}_i \overline{L}_{ii} + \overline{L}_{ii}^T \overline{\Xi}_i)/2$, $\overline{\Xi}_i = \text{diag}\{\overline{\xi}_{i1}, \overline{\xi}_{i2}, \cdots, \overline{\xi}_{iq_i}\}$, $\overline{\xi}_i = (\overline{\xi}_{i1}, \overline{\xi}_{i2}, \cdots, \overline{\xi}_{iq_i})^T > 0$ and $\overline{\xi}_i^T \overline{L}_i = 0$, $i = 2, 3, \cdots m$.

Lemma 8.8 *([149]) If the directed network $\mathscr{G}$ contains a directed spanning tree, then* $\min_{2 \leq i \leq m} \{a(L_{11}), c(\overline{L}_{ii})\} > 0$.

Theorem 8.9 *Suppose that the communication topology $\mathcal{G}$ contains a directed spanning tree. Then, second-order consensus in network (8.2) is achieved if there exists an infinite time sequence of uniformly bounded and non-overlapping time intervals $[t_k, t_{k+1})$, $k \in \mathbb{N}$, with $t_1 = 0$, such that for each time interval $[t_k, t_{k+1})$, $k \in \mathbb{N}$, the following conditions hold:*

(i) $\min_{2\le i\le m} \{a(\overline{L}_{11}), c(\overline{L}_{ii})\} > \alpha/\beta^2$,

(ii) $\delta_k > \omega_k \max_{2\le i\le m} \left\{ \frac{\gamma_4}{\gamma_3+\gamma_4}, \frac{\gamma_4^i}{\gamma_3^i+\gamma_4^i} \right\}$,

where δ_k *represents the Lebesgue measure of set* $\{t \mid t \in [t_k, t_{k+1}) \cap T\}$, $\omega_k = t_{k+1} - t_k$, $\gamma_3 = \frac{4\overline{\xi}_{1\min}\min\{\alpha^2 a(\overline{L}_{11}),(\beta^2 a(\overline{L}_{11})-\alpha)\}}{\overline{\xi}_{1\max}(2\alpha\beta b(\overline{L}_{11})+\sqrt{(2\alpha\beta b(\overline{L}_{11})-\beta)^2+4\alpha^2})}$, $\gamma_4 = 2\lambda_{\max}(Q^{-1}P_3)$, $Q = \begin{pmatrix} 2\alpha\beta a(\overline{L}_{11})\overline{\Xi}_1 & \alpha\overline{\Xi}_1 \\ \alpha\overline{\Xi}_1 & \beta\overline{\Xi}_1 \end{pmatrix}$, $P_3 = \begin{pmatrix} O_N & \frac{1}{2}\alpha\beta(\overline{\Xi}_1\overline{L}_{11} + \overline{L}_{11}^T\overline{\Xi}_1) \\ \frac{1}{2}\alpha\beta(\overline{\Xi}_1\overline{L}_{11} + \overline{L}_{11}^T\overline{\Xi}_1) & \alpha\overline{\Xi}_1 \end{pmatrix}$, $\gamma_3^i = \frac{4\overline{\xi}_{i\min}\min\{\alpha^2 c(\overline{L}_{ii}),(\beta^2 c(\overline{L}_{ii})-\alpha)\}}{\overline{\xi}_{i\max}(2\alpha\beta d(\overline{L}_{ii})+\sqrt{(2\alpha\beta d(\overline{L}_{ii})-\beta)^2+4\alpha^2})}$, $\gamma_4^i = 2\lambda_{\max}((Q^i)^{-1}P_3^i)$, $Q^i = \begin{pmatrix} 2\alpha\beta c(\overline{L}_{ii})\overline{\Xi}_i & \alpha\overline{\Xi}_i \\ \alpha\overline{\Xi}_i & \beta\overline{\Xi}_i \end{pmatrix}$,

$$P_3^i = \begin{pmatrix} O_N & \frac{1}{2}\alpha\beta(\overline{\Xi}_i\overline{L}_{ii} + \overline{L}_{ii}^T\overline{\Xi}_i) \\ \frac{1}{2}\alpha\beta(\overline{\Xi}_i\overline{L}_{ii} + \overline{L}_{ii}^T\overline{\Xi}_i) & \alpha\overline{\Xi}_i \end{pmatrix}, \; i = 2, 3, \cdots, m.$$

Proof. Obviously, the condensation network of $\mathcal{G}$, denoted by $\mathcal{G}^*$, is itself a directed spanning tree. The dynamics of the agents corresponding to the node set of the root of $\mathcal{G}^*$ will not affected by others and the local topology among them is strongly connected. According to conditions (i) and (ii), and by Theorem 8.5, the states of these agents will reach consensus with an exponential decay rate, i.e., there exists $\epsilon_1 > 0$, such that $x_i(t) = x_{con} + \mathcal{O}(e^{-\epsilon_1 t})$, $v_i(t) = v_{con}(t) + \mathcal{O}(e^{-\epsilon_1 t})$, where $i = 1, 2, \cdots, q_1$, $x_{\text{con}} = \sum_{j=1}^{q_1} \xi_{1j}x_j(0) + t\sum_{j=1}^{q_1} \xi_{1j}v_j(0)$, and $v_{\text{con}} = \sum_{j=1}^{q_1} \xi_{1j}v_j(0)$.

Next, consider the dynamics of the agents, denoted by $v_{i_1}, v_{i_2}, \cdots, v_{i_{q_i}}$, $2 \le i \le m$, corresponding to the ith node in $\mathcal{G}^*$. It is only affected by these nodes, such that there exist directed paths from them to $v_{i_s}, s = 1, 2, \cdots, q_i$. Suppose that such agents excluding $v_{i_s}, s = 1, 2, \cdots, q_i$, are $v_{j_1}, v_{j_2}, \cdots, v_{j_{k_i}}$. Furthermore, assume that the states of agents $v_{j_1}, v_{j_2}, \cdots, v_{j_{k_i}}$ have already reached consensus, and the final consensus values of position and velocity states are x_{con} and v_{con}, respectively. Let $\hat{x}_{i_r}(t) = x_{i_r}(t) - x_{con}$ and $\hat{v}_{i_r}(t) = v_{i_r}(t) - v_{con}$, $r = 1, 2, \cdots, q_i$. Then,

$$\begin{cases} \dot{\hat{x}}_{i_r}(t) = \hat{v}_{i_r}(t) \\ \dot{\hat{v}}_{i_r}(t) = \alpha \sum\limits_{j=1}^{q_i} a_{i_r i_j}(\hat{x}_{i_j}(t) - \hat{x}_{i_r}(t)) + \beta \sum\limits_{j=1}^{q_i} a_{i_r i_j}(\hat{v}_{i_j}(t) - \hat{v}_{i_r}(t)) \\ \qquad\quad -\alpha \sum\limits_{p=1}^{k_i} a_{i_r j_p}\hat{x}_{i_r}(t) - \beta \sum\limits_{p=1}^{k_i} a_{i_r j_p}\hat{v}_{i_r}(t) + \mathcal{O}(e^{-\epsilon t}), \quad t \in T, \\ \dot{\hat{v}}_i(t) = 0, \quad t \in \overline{T}, r = 1, 2, \cdots, q_i, \end{cases}$$

for some $\epsilon > 0$. Let $\hat{x}(t) = (\hat{x}_{i_1}^T(t), \hat{x}_{i_2}^T(t), \cdots, \hat{x}_{q_i}^T(t))^T$, $\hat{v}(t) = (\hat{v}_{i_1}^T(t), \hat{v}_{i_2}^T(t), \cdots, \hat{v}_{q_i}^T(t))^T$ and $\hat{y}(t) = (\hat{x}^T(t), \hat{v}^T(t))^T$. Then, the above system can be rewritten as

$$\begin{cases} \dot{\hat{y}}(t) = (\overline{B}_1 \otimes I_n)\hat{y}(t) + (\overline{B}_2 \otimes I_n)\mathcal{O}(e^{-\epsilon t}), & t \in T, \\ \dot{\hat{y}}(t) = (\overline{B}_3 \otimes I_n)\hat{y}(t), & t \in \overline{T}, \end{cases} \tag{8.22}$$

where $\overline{B}_1 = \begin{pmatrix} O_{q_i} & I_{q_i} \\ -\alpha\overline{L}_{ii} & -\beta\overline{L}_{ii} \end{pmatrix}$, $\overline{B}_2 = \begin{pmatrix} O_{q_i} & O_{q_i} \\ I_{q_i} & I_{q_i} \end{pmatrix}$, $\overline{B}_3 = \begin{pmatrix} O_{q_i} & I_{q_i} \\ O_{q_i} & O_{q_i} \end{pmatrix}$. Construct the following Lyapunov function

$$V(t) = \frac{1}{2}\hat{y}^T(t)(P \otimes I_n)\hat{y}(t), \tag{8.23}$$

where $P = \begin{pmatrix} \alpha\beta(\overline{\Xi}_i\overline{L}_{ii} + \overline{L}_{ii}^T\overline{\Xi}_i) & \alpha\overline{\Xi}_i \\ \alpha\overline{\Xi}_i & \beta\overline{\Xi}_i \end{pmatrix}$. Then, according to conditions (i) and (ii), by following the proof of Theorem 1, one can show that the states of agents $v_{i_1}, v_{i_2}, \cdots, v_{i_{q_i}}$, $2 \le i \le m$, will reach consensus exponentially. Furthermore, the final consensus value of the position state is $x_{\text{con}} = \sum_{j=1}^{q_i} \overline{\xi}_{1j}x_j(0) + t\sum_{j=1}^{q_i} \overline{\xi}_{1j}v_j(0)$, and the final consensus value of the velocity state is $v_{\text{con}} = \sum_{j=1}^{q_i} \overline{\xi}_{1j}v_j(0)$. This completes the proof. □

8.4 Consensus of Second-Order Multi-agent Systems with Nonlinear Dynamics and Intermittent Communication

In this section, consensus of second-order multi-agent systems with nonlinear dynamics and missing control inputs is studied.

The model under consideration is

$$\begin{cases} \dot{x}_i(t) = v_i(t) \\ \dot{v}_i(t) = f(x_i(t), v_i(t), t) - \alpha \sum_{j=1}^{N} l_{ij}x_j(t) - \beta \sum_{j=1}^{N} l_{ij}v_j(t), t \in T, \\ \dot{v}_i(t) = f(x_i(t), v_i(t), t), \quad t \in T^c, \quad i = 1, 2, \cdots, N, \end{cases} \tag{8.24}$$

where $T \bigcup T^c = [0, +\infty)$.

Clearly, if consensus can be achieved, it is natural to require a consensus state $s(t) = (s_1^T(t), s_2^T(t))^T \in \mathbb{R}^{2n}$ of the system (8.24) to be a trajectory of an isolated node, satisfying

$$\begin{cases} \dot{s}_1(t) = s_2(t), \\ \dot{s}_2(t) = f(s_1(t), s_2(t), t). \end{cases} \tag{8.25}$$

Here, $s(t)$ may be an isolated equilibrium point, a periodic orbit, or even a chaotic orbit in some applications.

Assumption 8.10 *There exist nonnegative constants ρ_i, $i \in \{1, 2\}$, such that*

$$\| f(x_1, x_2, t) - f(y_1, y_2, t) \| \leq \sum_{i=1}^{2} \rho_i \| x_i - y_i \|,$$

$\forall x_i, y_i \in \mathbb{R}^n$, $i \in \{1, 2\}$, $t \geq 0$.

Let $\widetilde{x}_i(t) = x_i(t) - \sum_{j=1}^{N} \xi_j x_j(t)$ and $\widetilde{v}_i(t) = v_i(t) - \sum_{j=1}^{N} \xi_j v_j(t)$, where $\xi = (\xi_1, \xi_2, \cdots, \xi_N)^T$ is the positive left eigenvector of L associated with eigenvalue 0, satisfying $\xi^T \mathbf{1}_N = 1$. One can easily obtain the following error dynamical system:

$$\begin{cases} \dot{\widetilde{x}}_i(t) = \widetilde{v}_i(t), \\ \dot{\widetilde{v}}_i(t) = f(x_i(t), v_i(t), t) - \sum_{j=1}^{N} \xi_j f(x_j(t), v_j(t), t) - \alpha \sum_{j=1}^{N} l_{ij} \widetilde{x}_j(t) \\ \quad -\beta \sum_{j=1}^{N} l_{ij} \widetilde{v}_j(t), \quad t \in T, \\ \dot{\widetilde{v}}_i(t) = f(x_i(t), v_i(t), t) - \sum_{j=1}^{N} \xi_j f(x_j(t), v_j(t), t), \ t \in T^c, \ i = 1, \cdots, N. \end{cases} \tag{8.26}$$

Let

$$\widetilde{x}(t) = (\widetilde{x}_1^T(t), \cdots, \widetilde{x}_N^T(t))^T, \widetilde{v}(t) = (\widetilde{v}_1^T(t), \cdots, \widetilde{v}_N^T(t))^T,$$
$$f(x(t), v(t), t) = (f^T(x_1(t), v_1(t), t), \cdots, f^T(x_N(t), v_N(t), t))^T,$$

and $\widetilde{y}(t) = (\widetilde{x}^T(t), \widetilde{v}^T(t))^T$. Then, system (8.26) can be written as

$$\begin{cases} \dot{\widetilde{y}}(t) = F(x(t), v(t), t) + (B_1 \otimes I_n)\widetilde{y}(t), \quad t \in T, \\ \dot{\widetilde{y}}(t) = F(x(t), v(t), t) + (B_2 \otimes I_n)\widetilde{y}(t), \quad t \in T^c, \end{cases} \tag{8.27}$$

where

$$F(x(t), v(t), t) = \begin{pmatrix} 0_{Nn} \\ [(I_N - \mathbf{1}_N \xi^T) \otimes I_n] f(x(t), v(t), t) \end{pmatrix},$$

$$B_1 = \begin{pmatrix} O_N & I_N \\ -\alpha L & -\beta L \end{pmatrix}, B_2 = \begin{pmatrix} O_N & I_N \\ O_N & O_N \end{pmatrix}.$$

Theorem 8.11 *Suppose that the network $\mathcal{G}$ is strongly connected and Assumption 8.10 holds. Then, second-order consensus in system (8.24) is achieved if there exists an infinite time sequence of uniformly bounded and nonoverlapping time intervals $[t_k, t_{k+1})$, $k \in \mathbb{N}$, with $t_1 = 0$, such that on each time interval $[t_k, t_{k+1})$, $k \in \mathbb{N}$, the following conditions hold:*

(*i*) $a(L) > \frac{1}{2}\left(\frac{\rho_1}{\alpha} + \frac{\alpha}{\beta^2} + \frac{\rho_2}{\beta} + \sqrt{\left(\frac{\rho_1}{\alpha} - \frac{\alpha}{\beta^2} - \frac{\rho_2}{\beta}\right)^2 + \frac{(\alpha\rho_2+\beta\rho_1)^2}{\alpha^2\beta^2}}\right)$,

(*ii*) $\delta_k > \frac{\gamma_2}{\gamma_1+\gamma_2}\omega_k$,

where δ_k *represents the Lebesgue measure of the set* $\{t \mid t \in [t_k, t_{k+1}) \cap T\}$, $\omega_k = t_{k+1} - t_k$, $\gamma_1 = (\lambda_{\min}(P_1)\xi_{\min})/(\lambda_{\max}(P_2)\xi_{\max})$, $P_1 = \begin{pmatrix} \alpha^2 a(L) - \alpha\rho_1 & -\frac{1}{2}(\alpha\rho_2+\beta\rho_1) \\ -\frac{1}{2}(\alpha\rho_2+\beta\rho_1) & \beta^2 a(L) - \alpha - \beta\rho_2 \end{pmatrix}$, $P_2 = \begin{pmatrix} \alpha\beta b(L) & \alpha/2 \\ \alpha/2 & \beta/2 \end{pmatrix}$, $\gamma_2 = 2\lambda_{\max}(Q^{-1}P_3)$, $Q = \begin{pmatrix} 2\alpha\beta a(L)\Xi & \alpha\Xi \\ \alpha\Xi & \beta\Xi \end{pmatrix}$, *and* $P_3 = \begin{pmatrix} \alpha\rho_1\Xi & \frac{(\alpha\rho_2+\beta\rho_1)\Xi+\alpha\beta(\Xi L+L^T\Xi)}{2} \\ \frac{(\alpha\rho_2+\beta\rho_1)\Xi+\alpha\beta(\Xi L+L^T\Xi)}{2} & (\beta\rho_2+\alpha)\Xi \end{pmatrix}$.

Proof. Construct the Lyapunov function candidate

$$V(t) = \frac{1}{2}\widetilde{y}^T(t)(P \otimes I_n)\widetilde{y}(t), \tag{8.28}$$

where $P = \begin{pmatrix} \alpha\beta(\Xi L + L^T\Xi) & \alpha\Xi \\ \alpha\Xi & \beta\Xi \end{pmatrix}$ and $\Xi = \text{diag}(\xi_1, \xi_2, \cdots, \xi_N)$. According to Lemma 2.25,

$$V(t) \geq \frac{1}{2}\widetilde{y}^T(t)(Q \otimes I_n)\widetilde{y}(t), \tag{8.29}$$

where $Q = \begin{pmatrix} 2\alpha\beta a(L)\Xi & \alpha\Xi \\ \alpha\Xi & \beta\Xi \end{pmatrix}$. By the Schur complement argument in Lemma 3.17, $Q > 0$ is equivalent to both $\beta > 0$ and $a(L) > \alpha/(2\beta^2)$. From condition (i), one obtains $Q > 0$, $V(t) \geq 0$ and $V(t) = 0$ if and only if $\widetilde{y}(t) = 0_{2Nn}$.

Let $\overline{x}(t) = \Sigma_{j=1}^N \xi_j(0)x_j(t)$ and $\overline{v}(t) = \Sigma_{j=1}^N \xi_j(0)v_j(0)$. For $t \in [t_k, t_{k+1}) \cap T$, taking the time derivative of $V(t)$ along the trajectories of (8.27) gives

$$\begin{aligned}
\dot{V}(t) =& \widetilde{y}^T(t)(P \otimes I_n)[F(x(t), v(t), t) + (B_1 \otimes I_n)\widetilde{y}(t)] \\
=& (\alpha\widetilde{x}^T(t) + \beta\widetilde{v}^T(t))[\Xi \otimes I_n][f(x(t), v(t), t) - 1_N \otimes f(\overline{x}(t), \overline{v}(t), t)] \\
& -\frac{\alpha^2}{2}\widetilde{x}^T(t)[(\Xi L + L^T\Xi) \otimes I_n]\widetilde{x}(t) - \frac{\beta^2}{2}\widetilde{v}^T(t)[(\Xi L + L^T\Xi) \otimes I_n]\widetilde{v}(t) \\
& +\alpha\widetilde{v}^T(t)(\Xi \otimes I_n)\widetilde{v}(t).
\end{aligned} \tag{8.30}$$

By Assumption 8.10,

$$\widetilde{x}^T(t)[\Xi \otimes I_n][f(x(t), v(t), t) - 1_N \otimes f(\overline{x}(t), \overline{v}(t), t)]$$

$$\leq \sum_{i=1}^{N} \| \widetilde{x}_i(t) \| \xi_i(\rho_1 \| \widetilde{x}_i(t) \| + \rho_2 \| \widetilde{v}_i(t) \|)$$

$$= \rho_1 \sum_{i=1}^{N} \xi_i \| \widetilde{x}_i(t)\|^2 + \rho_2 \sum_{i=1}^{N} \xi_i \| \widetilde{x}_i(t) \| \| \widetilde{v}_i(t) \|, \tag{8.31}$$

and

$$\widetilde{v}^T(t)[\Xi \otimes I_n][f(x(t), v(t), t) - 1_N \otimes f(\overline{x}(t), \overline{v}(t), t)]$$

$$\leq \rho_2 \sum_{i=1}^{N} \| \widetilde{v}_i(t)\|^2 + \rho_1 \sum_{i=1}^{N} \xi_i \| \widetilde{x}_i(t) \| \| \widetilde{v}_i(t) \|. \tag{8.32}$$

Combining (8.30 –8.32) gives

$$\dot{V}(t) \leq - \| \widetilde{y}(t)\|^T (P_1 \otimes \Xi) \| \widetilde{y}(t) \|, \tag{8.33}$$

where $\| \widetilde{x}(t) \| = (\| \widetilde{x}_1(t) \|, \cdots, \| \widetilde{x}_N(t) \|)^T$, $\| \widetilde{v}(t) \| = (\| \widetilde{v}_1(t) \|, \cdots, \| \widetilde{v}_N(t) \|)^T$, $\| \widetilde{y}(t) \| = (\| \widetilde{x}(t)\|^T, \| \widetilde{v}(t)\|^T)^T$, $P_1 = \begin{pmatrix} \alpha^2 a(L) - \alpha\rho_1 & -\frac{1}{2}(\alpha\rho_2 + \beta\rho_1) \\ -\frac{1}{2}(\alpha\rho_2 + \beta\rho_1) & \beta^2 a(L) - \alpha - \beta\rho_2 \end{pmatrix}$. By condition (i), P_1 is a positive-definite matrix. According to (8.33), one has

$$\dot{V}(t) \leq -\lambda_{\min}(P_1)\xi_{\min}\widetilde{y}^T(t)\widetilde{y}(t). \tag{8.34}$$

On the other hand,

$$\begin{aligned} V(t) &= \frac{1}{2}\widetilde{y}^T(t)(P \otimes I_n)\widetilde{y}(t) \\ &\leq \widetilde{y}^T(t)[(P_2 \otimes \Xi) \otimes I_n]\widetilde{y}(t), \end{aligned} \tag{8.35}$$

where the positive matrix $P_2 = \begin{pmatrix} \alpha\beta b(L) & \alpha/2 \\ \alpha/2 & \beta/2 \end{pmatrix}$. Thus,

$$V(t) \leq \lambda_{\max}(P_2)\xi_{\max}\widetilde{y}^T(t)\widetilde{y}(t). \tag{8.36}$$

Consequently,

$$\dot{V}(t) \leq -\gamma_1 V(t), \tag{8.37}$$

where $\gamma_1 = (\lambda_{\min}(P_1)\xi_{\min}) / (\lambda_{\max}(P_2)\xi_{\max})$.

For $t \in \{[t_k, t_{k+1}) \cap T\}$, taking the time derivative of $V(t)$ along the trajectories of (8.27) gives

$$\begin{aligned} \dot{V}(t) &= \widetilde{y}^T(t)(P \otimes I_n)[F(x(t), v(t), t) + (B_2 \otimes I_n)\widetilde{y}(t)] \\ &= \widetilde{y}^T(t)(P_3 \otimes I_n)\widetilde{y}(t), \end{aligned} \tag{8.38}$$

where

$$P_3 = \begin{pmatrix} \alpha\rho_1\Xi & \frac{(\alpha\rho_2+\beta\rho_1)\Xi+\alpha\beta(\Xi L+L^T\Xi)}{2} \\ \frac{(\alpha\rho_2+\beta\rho_1)\Xi+\alpha\beta(\Xi L+L^T\Xi)}{2} & (\beta\rho_2+\alpha)\Xi \end{pmatrix}.$$

It follows from Lemma 8.3 and inequality (8.29) that

$$\begin{aligned} \dot{V}(t) &\leq \lambda_{\max}((Q\otimes I_n)^{-1}(P_3\otimes I_n))\widetilde{y}^T(t)(Q\otimes I_n)\widetilde{y}(t) \\ &= \lambda_{\max}(Q^{-1}P_3)\widetilde{y}^T(t)(Q\otimes I_n)\widetilde{y}(t) \\ &\leq \gamma_2 V(t), \end{aligned} \tag{8.39}$$

where $\gamma_2 = 2\lambda_{\max}(Q^{-1}P_3)$.

Based on the above analysis,

$$V(t_2) \leq V(0)e^{-\Delta_1}, \tag{8.40}$$

where $\Delta_1 = \gamma_1\delta_1 - \gamma_2(\omega_1 - \delta_1)$. Then, according to condition (ii), $\Delta_1 > 0$. By recursion, for any positive integer k,

$$V(t_{k+1}) \leq V(0)e^{-\sum_{j=1}^{k}\Delta_j}, \tag{8.41}$$

where $\Delta_j = \gamma_1\delta_j - \gamma_2(\omega_j - \delta_j) > 0$, $j = 1, 2, \cdots, k$. For arbitrary $t > 0$, there exists a positive integer s such that $t_{s+1} < t \leq t_{s+2}$. Furthermore, since $[t_k, t_{k+1})$, $k \in \mathbb{N}$, is a uniformly bounded and non-overlapping time sequence, one can let $\omega_{\max} = \max_{i\in\mathbb{N}} \omega_i$ and $\kappa = \min_{i\in\mathbb{N}} \Delta_i > 0$. Thus, it follows that

$$\begin{aligned} V(t) &\leq V(t_{s+1})e^{\omega_{\max}\gamma_2} \\ &\leq e^{\omega_{\max}\gamma_2}V(0)e^{-\sum_{j=1}^{s}\Delta_j} \\ &\leq e^{\omega_{\max}\gamma_2}V(0)e^{-(\kappa/\omega_{\max})t} \\ &\leq K_0 e^{-K_1 t}, \quad \text{for all } t > 0, \end{aligned} \tag{8.42}$$

where $K_0 = e^{\gamma_2}V(0)$ and $K_1 = \frac{\kappa}{\omega_{\max}}$, which indicates that the states of agents converge exponentially, thereby achieving consensus. This completes the proof. □

Example 8.1 *Consider a multi-agent system consisting of four nonlinearly dynamical agents, where the communication topology is shown in Fig. 8.1 with weighting on the edges. The nonlinear function f is described by Chua's circuit [24]: $f(x_i(t), v_i(t), t) = (-v_{i1} + v_{i2} - l(v_{i1}), v_{i1} - v_{i2} + v_{i3}, -\varsigma v_{i2})^T$, $i = 1, 2, 3$, where $l(v_{i1}) = bv_{i1} + 0.5(a-b)(|v_{i1}+1| - |v_{i1}-1|)$, $x_i = (x_{i1}, x_{i2}, x_{i3})^T$, $v_i = (v_{i1}, v_{i2}, v_{i3})^T$. The isolated system is chaotic when $\mu = 10, \varsigma = 18, a = -4/3$, and $b = -3/4$, as shown in [149]. In view of Assumption 8.10, one has $\rho_1 = 4.3871$, $\rho_2 = 0$. It is assumed that the four agents communicate with their neighbors only when*

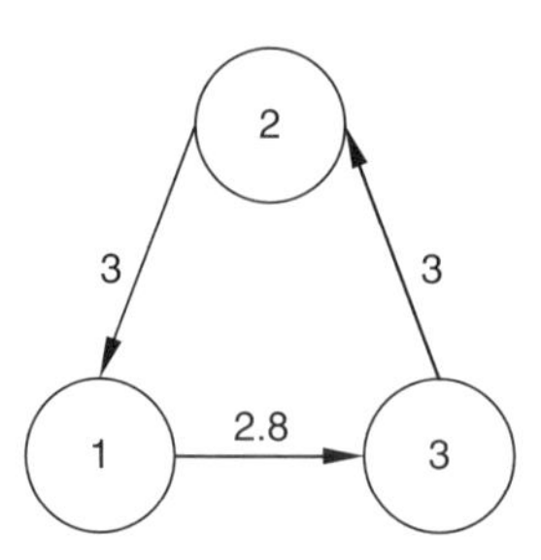

Figure 8.1 The network $\mathscr{G}(\mathscr{A})$ in Example 8.1

$t \in \bigcup_{k\in\mathbb{N}}[k, k+0.95)$. *Furthermore, let* $\alpha = 5$ *and* $\beta = 6$ *in system (8.24). From Fig. 8.1, it is easy to see that the network* $\mathscr{G}$ *is strongly connected. By Theorem 8.11, one has* $a(L) = 4.2997 > \frac{1}{2}\left(\frac{\rho_1}{\alpha} + \frac{\alpha}{\beta^2} + \frac{\rho_2}{\beta} + \sqrt{\left(\frac{\rho_1}{\alpha} - \frac{\alpha}{\beta^2} - \frac{\rho_2}{\beta}\right)^2 + \frac{(\alpha\rho_2+\beta\rho_1)^2}{\alpha^2\beta^2}}\right) = 1.0816$, *and* $\delta_k = 0.95 > \frac{\gamma_2}{\gamma_1+\gamma_2}\omega_k = 0.9293$, *for all* $k \in \mathbb{N}$. *Therefore, second-order consensus can be achieved in system (8.24). The position and velocity states of all agents are shown in Figs. 8.2 and 8.3, respectively, with initial conditions* $x_1(0) = (0.15, -0.25, 0.14)^T$, $x_2(0) = (-1.6, 1.8, 2.0)^T$, $x_3(0) = (-1, -1.2, 1.8)^T$, $v_1(0) = (3.2, 2.4, 0.8)^T$, $v_2(0) = (2.4, 3.0, 1.4)^T$, $v_3(0) = (-1.5, -1.2, 1.4)^T$. *Simulation results verify the theoretical analysis very well.*

8.5 Notes

The results provided in Sections 8.2 and 8.3 are based on [126], while the results in Section 8.4 are based on [124]. For further results on consensus of multi-agent systems

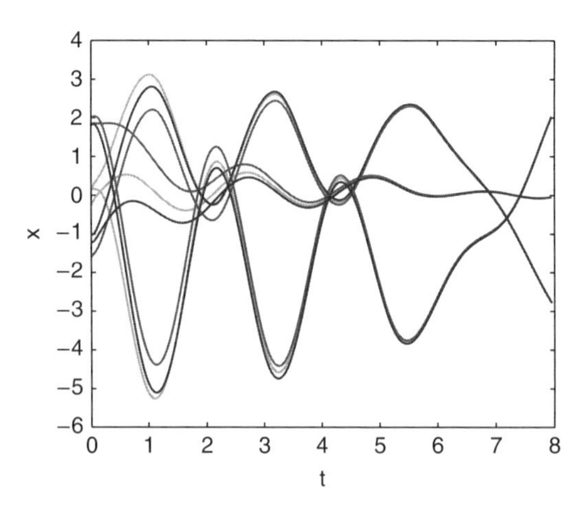

Figure 8.2 Consensus of state trajectories of multiple agents

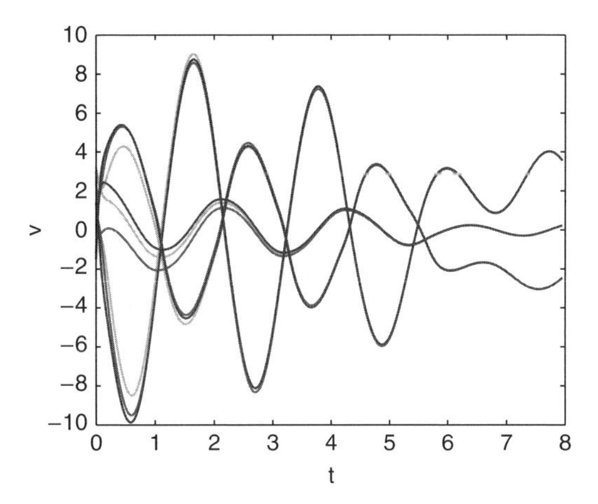

Figure 8.3 Consensus of velocity trajectories of multiple agents

with general linear node dynamics, refer to [125, 129]. In particular, consensus for multi-agent systems with general linear node dynamics and a fixed directed communication topology was studied in [125]. In [129], Consensus tracking in multi-agent systems with general linear node dynamics and switching directed topologies was studied by using tools from M-matrix theory and Lyapunov stability analysis for switched systems. It should be noted that much research on consensus of multi-agent systems with intermittent communication is still ongoing today, such as consensus of second-order multi-agent systems with delayed nonlinear dynamics and intermittent communication [128], semi-global consensus of multi-agent systems with intermittent communications [121], motion coordination of thrust-propelled underactuated vehicles in the presence of intermittent communication [1], solving the consensus problem for linear multi-agent systems with intermittent information transmissions via time-scale theory [113], to name only a few.

It is worth noting that there are still many unsolved interesting problems on cooperative control of multi-agent systems with intermittent communication, thus deserving further investigations. For examples, it is important yet challenging to study nonlinear multi-agent systems with asynchronously intermittent communication, finite-time consensus for multi-agent systems with intermittent communication, and consensus of multi-agent systems with intermittently quantized communication.

9

Distributed Adaptive Control of Multi-agent Systems

In this chapter 9, distributed adaptive control of multi-agent systems is studied [156, 157].

First, distributed adaptive control of synchronization in complex networks is discussed [156]. An effective distributed adaptive strategy to tune the coupling weights of a network is designed, based on local information of node dynamics. The analysis is then extended to the case where only a small fraction of coupling weights are adjusted. A general criterion is derived and it is found that synchronization can be reached if the subgraph consisting of the edges and nodes corresponding to the updated coupling weights contains a spanning tree.

Then, pinning control in complex networks is investigated. There exist some fundamental and yet challenging problems in pinning control of complex networks: (1) What types of pinning schemes may be chosen for a given complex network to realize synchronization? (2) What kinds of controllers may be designed to ensure network synchronization? (3) How large should the coupling strength be when being used in a given complex network to achieve synchronization? The second section addresses these technical questions. Surprisingly, it is found that a network under a typical framework can realize synchronization subject to any linear feedback pinning scheme by adaptively tuning the coupling strength. In addition, it is found that the nodes with low degrees should be pinned first when the coupling strength is small, which is contrary to the common view that the most highly connected nodes should be pinned first. Furthermore, it is interesting to find that the derived pinning condition with controllers given in a higher-dimensional setting can be reduced to a lower-dimensional condition without involving pinning controllers.

This chapter further discusses the design of distributed control gains for consensus in multi-agent systems with second-order nonlinear dynamics [157]. First, an effective distributed adaptive gain-design strategy is derived based only on local information of the network structure. Then, a leader-follower consensus problem in multi-agent

Distributed Cooperative Control of Multi-agent Systems, First Edition.
Wenwu Yu, Guanghui Wen, Guanrong Chen, and Jinde Cao.

systems with updated control gains is studied. A distributed adaptive law is then derived for each follower based on local information of neighboring agents and of the leader if this follower is an informed agent. Furthermore, a distributed leader-follower consensus problem in multi-agent systems with unknown nonlinear dynamics is investigated by combining the variable structure approach and the adaptive method.

9.1 Distributed Adaptive Control in Complex Networks

9.1.1 Preliminaries

Consider the complex network

$$\dot{x}_i(t) = f(x_i(t), t) - c\sum_{j=1}^{N} L_{ij}(t)\Gamma x_j(t), i = 1, 2, \cdots, N, \tag{9.1}$$

where $x_i = (x_{i1}, x_{i2}, \cdots, x_{in})^T \in \mathbb{R}^n$ is the state vector of the i-th node, $f : \mathbb{R}^n \times \mathbb{R}^+ \to \mathbb{R}^n$ is a nonlinear vector function, c is the constant overall coupling strength, and $\Gamma = \text{diag}(\gamma_1, \ldots, \gamma_n) \in \mathbb{R}^{n\times n}$ is a positive semi-definite inner coupling matrix, where $\gamma_j > 0$ if two nodes can communicate through the j-th state, and $\gamma_j = 0$ otherwise.

The objective of control here is to find some adaptive laws acting on $L_{ij}(t)$ under a fixed network topology such that the solutions of the controlled network (9.1) can achieve synchronization:

$$\lim_{t\to\infty} \| x_i(t) - x_j(t) \| = 0, \forall i, j = 1, 2, \cdots, N. \tag{9.2}$$

Note that the network structure is fixed and only the coupling weights can be time-varying, but if there is no connection between nodes i and j, then $G_{ij}(t) = G_{ji}(t) = 0$ for all t.

Assumption 9.1 *There exist a constant diagonal matrix* $\Delta = \text{diag}(\delta_1, \ldots, \delta_n)$ *and an* $\epsilon > 0$ *such that*

$$\begin{aligned} &(x-y)^T(f(x,t) - f(y,t)) - (x-y)^T\Delta(x-y) \\ &\leq -\epsilon(x-y)^T(x-y), \forall x, y \in \mathbb{R}^n, \forall t \in \mathbb{R}^+. \end{aligned} \tag{9.3}$$

Assumption 9.1 is the so-called QUAD condition on vector fields [23]. In what follows, assume $\Delta = H\Gamma$, where H is a diagonal matrix and Γ is the inner coupling matrix defined in (9.1). Note that this assumption is very mild. For example, all the linear and piecewise-linear continuous functions satisfy this condition. In addition, the condition is satisfied if $\partial f_i/\partial x_j$ $(i, j = 1, 2, \ldots, n)$ are uniformly bounded and Γ is positive definite, which includes many well-known systems.

Without loss of generality, only connected networks are considered throughout the chapter; otherwise, one can consider the synchronization on each connected component of the network separately.

9.1.2 Distributed Adaptive Control in Complex Networks

Following [29], some distributed adaptive laws on the weights $L_{ij}(t)$ with $i \neq j$ are proposed in this subsection, which result in corresponding adaptive laws on G_{ij} since $L_{ij} = -G_{ij}$, $i \neq j$.

Let $\bar{x} = \frac{1}{N}\sum_{j=1}^{N} x_j$. Then, one has

$$\dot{\bar{x}}(t) = \frac{1}{N}\sum_{j=1}^{N} f(x_j(t), t). \tag{9.4}$$

Subtracting (9.4) from (9.1) yields the following error dynamical network:

$$\dot{e}_i(t) = f(x_i(t), t) - \frac{1}{N}\sum_{j=1}^{N} f(x_j(t), t) - c\sum_{j=1}^{N} L_{ij}(t)\Gamma e_j(t), i = 1, 2, \ldots, N, \tag{9.5}$$

where $e_i = x_i - \bar{x}$, $i = 1, \ldots, N$.

Theorem 9.2 *Suppose that Assumption 9.1 holds and Γ is positive definite. Then, network (9.1) achieves synchronization under the following distributed adaptive law:*

$$\dot{L}_{ij}(t) = -\alpha_{ij}(x_i - x_j)^T\Gamma(x_i - x_j), \quad L_{ij}(0) = L_{ji}(0) \leq 0, \quad \forall (i,j) \in \mathscr{E}, \tag{9.6}$$

where $\alpha_{ij} = \alpha_{ji}$ are positive constants.

Proof. Consider the Lyapunov function candidate

$$V(t) = \frac{1}{2}\sum_{i=1}^{N} e_i^T(t)e_i(t) + \sum_{i=1}^{N}\sum_{j=1, j\neq i}^{N} \frac{c}{4\alpha_{ij}}(L_{ij}(t) + c_{ij})^2, \tag{9.7}$$

where $c_{ij} = c_{ji}$ are nonnegative constants, and $c_{ij} = 0$ if and only if $L_{ij} = 0$.

The derivative of $V(t)$ along the trajectories of (9.5) gives

$$\begin{aligned}
\dot{V} &= \sum_{i=1}^{N} e_i^T(t)\dot{e}_i(t) + \sum_{i=1}^{N}\sum_{j=1, j\neq i}^{N} \frac{c}{2\alpha_{ij}}(L_{ij}(t) + c_{ij})\dot{L}_{ij}(t) \\
&= \sum_{i=1}^{N} e_i^T(t)\left[f(x_i(t), t) - \frac{1}{N}\sum_{j=1}^{N} f(x_j(t), t) - c\sum_{j=1}^{N} L_{ij}(t)\Gamma e_j(t)\right] \\
&\quad - \frac{c}{2}\sum_{i=1}^{N}\sum_{j=1, j\neq i}^{N} (L_{ij}(t) + c_{ij})(x_i - x_j)^T\Gamma(x_i - x_j) \\
&= \sum_{i=1}^{N} e_i^T(t)\left[f(x_i(t), t) - f(\bar{x}, t) + f(\bar{x}, t) - \frac{1}{N}\sum_{j=1}^{N} f(x_j(t), t) - c\sum_{j=1}^{N} L_{ij}(t)\Gamma e_j(t)\right] \\
&\quad - \frac{c}{2}\sum_{i=1}^{N}\sum_{j=1, j\neq i}^{N} (L_{ij}(t) + c_{ij})(e_i - e_j)^T\Gamma(e_i - e_j).
\end{aligned} \tag{9.8}$$

Since $\sum_{i=1}^{N} e_i^T(t) = 0$, one has $\sum_{i=1}^{N} e_i^T(t)\left[f(\bar{x},t) - \frac{1}{N}\sum_{j=1}^{N} f(x_j(t),t)\right] = 0$. From Assumption 9.1, it follows that

$$\sum_{i=1}^{N} e_i^T(t)[f(x_i(t),t) - f(\bar{x},t)] \le -\epsilon \sum_{i=1}^{N} e_i^T(t)e_i(t) + \sum_{i=1}^{N} e_i^T(t)H\Gamma e_i(t). \tag{9.9}$$

Define the Laplacian matrix $\Sigma = (\widetilde{\sigma}_{ij})_{N\times N}$, where $\widetilde{\sigma}_{ij} = -c_{ij}, i \neq j$; $\widetilde{\sigma}_{ii} = -\sum_{j=1, j\neq i}^{N} \widetilde{\sigma}_{ij}$. In view of Lemma 2.6, one obtains

$$\begin{aligned} & c\sum_{i=1}^{N}\sum_{j=1, j\neq i}^{N} (L_{ij}(t) + \sigma_{ij})(e_i - e_j)^T\Gamma(e_i - e_j) \\ & = -2c\sum_{i=1}^{N}\sum_{j=1}^{N} L_{ij}(t)e_i^T\Gamma e_j + 2c\sum_{i=1}^{N}\sum_{j=1}^{N} \widetilde{\sigma}_{ij}(t)e_i^T\Gamma e_j. \end{aligned} \tag{9.10}$$

Therefore, combining (9.8)–(9.10) and by using Lemma 2.6, one gets

$$\begin{aligned} \dot{V} & \le -\epsilon\sum_{i=1}^{N} e_i^T(t)e_i(t) + \sum_{i=1}^{N} e_i^T(t)H\Gamma e_i(t) - c\sum_{i=1}^{N}\sum_{j=1}^{N} \widetilde{\sigma}_{ij}(t)e_i^T\Gamma e_j \\ & = e^T(t)[-\epsilon(I_N \otimes I_n) + (I_N \otimes H\Gamma) - c(\Sigma \otimes \Gamma)]e(t). \end{aligned} \tag{9.11}$$

Let Λ be the diagonal matrix associated with Σ, that is, there exists a unitary matrix $P = (p_1, \dots, p_N)$ such that $P^T\Sigma P = \Lambda$. Let $y(t) = (P^T \otimes I_N)e(t)$. Since $p_1 = \frac{1}{\sqrt{N}}(1, \dots, 1)^T$, one has $y_1 = (p_1^T \otimes I_N)e(t) = 0$. Then, it follows that

$$\begin{aligned} \dot{V} & \le e^T(t)[-\epsilon(I_N \otimes I_n) + (I_N \otimes H\Gamma) - c(P \otimes I_N)(\Lambda \otimes \Gamma)(P^T \otimes I_N)]e(t) \\ & = e^T(t)[-\epsilon(I_N \otimes I_n) + (I_N \otimes H\Gamma)]e(t) - cy^T(t)(\Lambda \otimes \Gamma)y(t). \end{aligned} \tag{9.12}$$

By Lemma 2.6 and since Γ is positive definite, one has $y^T(t)(\Lambda \otimes \Gamma)y(t) \ge \lambda_2(\Sigma)y^T(t)(I_N \otimes \Gamma)y(t)$. Hence,

$$\begin{aligned} \dot{V} & \le e^T(t)[-\epsilon(I_N \otimes I_n) + (I_N \otimes H\Gamma)]e(t) - c\lambda_2(\Sigma)y^T(t)(I_N \otimes \Gamma)y(t) \\ & = e^T(t)[-\epsilon(I_N \otimes I_n) + (I_N \otimes H\Gamma)]e(t) \\ & \quad -c\lambda_2(\Sigma)e^T(t)(P \otimes I_N)(I_N \otimes \Gamma)(P^T \otimes I_N)e(t) \\ & = e^T(t)[-\epsilon(I_N \otimes I_n) + (I_N \otimes H\Gamma) - c\lambda_2(\Sigma)(I_N \otimes \Gamma)]e(t). \end{aligned} \tag{9.13}$$

By choosing c_{ij} sufficiently large, such that $c\lambda_2(\Sigma) > \max_j(h_j)$, one obtains $(I_N \otimes H\Gamma) - c\lambda_2(\Sigma)(I_N \otimes \Gamma) \le 0$. Therefore,

$$\dot{V} \le -\epsilon e^T(t)e(t), \tag{9.14}$$

where $e(t) = (e_1^T(t), e_2^T(t), \dots, e_N^T(t))^T$.

Thus, it follows that function (9.7) is not increasing, and so each term of (9.7) is bounded. Consequently, both the error vector e and each of the coupling gains L_{ij} are bounded. Since L_{ij} is monotonically decreasing (see (9.6)), one concludes that each gain asymptotically converges to some finite negative value. The convergence of all L_{ij} implies, due to (9.6) and since Γ is positive definite, that the error vector asymptotically approaches zero. □

Corollary 9.3 *Suppose that Assumption 9.1 holds, Γ is positive semi-definite, and system (9.1) is autonomous (i.e., $f(\cdot, t)$ is independent of time t). Then, network (9.1) achieve synchronization under the distributed adaptive law (9.6).*

Proof. The proof can be completed by applying the LaSalle invariance principle ([49], Chapter 5, Theorem 2.1). □

When system (9.1) is nonautonomous, a similar result is given as follows.

Theorem 9.4 *Suppose that Assumption 9.1 holds, the vector field f is continuously differentiable, and Γ is positive semi-definite. Then, network (9.1) achieves synchronization under the distributed adaptive law (9.6).*

Proof. Proceeding as in the proof of Theorem 9.2, one can show that e and each L_{ij} are bounded by (9.14). Since f is differentiable and, from (9.8), $\dot{V}$ is continuous, the boundedness of e and of each L_{ij} implies that $\ddot{V}$ is itself bounded. Therefore, $\dot{V}$ is uniformly continuous. So, as shown in [5] (pp. 199–208), one concludes that the error goes to zero asymptotically, and that, due to (9.6), all L_{ij} converge to a finite value. □

Remark 9.5 In retrospect, the dynamical coupling weights were adjusted, according to the local synchronization property between each couple of nodes, in [29, 30]. A detailed analysis of the families of adaptive strategies guaranteeing synchronization was given in [29]. However, those derived conditions are conservative, as shown in Subsection 9.1.4 therein. In Theorem 9.2 here, by designing a simple distributed adaptive law on the coupling weights, the conditions in [29] are significantly relaxed. As long as Assumption 9.1 is satisfied, the network can reach synchronization under the distributed adaptive law (9.6).

9.1.3 Pinning Edges Control

In Theorem 9.2, all the coupling weights are adjusted according to the distributed adaptive law (9.6). Though the algorithm is local and decentralized, it is literally impossible to effectively update all the coupling weights of the network in applications. Here, the aim is to update only a small fraction of the coupling weights such that synchronization can be reached. This procedure is called *pinning edges control.* For pinning nodes control, refer to [23, 119, 151].

Assume that the pinning strategy is applied on a small fraction δ $(0 < \delta < 1)$ of the coupling weights in network (9.1). The total number of selected edges will be $l = \lfloor \delta N \rfloor$, the integer part of the real number δN.

Let $\widetilde{\mathscr{G}} = (\widetilde{\mathscr{V}}, \widetilde{\mathscr{E}}, \widetilde{G})$ be a subgraph of $\mathscr{G}$, with a set of l selected undirected edges $\widetilde{\mathscr{E}}$, the corresponding set of nodes $\widetilde{\mathscr{V}}$, and a weighted adjacency matrix $\widetilde{G} = (\widetilde{G}_{ij})_{N\times N}$, where $\widetilde{\mathscr{V}} = \{v_{k_1}, v_{k_2}, \cdots, v_{k_l}, v_{m_1}, v_{m_2}, \cdots, v_{m_l}\}$, $\widetilde{e}_{k_i m_i} = (\widetilde{v}_{k_i}, \widetilde{v}_{m_i}) \in \widetilde{\mathscr{E}}$, and $\widetilde{G}_{k_i m_i}(t) = \widetilde{G}_{m_i k_i}(t) > 0$, $i = 1, \ldots, l$. Let $\widetilde{L}$ be the corresponding Laplacian matrix of the adjacency matrix $\widetilde{G}$.

Theorem 9.6 *Suppose that Assumption 9.1 holds and Γ is positive definite. Under the distributed adaptive law*

$$\dot{L}_{ij}(t) = \dot{L}_{ji}(t) = -\alpha_{ij}(x_i - x_j)^T \Gamma (x_i - x_j), \quad (i,j) \in \widetilde{\mathscr{E}}, \tag{9.15}$$

where $\alpha_{k_i m_i} = \alpha_{m_i k_i}$ are positive constants, network (9.1) achieves synchronization if there are positive constants $\widetilde{\sigma}_{m_i k_i}$ such that

$$(I_N \otimes H\Gamma) - c\lambda_2(\underline{L})(I_N \otimes \Gamma) \leq 0, \tag{9.16}$$

where $\underline{L} = (\underline{L}_{js})_{N\times N}$ is a Laplacian matrix with $\underline{L}_{js} = \underline{L}_{sj} = -\widetilde{\sigma}_{m_i k_i}$ when $j = m_i$ and $s = k_i$; otherwise, $\underline{L}_{js} = \underline{L}_{sj} = L_{js}$, $i = 1, \ldots, l$, $1 \leq j \neq s \leq N$.

Proof. Consider the Lyapunov function candidate

$$V(t) = \frac{1}{2}\sum_{i=1}^{N} e_i^T(t) e_i(t) + \sum_{i=1}^{l} \frac{c}{2\alpha_{k_i m_i}} (L_{m_i k_i}(t) + \widetilde{\sigma}_{m_i k_i})^2, \tag{9.17}$$

where $\widetilde{\sigma}_{m_i k_i} = \widetilde{\sigma}_{k_i m_i}$ are positive constants to be determined, $i = 1, 2, \ldots, l$.

The derivative of $V(t)$ along the trajectories of (9.17) gives

$$\begin{aligned} \dot{V} = & \sum_{i=1}^{N} e_i^T(t) \left[f(x_i(t), t) - f(\overline{x}, t) - c\sum_{j=1}^{N} L_{ij}(t) \Gamma e_j(t) \right] \\ & - c\sum_{i=1}^{l} (L_{m_i k_i}(t) + \widetilde{\sigma}_{m_i k_i})(e_{k_i} - e_{m_i})^T \Gamma (e_{k_i} - e_{m_i}) \\ \leq & \, e^T(t)[-\epsilon(I_N \otimes I_n) + (I_N \otimes H\Gamma) - c(\underline{L} \otimes \Gamma)] e(t). \end{aligned} \tag{9.18}$$

The proof can be completed by following the same process as in the proof of Theorem 9.2 above. □

In Theorem 9.6, a general criterion is given for reaching synchronization by using the designed adaptive law (9.15). Clearly, one can solve (9.16) by using an LMI approach. However, the condition in (9.16) may be difficult to apply to a very large-scale network. Therefore, the next objective is to simplify the condition in (9.16).

Corollary 9.7 *Suppose that Assumption 9.1 holds and Γ is positive definite. If the subgraph $\widetilde{\mathcal{G}}$ contains a spanning tree of $\mathcal{G}$, then network (9.1) achieves synchronization under the following distributed adaptive law:*

$$\dot{L}_{ij}(t) = \dot{L}_{ji}(t) = -\alpha_{ij}(x_i - x_j)^T\Gamma(x_i - x_j), \quad (i,j) \in \widetilde{\mathcal{E}}, \tag{9.19}$$

where $\alpha_{ij} = \alpha_{ji}$ are positive constants.

Proof. Consider the same Lyapunov function candidate given in (9.17) and take the derivative of $V(t)$ along the trajectories of (9.17):

$$\dot{V} \leq e^T(t)[-\epsilon(I_N \otimes I_n) + (I_N \otimes H\Gamma) - c(\overline{L} \otimes \Gamma) - c(\widetilde{\Sigma} \otimes \Gamma)]e(t), \tag{9.20}$$

where $\overline{L}$ and $\widetilde{\Sigma}$ are Laplacian matrices defined by $\overline{L} = L - \widetilde{L}$ and $\widetilde{\Sigma} = (-\widetilde{\sigma}_{ij})_{N\times N}$, with $\widetilde{\sigma}_{m_pk_p} > 0, p = 1, \ldots, l$ and $\widetilde{\sigma}_{ij} = 0$; otherwise, $i, \neq j, i,j = 1, \ldots, N$. It is easy to see that each off-diagonal entry in $\overline{L}$ is the same as in L, except that $\overline{L}_{ij} = \overline{L}_{ji} = 0, \forall (i,j) \in \widetilde{\mathcal{E}}$. Since $\widetilde{\mathcal{G}}$ contains a spanning tree of $\mathcal{G}$, $\lambda_2(\widetilde{\Sigma})$ can be sufficiently large if one chooses sufficiently large positive constants $\widetilde{\sigma}_{k_im_i}$, $i = 1, \ldots, N$. Therefore,

$$\dot{V} \leq e^T(t)[-\epsilon(I_N \otimes I_n) + (I_N \otimes H\Gamma) - c\lambda_2(\widetilde{\Sigma})(I_N \otimes \Gamma)]e(t). \tag{9.21}$$

This completes the proof. □

Remark 9.8 In Corollary 9.7, if the subgraph consisting of the edges and nodes corresponding to the updated coupling weights contains a spanning tree of $\mathcal{G}$, then synchronization in network (9.1) can be reached. Therefore, the minimal number of updated coupling weights under this scheme is $N - 1$, which is much less than number of all the coupling weights which need updating.

As in Subsection 9.1.2, the analysis is extended to the case where the network nodes are connected through only a subset of the state vectors, that is, Γ is positive semi-definite.

Corollary 9.9 *Suppose that Assumption 9.1 holds, Γ is positive semi-definite, and system (9.1) is autonomous (i.e., $f(\cdot, t)$ is independent of time t). Then, network (9.1) achieves synchronization under the distributed adaptive law (9.19).*

Proof. The proof can be completed by applying the LaSalle invariance principle ([49], Chapter 5, Theorem 2.1). □

Theorem 9.10 *Suppose that Assumption 9.1 holds, Γ is positive semi-definite and f belongs to class C^1. Then, network (9.1) achieves synchronization under the distributed adaptive law (9.19).*

Proof. Proceeding as in Corollary 9.7, one has equation (9.21). Then, with the same arguments as in Theorem 9.4, the conclusion follows. □

9.1.4 Simulation Examples

In this subsection, some simulation examples are provided to validate the theoretical analysis.

Example 1. A network of Chua's circuits Let the nonlinear dynamical system $\dot{s}(t) = f(s(t), t)$ be Chua's circuit

$$\begin{cases} \dot{s}_1 = \eta(-s_1 + s_2 - l(s_1)), \\ \dot{s}_2 = s_1 - s_2 + s_3, \\ \dot{s}_3 = -\beta s_2, \end{cases} \tag{9.22}$$

where $l(x_1) = bx_1 + 0.5(a - b)(|x_1 + 1| - |x_1 - 1|)$. System (9.22) is chaotic when $\eta = 10$, $\beta = 18$, $a = -4/3$, and $b = -3/4$. It is easy to verify that Assumption 9.1 is satisfied.

Consider an ER random network [35] of $N = 100$ nodes, generated with probability $p = 0.15$, which contains about $pN(N - 1)/2 \approx 724$ edges. If there is a connection between nodes i and j, then $G_{ij}(0) = G_{ji}(0) = 1$ $(i \neq j)$, $i, j = 1, 2, \ldots, 100$, and the coupling strength $c = 0.2$. A simulation-based analysis on the random network is performed by using a random-edge pinning scheme, where one randomly selects $m = \lfloor \delta pN(N - 1)/2 \rfloor$ edges and only pins a small fraction with $\delta = 0.35$, i.e., 35% of them ($m = 259$).

By Corollary 9.7, since the subgraph $\widetilde{\mathcal{G}}$ contains a spanning tree of $\mathcal{G}$, the network is synchronized under the designed distributed adaptive law. The orbits of the states x_{ij} are shown in Fig. 9.1, for $i = 1, 2, \ldots, 100$, and $j = 1, 2, 3$, and the updated weights G_{ij} are illustrated in Fig. 9.2.

From Fig. 9.2, it is easy to see that the coupling weights are varying slightly above 1, which is very hard to check by using other criteria given in the literature for a very small coupling strength c. Here, by using the proposed distributed adaptive pinning edges control scheme, synchronization can be reached in the network.

It is worth remarking that, as proved in [29], Chua's circuit is QUAD with $\Gamma \geq 0$, and all the assumptions of Corollary 9.7 are satisfied. The numerical simulations therefore agree with the theoretical analysis of stability.

Example 2. A network of Lorenz oscillators As another example, consider a network of 100 Lorenz oscillators. In this case, the state has three components $x = (p, q, r)^T$ and $f(x) = (10(q - p), -pr + 28z - q, pq - \frac{8}{3}r)^T$. In what follows, consider the network

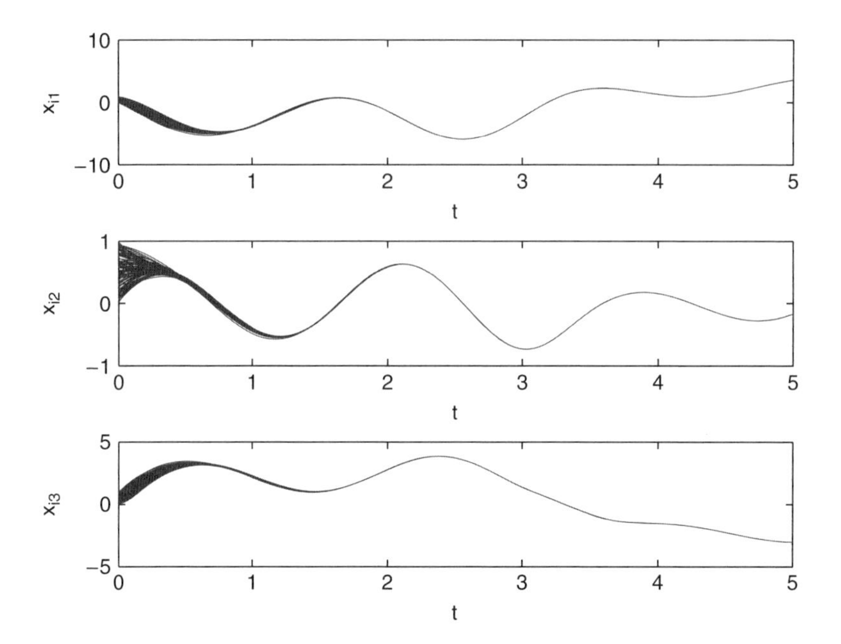

Figure 9.1 Orbits of the states x_{ij}, $i = 1, 2, \ldots, 100$; $j = 1, 2, 3$

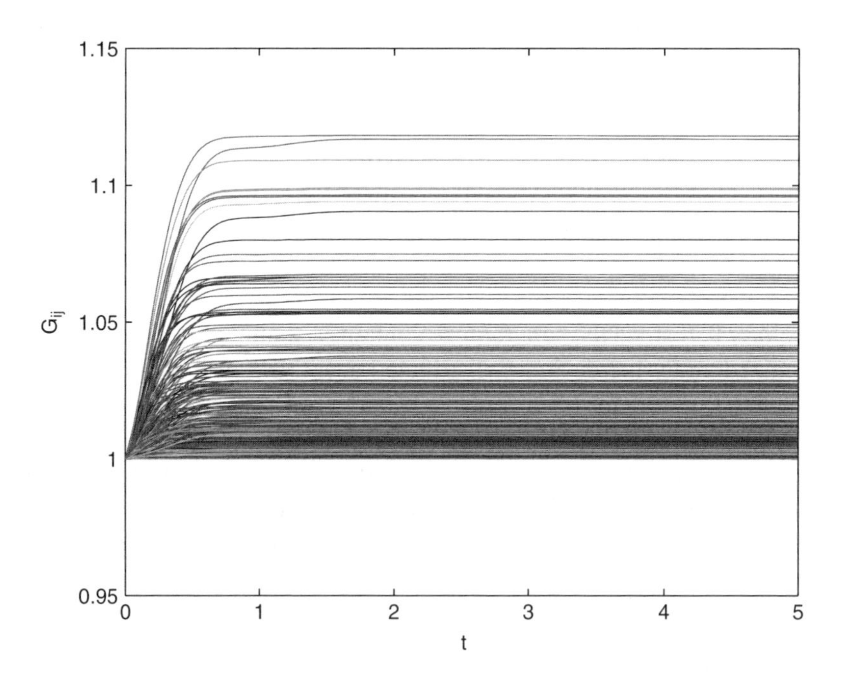

Figure 9.2 Orbits of the weights G_{ij}, $i, j = 1, 2, \ldots, 100$

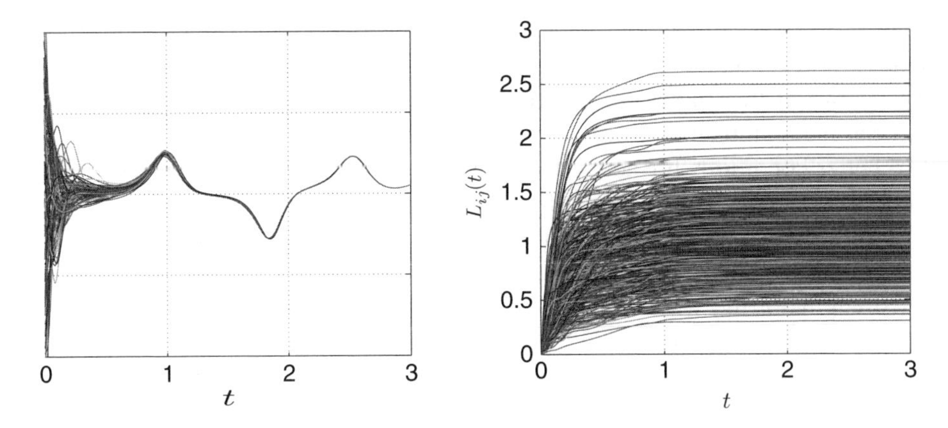

Figure 9.3 Network of 100 Lorenz oscillators. Evolution of the state variables p_i (left) and the adaptive gains L_{ij}

nodes coupled through the decentralized adaptive mechanism (9.6) only on the first two state variables, that is,

$$\Gamma = \begin{bmatrix} 1 & 0 & 0 \\ 0 & 1 & 0 \\ 0 & 0 & 0 \end{bmatrix}.$$

The simulation starts from null initial coupling gains and initial states of a normal distribution with standard deviation 40. Set $\alpha_{ij} = \alpha_{ji} = 0.1, \forall (i,j) \in \mathcal{E}$. As depicted in Fig. 9.3, synchronization is asymptotically achieved and the adaptive gains asymptotically converge to constant values.

In fact, the Lorenz system is QUAD [67] and clearly continuously differentiable. Moreover, $\Gamma \geq 0$. So all the assumptions of Corollary 9.9 are satisfied.

The stability of a network of the Lorenz system in the presence of adaptive couplings has never been proved analytically before. In fact, in [29], additional assumptions were required, which were not satisfied by any continuously differentiable chaotic system.

9.2 Distributed Control Gains Design for Second-Order Consensus in Nonlinear Multi-agent Systems

As discussed in [50, 95, 146, 149], in order to reach second-order consensus, some additional global conditions in terms of the spectrum of the Laplacian matrix must be satisfied even if the multi-agent system is linear. Note that a typical consensus protocol in multi-agent systems with second-order dynamics is distributed, and yet some centralized information depending on the spectrum of the Laplacian matrix is required a priori for designing the control gains, which actually did not take full advantage of the powerful distributed protocol. This is a serious drawback for the so-called local distributed protocols developed earlier in the literature for multi-agent systems. To overcome this problem and design a fully distributed protocol, adaptive control represents a good approach.

By contrast, in the studies of synchronization in complex networks [4], great efforts have been made in using adaptive strategies to adjust network parameters aiming to derive better conditions for network synchronization. For example, in [70, 164], an algorithm was proposed for updating the coupling strengths for network synchronization. In particular, synchronization may be achieved by using high-gain output feedback. For instance, in [23, 151], adaptive schemes were designed for the coupling strengths of the whole network by using the states information of all agents, which are therefore centralized algorithms by their nature. Recently, a general distributed adaptive strategy on the coupling weights was proposed and some theoretical conditions were derived for reaching network synchronization in [29, 156].

It has been observed that many derived conditions for reaching second-order consensus in multi-agent systems were only sufficient and somewhat conservative [50, 95, 149]. Very recently, some research focus has been directed to the adaptive control for reaching cooperative behaviors in coupled Euler-Lagrange systems and multi-agent systems with second-order dynamics [20, 27], originated from synchronization in complex networks. Some distributed adaptive laws were designed on the weights of the network and uniform ultimate boundedness was achieved for multi-agent systems with second-order dynamics [20, 27]. However, it is still a challenging problem to design a distributed local protocol without utilizing the global information for reaching second-order consensus in multi-agent systems, especially those with nonlinear dynamics. This motivates the study in the present section, which designs a fully distributed consensus protocol without using global spectrum information for multi-agent systems with second-order nonlinear dynamics to reach global consensus.

9.2.1 Preliminaries

The second-order consensus protocol in multi-agent dynamical systems is described by [95, 146]

$$\begin{aligned}
\dot{x}_i(t) &= v_i(t), \\
\dot{v}_i(t) &= u_i(t), \\
u_i(t) &= \widetilde{\alpha} \sum_{j=1,\, j\neq i}^{N} G_{ij}(x_j(t) - x_i(t)) + \widetilde{\beta} \sum_{j=1,\, j\neq i}^{N} G_{ij}(v_j(t) - v_i(t)), \\
& i = 1, 2, \cdots, N,
\end{aligned} \tag{9.23}$$

where $x_i \in R^n$ and $v_i \in R^n$ (R^n is the n-dimensional real space) are the position and velocity states of the ith agent (node), respectively, u_i is the control input, $\widetilde{\alpha} > 0$ and $\widetilde{\beta} > 0$ are the coupling strengths (gains), and $G = (G_{ij})_{N\times N}$ is the weighted adjacency matrix of the network.

The second-order consensus protocol in multi-agent system (9.23) is distributed since each agent only uses local information of neighboring agents in the control input. In most cases, second-order consensus can be reached if the coupling gains and the spectrum of the Laplacian matrix satisfy some additional conditions [50, 95, 146, 149, 161]; for example, second-order consensus in multi-agent system (9.23) can be reached if and only if $\frac{\widetilde{\beta}^2}{\widetilde{\alpha}} > \max_{2\leq i\leq N} \frac{\mathcal{I}^2(\lambda_i)}{\mathcal{R}(\lambda_i)[\mathcal{R}^2(\lambda_i)+\mathcal{I}^2(\lambda_i)]}$ [146]. To satisfy this kind of condition, a centralized information scheme depending on the spectrum of the Laplacian matrix must be known a priori for the control gains design. One might question why not apply a sufficiently large value $\widetilde{\beta}^2/\widetilde{\alpha}$ without using the global spectral information for solving this problem? It is noticed that a sufficiently large coupling gain $\widetilde{\beta}$ can induce a high cost for the controller design while a small $\widetilde{\alpha}$ can result in a low convergent speed. Thus, it is not practical to choose a very large value $\widetilde{\beta}^2/\widetilde{\alpha}$ without using prior knowledge of the global spectral information.

In order to design a fully distributed consensus protocol, the following consensus protocol with updated coupling control gains is considered, even for a network with nonlinear dynamics:

$$\begin{aligned}
\dot{x}_i(t) &= v_i(t), \\
\dot{v}_i(t) &= f(x_i(t), v_i(t), t) + \alpha c_i(t) \sum_{j=1,\, j\neq i}^{N} G_{ij}(x_j(t) - x_i(t)) \\
&\quad + \beta c_i(t) \sum_{j=1,\, j\neq i}^{N} G_{ij}(v_j(t) - v_i(t)), i = 1, 2, \cdots, N,
\end{aligned} \tag{9.24}$$

where $f : R^n \times R^n \times R^+ \to R^n$ is a continuously differentiable vector-valued nonlinear function, $\alpha > 0$ and $\beta > 0$ are coupling strengths, and $c_i(t)$ is the time-varying distributed control gain of the ith agent. The objective of introducing the time-varying control gains here is to find some adaptive distributed laws acting on the control gains such that second-order consensus can be reached in the multi-agent system (9.24) without requiring the knowledge of the spectrum of the Laplacian matrix of the graph.

Clearly, since $\sum_{j=1}^{N} L_{ij} = 0$, if consensus can be achieved, any solution $(x_0^T(t), v_0^T(t))^T \in R^{2n}$ of system (9.24) is a trajectory of an isolated node satisfying

$$\begin{aligned}
\dot{x}_0(t) &= v_0(t), \\
\dot{v}_0(t) &= f(x_0(t), v_0(t), t).
\end{aligned} \tag{9.25}$$

Here, $(x_0^T, v_0^T)^T \in R^{2n}$ is used to describe a group leader, followed by agents satisfying (9.24).

By simple calculation, system (9.24) can be equivalently rewritten as

$$\dot{x}_i(t) = v_i(t),$$

$$\dot{v}_i(t) = f(x_i(t), v_i(t), t) - \alpha c_i(t) \sum_{j=1}^{N} L_{ij} x_j(t) - \beta c_i(t) \sum_{j=1}^{N} L_{ij} v_j(t),$$
$$i = 1, 2, \cdots, N. \tag{9.26}$$

In order to derive the main results of this chapter, the following lemmas are needed.

Lemma 9.11

(i) Let L be the Laplacian matrix of the undirected graph $\mathcal{G}$. Then, the matrix L^2 is semi-positive definite. It has a simple eigenvalue 0, and all the other eigenvalues satisfy $0 < \lambda_2^2 \leq \cdots \leq \lambda_N^2$ if and only if the graph $\mathcal{G}$ is connected.
(ii) The second smallest eigenvalue $\lambda_2(L^2)$ of the matrix L^2 satisfies $\lambda_2(L^2) = \lambda_2^2(L) = \min_{x^T 1_N = 0, x \neq 0} \frac{x^T L^2 x}{x^T x}$.

Proof. Since $\lambda_i(L^2) = \lambda_i^2(L)$ for $i = 1, 2, \ldots, N$, the proof can be completed by using Lemma 2.6 and matrix calculations. □

Lemma 9.12 *[23] If L is irreducible, $L_{ij} = L_{ji} \leq 0$ for $i \neq j$, and $\sum_{j=1}^{N} L_{ij} = 0$, $i = 1, 2, \ldots, N$, then, for any constant $\epsilon > 0$, all eigenvalues of the matrix $L + \Xi$ are positive, where $\Xi = \text{diag}(\epsilon, 0, \ldots, 0)$.*

Lemma 9.13 *[107] (Barbalat's lemma) If $V(x, t)$ satisfies the following conditions:*

(i) $V(x, t)$ is lower bounded;
(ii) $\dot{V}(x, t)$ is negative semi-definite;
(iii) $\dot{V}(x, t)$ is uniformly continuous in time or $\dot{V}(x, t)$ is bounded,
then $\ddot{V}(x, t) \to 0$ as $t \to \infty$.

Next, an assumption on the nonlinear function of multi-agent system (9.24) is made.

Assumption 9.14 *There exist nonnegative constants ρ_1 and ρ_2 such that*

$$\| f(x, v, t) - f(y, z, t) \| \leq \rho_1 \| x - y \| + \rho_2 \| v - z \|, \forall x, y, v, z \in R^n.$$

Assumption 9.14 is the so-called QUAD condition on vector fields [23]. Note that this assumption is very mild. For example, all the linear and piecewise-linear continuous functions satisfy this condition. In addition, the condition is satisfied if $\partial f_i / \partial x_j (i, j = 1, 2, \ldots, n)$ are uniformly bounded, which includes many well-known systems.

9.2.2 Distributed Control Gains Design: Leaderless Case

In this subsection, distributed control gains design for system (9.26) is considered.

Let $\overline{x}(t) = \frac{1}{N}\sum_{j=1}^{N} x_j(t)$ and $\overline{v}(t) = \frac{1}{N}\sum_{j=1}^{N} v_j(t)$ be the average position and velocity states of all agents, and $\widehat{x}_i(t) = x_i(t) - \overline{x}(t)$ and $\widehat{v}_i(t) = v_i(t) - \overline{v}(t)$ represent the position and velocity vectors relative to the average position and velocity of agents in system (9.26), respectively. Then, the following error dynamical system can be obtained by a simple calculation:

$$
\begin{aligned}
\dot{\widehat{x}}_i(t) =& \widehat{v}_i(t),\\
\dot{\widehat{v}}_i(t) =& f(x_i(t), v_i(t), t) - \frac{1}{N}\sum_{k=1}^{N} f(x_k(t), v_k(t), t) - \alpha c_i(t)\sum_{j=1}^{N} L_{ij}x_j(t)\\
&-\beta c_i(t)\sum_{j=1}^{N} L_{ij}v_j(t) + \alpha\frac{1}{N}\sum_{k=1}^{N} c_k \sum_{j=1}^{N} L_{kj}x_j(t)\\
&+\beta\frac{1}{N}\sum_{k=1}^{N} c_k \sum_{j=1}^{N} L_{kj}v_j(t), i = 1, 2, \cdots, N.
\end{aligned}
\tag{9.27}
$$

Let $f(x, v, t) = (f^T(x_1, v_1, t), \ldots, f^T(x_N, v_N, t))^T$, $\widehat{x} = (\widehat{x}_1^T, \widehat{x}_2^T, \ldots, \widehat{x}_N^T)^T$, $\widehat{v} = (\widehat{v}_1^T, \widehat{v}_2^T, \ldots, \widehat{v}_N^T)^T$, and $\widehat{y} = (\widehat{x}^T, \widehat{v}^T)^T$. Then, since $\sum_{j=1}^{N} L_{ij} = 0$, system (9.27) can be recast into a compact matrix form as follows:

$$
\dot{\widehat{y}}(t) = F(x, v, t) + (\widehat{L} \otimes I_n)\widehat{y}(t) + G(x, v), \tag{9.28}
$$

where $F(x, v, t) = \begin{pmatrix} 0_{Nn} \\ ((I_N - 1_N 1_N^T/N) \otimes I_n)f(x, v, t) \end{pmatrix}$, $\widehat{L} = \begin{pmatrix} O_N & I_N \\ -\alpha C(t)L & -\beta C(t)L \end{pmatrix}$, $G(x, v) = \begin{pmatrix} 0_{Nn} \\ 1_N \otimes g(x, v) \end{pmatrix}$, $C(t) = \mathrm{diag}(c_1(t), c_2(t), \ldots, c_N(t))$, and $g(x, v) = \alpha\frac{1}{N}\sum_{k=1}^{N} c_k(t)\sum_{j=1}^{N} L_{kj}x_j(t) + \beta\frac{1}{N}\sum_{k=1}^{N} c_k(t)\sum_{j=1}^{N} L_{kj}v_j(t)$.

Theorem 9.15 *Suppose that the graph $\mathcal{G}$ is connected and Assumption 9.14 holds. Then, second-order consensus in system (9.24) can be reached under the following distributed adaptive laws:*

$$
\begin{aligned}
\dot{c}_i(t) = \xi_i \Bigg[& \alpha\left(\sum_{j=1}^{N} L_{ij}x_j\right)^T\left(\sum_{j=1}^{N} L_{ij}x_j\right) + \beta\gamma\left(\sum_{j=1}^{N} L_{ij}v_j\right)^T\left(\sum_{j=1}^{N} L_{ij}v_j\right)\\
&+(\beta + \alpha\gamma)\left(\sum_{j=1}^{N} L_{ij}x_j\right)^T\left(\sum_{j=1}^{N} L_{ij}v_j\right)\Bigg], i = 1, 2, \ldots, N,
\end{aligned}
\tag{9.29}
$$

where $\xi_i > 0$ and $\gamma > 0$ are constants.

Proof. Consider the following Lyapunov function candidate:

$$V(\hat{y}, c, t) = \frac{1}{2}\hat{y}^T(t)(\Omega \otimes I_n)\hat{y}(t) + \sum_{i=1}^{N} \frac{\epsilon}{2\xi_i}(c_i(t) - \hat{c}_i)^2, \tag{9.30}$$

where $\Omega = \begin{pmatrix} \mu LL & \epsilon L \\ \epsilon L & \eta L \end{pmatrix}$, $c(t) = (c_1(t), \dots, c_N(t))^T$, $\hat{c}_i > 0$ are constants to be determined, $\mu \gg \epsilon > 0$, and $\eta > 0$. It will be shown that $V(\hat{y}, c, t) \geq 0$ and $V(\hat{y}, c, t) = 0$ if and only if $\hat{y}(t) = 0$ and all $c_i(t) = \hat{c}_i$. From Lemmas 2.6 and 9.11,

$$V(\hat{y}, c, t) \geq \frac{1}{2}\hat{y}^T(t)(\hat{\Omega} \otimes I_n)\hat{y}(t) + \sum_{i=1}^{N} \frac{\epsilon}{2\xi_i}(c_i(t) - \hat{c}_i)^2, \tag{9.31}$$

where $\hat{\Omega} = \begin{pmatrix} \mu\lambda_2^2(L)I_N & \epsilon L \\ \epsilon L & \eta\lambda_2(L)I_N \end{pmatrix}$. By Lemma 3.17, $\hat{\Omega} > 0$ is equivalent to that $\eta\lambda_2(L) > 0$ and $\mu\lambda_2^2(L)I_N - \frac{\epsilon^2}{\eta\lambda_2(L)}L^2 > 0$, which are satisfied since $\mu \gg \epsilon > 0$. Therefore, $V(\hat{y}, c, t) \geq 0$ and $V(\hat{y}, c, t) = 0$ if and only if $\hat{y}(t) = 0$ and $c_i(t) = \hat{c}_i$, $i = 1, 2, \dots, N$.

By Lemma 2.8, taking the derivative of $V(t)$ along the trajectories of (9.28) yields

$$\begin{aligned}
&\dot{V}(\hat{y}, c, t) \\
&= \hat{y}^T(t)(\Omega \otimes I_n)[F(x, v, t) + (\hat{L} \otimes I_n)\hat{y}(t) + G(x, v)] \\
&\quad + \sum_{i=1}^{N} \frac{\epsilon}{\xi_i}(c_i(t) - \hat{c}_i)\dot{c}_i(t) \\
&= (\epsilon\hat{x}^T(t) + \eta\hat{v}^T(t))[(L(I_N - 1_N 1_N^T/N)) \otimes I_n]f(x, v, t) \\
&\quad + \hat{y}^T(t)[(\Omega\hat{L}) \otimes I_n]\hat{y} + \sum_{i=1}^{N} \frac{\epsilon}{\xi_i}(c_i(t) - \hat{c}_i)\dot{c}_i(t) \\
&\quad + (\epsilon\hat{x}^T(t) + \eta\hat{v}^T(t))[(L1_N) \otimes g(x, v)] \\
&= [\epsilon\hat{x}^T(t) + \eta\hat{v}^T(t)](L \otimes I_n)[f(x, v, t) - 1_N \otimes f(\bar{x}, \bar{v}, t)] \\
&\quad + \frac{1}{2}\hat{y}^T(t)[(\Omega\hat{L} + \hat{L}^T\Omega) \otimes I_n]\hat{y}(t) + \sum_{i=1}^{N} \frac{\epsilon}{\xi_i}(c_i(t) - \hat{c}_i)\dot{c}_i(t),
\end{aligned} \tag{9.32}$$

where $[\epsilon\hat{x}^T(t) + \eta\hat{v}^T(t)](L \otimes I_n)[1_N \otimes f(\bar{x}, \bar{v}, t)] = [\epsilon\hat{x}^T(t) + \eta\hat{v}^T(t)](L1_N) \otimes [I_n f(\bar{x}, \bar{v}, t)] = 0$ and $(L1_N) \otimes g(x, v) = 0$ from Lemma 2.8 since $L1_N = 0$.

By simple calculation,

$$\begin{aligned}
&(\Omega\hat{L} + \hat{L}^T\Omega) \\
&= \begin{pmatrix} -2\alpha\epsilon LC(t)L & \mu L^2 - (\beta\epsilon + \alpha\eta)LC(t)L \\ \mu L^2 - (\beta\epsilon + \alpha\eta)LC(t)L & 2\epsilon L - 2\beta\eta LC(t)L \end{pmatrix}.
\end{aligned} \tag{9.33}$$

From Assumption 9.14,

$$
\begin{aligned}
&\hat{x}^T(t)(L \otimes I_n)[f(x, v, t) - 1_N \otimes f(\bar{x}, \bar{v}, t)] \\
&= \sum_{i=1}^{N} \sum_{j=1}^{N} L_{ij}(x_i - \bar{x})^T [f(x_j, v_j, t) - f(\bar{x}, \bar{v}, t)] \\
&\leq \sum_{i=1}^{N} \sum_{j=1}^{N} |L_{ij}| \parallel \hat{x}_i \parallel (\rho_1 \parallel \hat{x}_j \parallel + \rho_2 \parallel \hat{v}_j \parallel) \\
&\leq \rho_1 L_{\max} N \sum_{i=1}^{N} \parallel \hat{x}_i \parallel^2 + \frac{\rho_2 L_{\max} N}{2} \sum_{i=1}^{N} (\parallel \hat{x}_i \parallel^2 + \parallel \hat{v}_i \parallel^2),
\end{aligned}
\tag{9.34}
$$

and

$$
\begin{aligned}
&\hat{v}^T(t)(L \otimes I_n)[f(x, v, t) - 1_N \otimes f(\bar{x}, \bar{v}, t)] \\
&\leq \rho_2 L_{\max} N \sum_{i=1}^{N} \parallel \hat{v}_i \parallel^2 + \frac{\rho_1 L_{\max} N}{2} \sum_{i=1}^{N} (\parallel \hat{x}_i \parallel^2 + \parallel \hat{v}_i \parallel^2),
\end{aligned}
\tag{9.35}
$$

where $L_{\max} = \max_{i,j=1,\ldots,N} |L_{ij}|$.

Since $\sum_{j=1}^{N} L_{ij} = 0$, one obtains $\sum_{i=1}^{N} L_{ij} \hat{x}_j = \sum_{i=1}^{N} L_{ij} x_j$ and $\sum_{i=1}^{N} L_{ij} \hat{v}_j = \sum_{i=1}^{N} L_{ij} v_j$. Combining (9.32)–(9.35) gives

$$
\begin{aligned}
\dot{V} \leq\; & (\epsilon \rho_1 + (\eta \rho_1 + \epsilon \rho_2)/2) L_{\max} N \sum_{i=1}^{N} \parallel \hat{x}_i \parallel^2 \\
& + (\eta \rho_2 + (\eta \rho_1 + \epsilon \rho_2)/2) L_{\max} N \sum_{i=1}^{N} \parallel \hat{v}_i \parallel^2 - \alpha \epsilon \hat{x}^T ((L C(t) L) \otimes I_n) \hat{x} \\
& + \hat{x}^T ((\mu L^2 - (\beta \epsilon + \alpha \eta) L C(t) L) \otimes I_n) \hat{v} \\
& + \hat{v}^T ((\epsilon L - \beta \eta L C(t) L) \otimes I_n) \hat{v} + \sum_{i=1}^{N} \frac{\epsilon}{\xi_i} (c_i(t) - \hat{c}_i) \dot{c}_i(t).
\end{aligned}
$$

From the adaptive laws in (9.29), letting $\gamma = \eta / \epsilon$ gives

$$
\begin{aligned}
\dot{V} \leq\; & (\epsilon \rho_1 + (\eta \rho_1 + \epsilon \rho_2)/2) L_{\max} N \sum_{i=1}^{N} \parallel \hat{x}_i \parallel^2 \\
& + (\eta \rho_2 + (\eta \rho_1 + \epsilon \rho_2)/2) L_{\max} N \sum_{i=1}^{N} \parallel \hat{v}_i \parallel^2 \\
& - \alpha \epsilon \hat{x}^T ((L \hat{C} L) \otimes I_n) \hat{x} + \hat{v}^T ((\epsilon L - \beta \eta L \hat{C} L) \otimes I_n) \hat{v} \\
& + \hat{x}^T ((\mu L^2 - (\beta \epsilon + \alpha \eta) L \hat{C} L) \otimes I_n) \hat{v},
\end{aligned}
\tag{9.36}
$$

where $\widehat{C} = \text{diag}(\widehat{c}_1, \widehat{c}_2, \dots, \widehat{c}_N)$. Let $\widehat{c}_i = \overline{c}$ and $\mu = (\beta\epsilon + \alpha\eta)\overline{c}$. By Lemma 9.11,

$$\begin{aligned}
&\dot{V}(\widehat{y}, c, t) \\
&\leq -(\alpha\epsilon\overline{c}\lambda_2^2(L) - \epsilon\rho_1 L_{\max}N - (\eta\rho_1 + \epsilon\rho_2)L_{\max}N/2)\widehat{x}^T\widehat{x} \\
&\quad -(\beta\eta\overline{c}\lambda_2^2(L) - \eta\rho_2 L_{\max}N - \epsilon\lambda_N - (\eta\rho_1 + \epsilon\rho_2)L_{\max}N/2)\widehat{v}^T\widehat{v}. \qquad (9.37)
\end{aligned}$$

Choose $\overline{c}$ sufficiently large such that $\alpha\epsilon\overline{c}\lambda_2^2(L) > \epsilon\rho_1 L_{\max}N + (\eta\rho_1 + \epsilon\rho_2)L_{\max}N/2 + 1$ and $\beta\eta\overline{c}\lambda_2^2(L) > \eta\rho_2 L_{\max}N + \epsilon\lambda_N + (\eta\rho_1 + \epsilon\rho_2)L_{\max}N/2 + 1$. Then,

$$\dot{V}(\widehat{y}, c, t) \leq -\widehat{x}^T\widehat{x} - \widehat{v}^T\widehat{v}. \qquad (9.38)$$

It is easy to see that $V(\widehat{y}, c, t)$ satisfies conditions (i) and (ii) in Lemma 9.13. From (9.38), one obtains that $\widehat{y}$ and c are bounded, and thus from (9.28), (9.29), and Assumption 9.14, $\dot{\widehat{y}}$ and $\dot{c}$ are bounded. From $\dot{V}(\widehat{y}, c, t)$ in (9.32), one finally gets that $\ddot{V}(\widehat{y}, c, t)$ is bounded. Thus, the proof can be completed by using Lemma 9.13. □

Remark 9.16 Second-order consensus in system (9.24) can be reached under the distributed adaptive laws given in (9.29) without requiring any additional centralized conditions, unlike most results in the literature [50, 95, 146, 149, 161]. In the design of distributed adaptive laws for the control gains here, each agent only uses local information of neighboring agents.

9.2.3 Distributed Control Gains Design: Leader-Follower Case

In this section, leader-follower consensus in second-order multi-agent system with distributed control gains and nonlinear dynamics is investigated, where the leader evolves according to system (9.25) and all followers are governed by the following multi-agent system:

$$\begin{aligned}
\dot{x}_i(t) &= v_i(t), \\
\dot{v}_i(t) &= f(x_i(t), v_i(t), t) + \alpha c_i(t)\left(-\sum_{j=1}^{N} L_{ij}x_j(t) - k_i(x_i(t) - x_0(t))\right) \\
&\quad +\beta c_i(t)\left(-\sum_{j=1}^{N} L_{ij}v_j(t) - k_i(v_i(t) - v_0(t))\right), \qquad (9.39)
\end{aligned}$$

in which $k_i \geq 0$ are the coupling strengths from the leader to agent i, $i = 1, 2, \dots, N$. If the follower i can measure leader's information, then $k_i > 0$; otherwise $k_i = 0$. Namely, only a fraction of agents, called informed agents, can measure leader's information.

Then, system (9.39) can be written as

$$\begin{aligned}\dot{x}_i(t) &= v_i(t),\\ \dot{v}_i(t) &= f(x_i(t), v_i(t), t) - \alpha c_i(t)\sum_{j=1}^{N} H_{ij}(x_j(t) - x_0(t))\\ &\quad - \beta c_i(t)\sum_{j=1}^{N} H_{ij}(v_j(t) - v_0(t)),\end{aligned} \tag{9.40}$$

where $H = (H_{ij})_{N\times N} = L + D$ and $D = \mathrm{diag}(k_1, k_2, \dots, k_N)$. From Lemma 9.12, H is positive definite if there is at least one informed agent, which is assumed throughout this section; otherwise, it is generally impossible to expect that the agents of the whole group can follow the leader.

Let $\widetilde{x}_i(t) = x_i(t) - x_0(t)$ and $\widetilde{v}_i(t) = v_i(t) - v_0(t)$. Then, one obtains the following error dynamical system:

$$\begin{aligned}\dot{\widetilde{x}}_i(t) &= \widetilde{v}_i(t),\\ \dot{\widetilde{v}}_i(t) &= f(x_i(t), v_i(t), t) - f(x_0(t), v_0(t)) - \alpha c_i(t)\sum_{j=1}^{N} H_{ij}\widetilde{x}_j(t)\\ &\quad - \beta c_i(t)\sum_{j=1}^{N} H_{ij}\widetilde{v}_j(t).\end{aligned} \tag{9.41}$$

Theorem 9.17 *Suppose that the graph $\mathcal{G}$ is connected and Assumption 9.14 holds. Then, all agents in system (9.39) can follow the leader described by system (9.25) under the following distributed adaptive laws:*

$$\begin{aligned}\dot{c}_i(t) = \xi_i \Bigg[& \alpha\left(\sum_{j=1}^{N} H_{ij}\widetilde{x}_j\right)^T\left(\sum_{j=1}^{N} H_{ij}\widetilde{x}_j\right) + \beta\gamma\left(\sum_{j=1}^{N} H_{ij}\widetilde{v}_j\right)^T\left(\sum_{j=1}^{N} H_{ij}\widetilde{v}_j\right)\\ & + (\beta + \alpha\gamma)\left(\sum_{j=1}^{N} H_{ij}\widetilde{x}_j\right)^T\left(\sum_{j=1}^{N} H_{ij}\widetilde{v}_j\right)\Bigg],\end{aligned} \tag{9.42}$$

where $\xi_i > 0$ and $\gamma > 0$ are constants, $i = 1, 2, \dots, N$.

Proof. Consider the following Lyapunov function candidate:

$$\widetilde{V}(t) = \frac{1}{2}\widetilde{y}^T(t)(\widetilde{\Omega} \otimes I_n)\widetilde{y}(t) + \sum_{i=1}^{N} \frac{\epsilon}{2\xi_i}(c_i(t) - \widehat{c}_i)^2, \tag{9.43}$$

where $\widetilde{y} = (\widetilde{x}_1^T(t), \ldots, \widetilde{x}_N^T(t), \widetilde{v}_1^T(t), \ldots, \widetilde{v}_N^T(t))^T$, $\widetilde{\Omega} = \begin{pmatrix} \mu HH & \epsilon H \\ \epsilon H & \eta H \end{pmatrix}$, $\widehat{c}_i > 0$ are constants to be determined, $\mu \gg \epsilon > 0$, and $\eta > 0$. By following a similar analysis as in Theorem 9.15, the proof can be completed. □

Remark 9.18 Since $\sum_{j=1}^{N} L_{ij} = 0$, it is easy to verify that $\sum_{j=1}^{N} H_{ij}\widetilde{x}_j = \sum_{j=1}^{N} L_{ij}(x_j - x_0) + k_i(x_i - x_0) = -\sum_{j=1}^{N} G_{ij}(x_j - x_i) + k_i(x_i - x_0)$ and $\sum_{j=1}^{N} H_{ij}\widetilde{v}_j = -\sum_{j=1}^{N} G_{ij}(x_j - x_i) + k_i(x_i - x_0)$. Therefore, the distributed adaptive laws (9.42), acting on the control gains, only use local information of neighboring agents. Specifically, the uninformed agents use the information of neighboring agents and the informed agents use the information of both neighboring agents and the leader.

Next, the following assumption on $f(x_0, v_0)$ is made.

Assumption 9.19 *There exists a nonnegative constant ρ_3 such that*

$$\| f(x_0(t), v_0(t), t)\|_\infty \leq \rho_3, \forall t > 0.$$

The above condition is very general and can be satisfied in many systems, for example, the multi-robot systems with uncertain dynamics [27] and stable or periodic linear systems [104].

Then, the followers satisfy

$$\begin{aligned} \dot{x}_i(t) &= v_i(t), \\ \dot{v}_i(t) &= -\alpha \sum_{j=1}^{N} H_{ij}(x_j(t) - x_0(t)) - \beta \sum_{j=1}^{N} H_{ij}(v_j(t) - v_0(t)) \\ &\quad - c_i(t)\mathrm{sgn}\left(\sum_{j=1}^{N} H_{ij}((x_j(t) - x_0(t)) + \gamma(v_j(t) - v_0(t))) \right), \\ i &= 1, 2, \cdots, N, \end{aligned} \tag{9.44}$$

where $\gamma > 0$ is a constant. Similarly, letting $\widetilde{x}_i(t) = x_i(t) - x_0(t)$ and $\widetilde{v}_i(t) = v_i(t) - v_0(t)$ yields the following error dynamical system in a compact matrix form:

$$\dot{\widetilde{y}}(t) = \widetilde{F}(x_0, v_0, t) + (\widetilde{H} \otimes I_n)\widetilde{y}(t) + \widetilde{G}(x, v), \tag{9.45}$$

where $\widetilde{x} = (\widetilde{x}_1^T, \widetilde{x}_2^T, \ldots, \widetilde{x}_N^T)^T$, $v = (\widetilde{v}_1^T, \widetilde{v}_2^T, \ldots, \widetilde{v}_N^T)^T$, $\widetilde{y} = (\widetilde{x}^T, \widetilde{v}^T)^T$, $\widetilde{F}(x_0, v_0, t) = \begin{pmatrix} 0_{Nn} \\ -1_N \otimes f(x_0, v_0, t) \end{pmatrix}$, $\widetilde{H} = \begin{pmatrix} O_N & I_N \\ -\alpha H & -\beta H \end{pmatrix}$, $C(t) = \mathrm{diag}(c_1(t), c_2(t), \ldots, c_N(t))$, and $\widetilde{G}(x, v) = \begin{pmatrix} 0_{Nn} \\ -(C(t) \otimes I_n)\mathrm{sgn}((H \otimes I_n)(\widetilde{x} + \gamma\widetilde{v})) \end{pmatrix}$.

Theorem 9.20 *Suppose that the graph $\mathcal{G}$ is connected and Assumption 9.19 holds. Then, all agents in system (9.44) can follow the leader described by system (9.25) under the following distributed adaptive laws:*

$$\dot{c}_i(t) = \xi_i \left\| \sum_{j=1}^{N} H_{ij}((x_j(t) - x_0(t)) + \gamma(v_j(t) - v_0(t))) \right\|_1 ,$$
$$i = 1, 2, \ldots, N, \tag{9.46}$$

where $\xi_i > 0$ and $\gamma > 0$ are constants.

Proof. Consider the same Lyapunov function candidate $\widetilde{V}(t)$ as in (9.43). Taking the derivative of $\widetilde{V}(t)$ along the trajectories of (9.45) yields

$$\begin{aligned}
&\dot{\widetilde{V}}(t) \\
&= (\epsilon\widetilde{x}^T(t) + \eta\widetilde{v}^T(t))(H \otimes I_n)(-1_N \otimes f(x_0, v_0, t)) + \widetilde{y}^T(t)[(\widetilde{\Omega}\widetilde{H}) \otimes I_n]\widetilde{y} \\
&\quad + \sum_{i=1}^{N} \frac{\epsilon}{\xi_i}(c_i(t) - \widehat{c}_i)\dot{c}_i(t) - \epsilon\Big((H \otimes I_n)\Big(\widetilde{x} + \frac{\eta}{\epsilon}\widetilde{v}\Big)\Big)^T (C(t) \otimes I_n) \\
&\quad \times \mathrm{sgn}((H \otimes I_n)(\widetilde{x} + \gamma\widetilde{v})) + \sum_{i=1}^{N} \frac{\epsilon}{\xi_i}(c_i(t) - \widehat{c}_i)\dot{c}_i(t).
\end{aligned} \tag{9.47}$$

Let $\mu = \beta\epsilon + \alpha\eta$, which indicates that $\mu \gg \epsilon > 0$ and $\eta \gg \epsilon > 0$. Then, by simple calculation,

$$\frac{1}{2}(\widetilde{\Omega}\widetilde{H} + \widetilde{H}^T\widetilde{\Omega}) = \begin{pmatrix} -\alpha\epsilon H^2 & O_N \\ O_N & \epsilon H - \beta\eta H^2 \end{pmatrix}, \tag{9.48}$$

which is negative definite since $\eta \gg \epsilon > 0$. Let $\widehat{c}_i = \overline{c}$ and $\eta/\epsilon = \gamma$, $i = 1, 2, \ldots, N$. By using the distributed adaptive laws (9.46) and Assumption 9.19, one obtains

$$\begin{aligned}
\dot{\widetilde{V}}(t) &= \epsilon(\widetilde{x}^T(t) + \gamma\widetilde{v}^T(t))(H \otimes I_n)(-1_N \otimes f(x_0, v_0, t)) \\
&\quad + \widetilde{y}^T(t)[(\widetilde{\Omega}\widetilde{H} + \widetilde{H}^T\widetilde{\Omega}) \otimes I_n]\widetilde{y}(t)/2 - \overline{c}\epsilon \left\| \sum_{j=1}^{N} H_{ij}(\widetilde{x}_j(t) + \gamma\widetilde{v}_j(t)) \right\|_1 \\
&\le \widetilde{y}^T(t)[(\widetilde{\Omega}\widetilde{H} + \widetilde{H}^T\widetilde{\Omega}) \otimes I_n]\widetilde{y}/2,
\end{aligned} \tag{9.49}$$

where $\overline{c}$ is chosen such that $\overline{c} > \rho_3$, and by (9.48), $(\widetilde{\Omega}\widetilde{H} + \widetilde{H}^T\widetilde{\Omega})$ is negative definite. By following a similar analysis as in Theorem 9.15, the proof can be completed. □

Remark 9.21 Since $\sum_{i=1}^{N} L_{ij} = 0$, one has $\sum_{j=1}^{N} H_{ij}(\widetilde{x}_j + \gamma\widetilde{v}_j) = -\sum_{j=1}^{N} G_{ij}((x_j - x_i) + \gamma(v_j - v_i)) + k_i((x_i - x_0) + \gamma(v_i - v_0))$. Therefore, under the distributed adaptive laws (9.46), each agent uses only local information of the network.

Remark 9.22 In Theorem 9.17, the nonlinear function satisfies the Lipschitz condition in Assumption 9.14, and simple linear followers in (9.40) are designed to reach the leader-follower consensus of multi-agent systems. However, in Theorem 9.20, the boundedness of the nonlinear function holds by Assumption 9.19, thus the complicated nonlinear followers are designed in (9.44). For these two cases, different followers are designed under different assumptions.

Remark 9.23 In [50, 95, 146, 149, 161], a centralized condition was given depending on the spectrum of the Laplacian matrix and network parameters for second-order consensus design, which is by nature not a distributed protocol. Under the distributed adaptive laws (9.42) and (9.46) here, all agents can follow the leader asymptotically if the graph $\mathcal{G}$ is connected and Assumption 9.14 or 9.19 is satisfied. Here, though some conditions on the nonlinear function f must be satisfied, the parameters ρ_i $(i = 1, 2, 3)$ can be unknown for the control gains design, as shown in (9.42) and (9.46).

9.2.4 Simulation Examples

Example 9.1 *Distributed control gains design for consensus in general multi-agent systems with second-order Chua's circuits*

Considering the second-order consensus protocol with time-varying velocities in system (9.24), where $\alpha = \beta = 1$, the Laplacian matrix of the graph $\mathcal{G}$ is given by $\begin{pmatrix} 5 & -2 & -3 & 0 \\ -2 & 6 & 0 & -4 \\ -3 & 0 & 3 & 0 \\ 0 & -4 & 0 & 4 \end{pmatrix}$, and the nonlinear function f is described by Chua's circuit [24]

$$f(x_i(t), v_i(t), t) = \begin{pmatrix} \varsigma(-v_{i1} + v_{i2} - l(v_{i1})), \\ v_{i1} - v_{i2} + v_{i3}, \\ -\rho v_{i2}, \end{pmatrix}, \tag{9.50}$$

with $l(v_{i1}) = bv_{i1} + 0.5(a - b)(|v_{i1} + 1| - |v_{i1} - 1|)$. System (9.50) is chaotic when $\varsigma = 10$, $\rho = 18$, $a = -4/3$, and $b = -3/4$, as shown in [142]. It is easy to verify that Assumption 9.14 is satisfied. Choose $\xi = 0.1$ and $\gamma = 1$. The distributed adaptive laws given in Theorem 9.15 are designed. Then, second-order consensus in system (9.24) can be reached under the distributed adaptive laws (9.29). The position and velocity states of all the agents are shown in Fig. 9.4. A distinct feature is that the designed distributed control gains are very small, as shown in Fig. 9.5, contrary to that many conditions in the previous works are only sufficient, which may be somewhat conservative.

Example 9.2 *Distributed control gains design for leader-follower consensus in multi-agent systems with second-order spacecraft formation under Assumption 9.14*

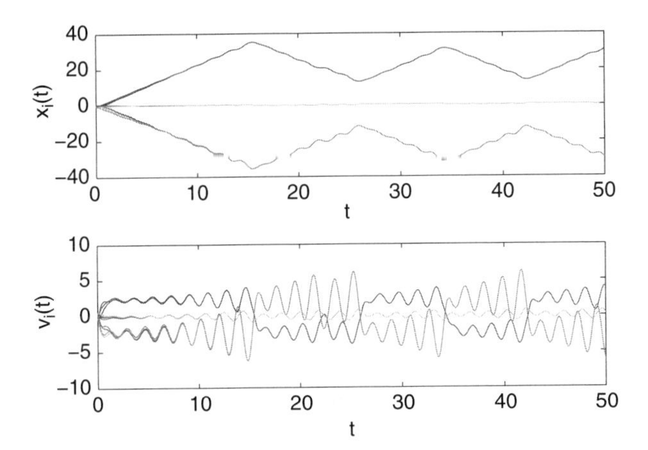

Figure 9.4 Position and velocity states of multiple agents in Example 9.1

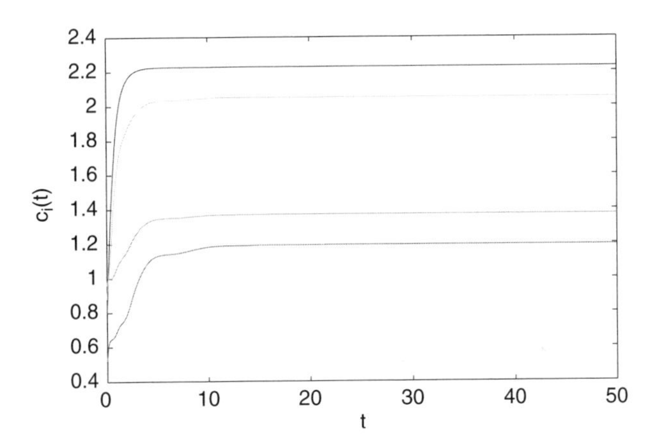

Figure 9.5 States of distributed control gains in Example 9.1

Consider the same model as in Example 9.1, except that the nonlinear system is the linearized equations of the relative dynamics of the satellite given by Hill's equations [62]

$$f(x_i(t), v_i(t), t) = \begin{pmatrix} 0 & 0 & 0 \\ 0 & 3w^2 & 0 \\ 0 & 0 & -w^2 \end{pmatrix} x_i(t) + \begin{pmatrix} 0 & 2w & 0 \\ -2w & 0 & 0 \\ 0 & 0 & 0 \end{pmatrix} v_i(t), \tag{9.51}$$

where $x_i(t)$ and $v_i(t)$ are the position and velocity components of the ith satellite in the rotating coordinate, and $w = 1$ denotes the angular rate of the satellite.

Assume $D = \text{diag}(3, 0, 0, 0)$, meaning that only the first agent can measure leader's information. As guaranteed by Theorem 9.17, all agents in system (9.39) can follow

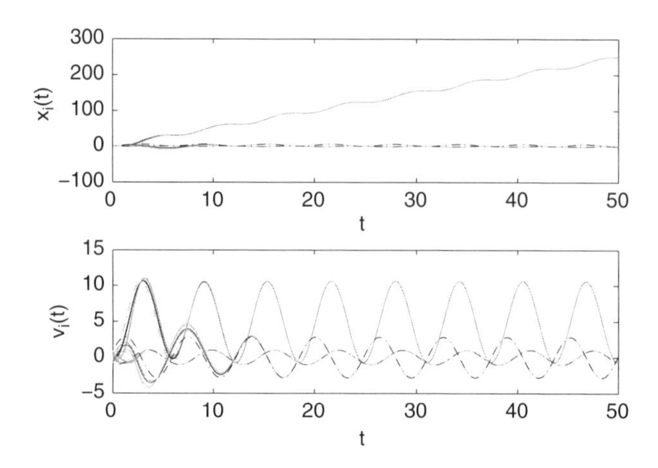

Figure 9.6 Position and velocity states of multiple agents in Example 9.2

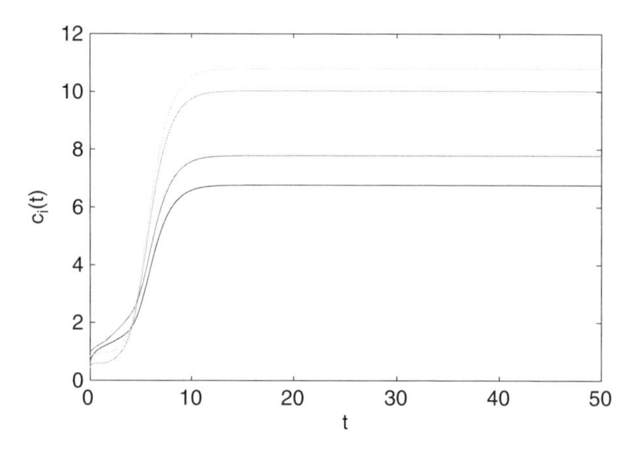

Figure 9.7 States of distributed control gains in Example 9.2

the leader described by (9.25) under the distributed adaptive laws (9.42). The position and velocity states of all the agents are shown in Fig. 9.6, and the designed distributed control gains are illustrated in Fig. 9.7. Here, only consensus is considered but one can also design the desired separation distance in the control input.

9.3 Notes

In this chapter, distributed adaptive control for synchronization in complex networks has been studied. Some distributed adaptive strategies have been designed on all the coupling weights of the network for reaching synchronization, which were then extended to the case where only a small fraction of coupling weights are adjusted. A general criterion has been given, and it has been found that synchronization can be reached if the corresponding edges and nodes with updated coupling weights form a spanning tree. Furthermore, a leader-follower control problem in complex networks has been considered. Some distributed adaptive pinning schemes have been proposed both on the coupling weights between the nodes and from the leader to the nodes. It has been found that synchronization can be reached if the network composed of the corresponding edges and nodes with controlled coupling weights contains a spanning tree, where the directions of the connections from the leader to the nodes were removed.

Then, distributed control gains have been designed for both leaderless consensus and leader-follower consensus in multi-agent systems with second-order nonlinear dynamics. Several distributed adaptive strategies for designing the control gains have been developed based only on local information. It has been found that consensus can be reached under the designed distributed control gains only if the undirected network is connected, which nevertheless is a very simple and natural condition.

Some novel distributed adaptive strategies on the coupling weights of the network have been proposed. However, it is still not understood what kind of pinning schemes may be chosen for a given complex network to realize synchronization if the number of the selected coupling weights is less than $N - 1$. The study of adaptive control on coupling weights of complex network is at its beginning, and many works about the detailed pinning weights strategies should be further investigated in the future. Also, it is still an unsolved and challenging problem to design distributed control gains in multi-agent systems with second-order dynamics and general directed topologies, leaving an important and interesting topic for future research.

10

Distributed Consensus Filtering in Sensor Networks

This chapter discusses distributed filtering in sensor networks by using consensus theory and techniques from multi-agent systems [154].

Sensor networks have attracted increasing attention from the scientific and engineering communities due to their wide applications in robotics, environment surveillance and monitoring, information collection, wireless networks, and so on. A sensor network consists of a large number of sensor nodes distributed over a spatial region. Each sensor has some level of communication, intelligence for signal processing, and data fusion, which together builds up a sensing network. Due to the limited energy, computational ability, and communication capability, a large number of sensor nodes should be used in a wider region so as to achieve higher accuracy in estimating the quantities of interests rather than using a single node. Each sensor node is equipped with a microelectronic device having limited power source, and it might not be possible to transmit messages over a large sensor network. In order to save energy, it is natural to carry out data fusion to reduce the communication overhead. Therefore, distributed estimation and tracking is one of the most important issues in large-scale sensor networks today.

From a network-theoretic point of view, a large-scale sensor network can be viewed as a complex network or a multi-agent system with each node representing a sensor and each edge carrying the information exchange between two sensors. It would be interesting to see if the techniques in synchronization of complex networks [70, 117, 118, 142] and consensus of multi-agent systems [57, 81, 98] can be used in the distributed consensus filtering design. In a complex network, each node communicates with its neighboring nodes to exchange information, so that all the states could reach synchronization or consensus asymptotically. Therefore, it is natural to use the synchronization fundamentals of complex networks and consensus in multi-agent systems as the theoretic basis for the distributed consensus filtering design.

Distributed Cooperative Control of Multi-agent Systems, First Edition.
Wenwu Yu, Guanghui Wen, Guanrong Chen, and Jinde Cao.

In the case where the whole network cannot synchronize by itself, some controllers may be designed and applied to force the network to synchronize or to be stabilized. However, it is practically impossible to add controllers to all nodes. To reduce the number of controlled nodes, some local feedback injections may be applied to a small fraction of the network nodes. Such an idea is known as pinning control [23, 119, 151]. Practically, it is very difficult to observe all the states of the target, so pinning observers may be designed in the case where the informed sensors can only measure partial states of the target. All such technical issues will be addressed in this chapter.

10.1 Preliminaries

Let the target be described by the following model:

$$ds(t) = f(s(t), t)dt + v(t)dv, \tag{10.1}$$

where $s(t) \in R^n$ is the state of the target, $v(t) \in R^n$ is an external noise intensity function, and $v(t)$ is a one-dimensional Brownian motion with expectation $\mathbf{E}\{dv(t)\} = 0$ and variance $\mathbf{D}\{dv(t)\} = 1$. The model is defined on a complete probability space $(\Omega, \mathcal{F}, \mathbb{P})$ with a natural filtration $\{\mathcal{F}_t\}_{t\geq 0}$ generated by $\{v(s) : 0 \leq s \leq t\}$, where Ω is associated with the canonical space generated by $v(t)$, and $\mathcal{F}$ is the associated σ-algebra generated by $\{v(t)\}$ with probability measure $\mathbb{P}$.

Consider a sensor network of size N and assume that if sensor i can measure the signal $s(t)$, then

$$y_i(t) = s(t) + \sigma_i(t)d\omega_i/dt, \tag{10.2}$$

where $y_i(t) \in R^n$ is the measurement of sensor i on target $s(t)$, $\sigma_i(t)$ is an external noise intensity function of agent i, and $\omega_i(t)$ is an independent one-dimensional Brownian motion with expectation $\mathbf{E}\{d\omega_i\} = 0$ and variance $\mathbf{D}\{d\omega_i\} = 1$, $i = 1, 2, \ldots, N$.

Note that there are a large number of sensors which can measure the target $s(t)$ from observations $y_i(t)$, $i = 1, 2, \ldots, N$. However, it is still a challenging problem to carry out the data fusion if there is not a centralized processor capable of collecting all the measurements from the sensors. The objective here, therefore, is to design a distributed filter to track the state $s(t)$ of the target.

Suppose that each sensor can only communicate with the neighboring sensors. Taking the measurement y_i as the input, the following filter is designed:

$$\dot{x}_i(t) = f(x_i(t), t) + c \sum_{j=1, j\neq i}^{N} a_{ij}(x_j(t) - x_i(t)) + u_i(t), i = 1, 2, \ldots, N, \tag{10.3}$$

where $x_i(t)$ is the estimation of the target $s(t)$ in sensor node i, c is the coupling strength, and $u_i(t)$ is the designed controller that is dependent on the measurement y_i. If sensor i is in the sensing range of sensor j, then there is a connection between sensor i and sensor j, i.e., $a_{ij} = a_{ji} > 0$ $(i \neq j)$; otherwise, $a_{ij} = a_{ji} = 0$. Let $a_{ii} = -\sum_{j=1, j\neq i}^{N} a_{ij}$ for $i = 1, 2, \ldots, N$, and $\mathcal{N}(i) = \{j \mid a_{ij} > 0\}$ denote the set of neighbors of sensor i. Then, (10.3) can be written as

$$\dot{x}_i(t) = f(x_i(t), t) + c \sum_{j=1}^{N} a_{ij} x_j(t) + u_i(t), i = 1, 2, \ldots, N. \quad (10.4)$$

From (10.4), it is easy to see that the sensor i can only receive estimated signals from its neighbors in $\mathcal{N}(i)$.

This section only considers the situation where the sensor network coupling matrix $A = (a_{ij})_{N\times N}$ is irreducible.

Assumption 10.1 *For all $x, y \in R^n$, there exists a constant θ such that*

$$(x-y)^T(f(x,t) - f(y,t)) \le \theta(x-y)^T(x-y), \forall t \in R. \quad (10.5)$$

Assumption 10.2 *Both $v(t) \in R^n$ and $\sigma_i(t) \in R^n$ belong to $L_\infty[0,\infty)$, i.e., $v(t)$ and $\sigma_i(t)$ are bounded vector functions satisfying*

$$v^T(t)v(t) \le \alpha, \forall t \in R, \quad (10.6)$$

$$\sigma_i^T(t)\sigma_i(t) \le \beta_i, \forall t \in R, \quad (10.7)$$

where α and β_i are positive constants, $i = 1, 2, \ldots, N$.

Assumption 10.3 *For all $x \in R^n$, there exists a constant $\gamma > 0$ such that*

$$\| f(x,t) \| \le \gamma, \forall t \in R. \quad (10.8)$$

Definition 10.4 The designed controllers u_i, $i = 1, 2, \ldots, N$, are said to be *distributed bounded consensus controllers* if there exist constants $\phi > 0$, $\eta_i > 0$ and $\mu > 0$, such that

$$\lim_{t\to\infty} \frac{1}{N}\mathbf{E}\left(\sum_{i=1}^{N} \| x_i(t) - s(t)\|^2\right) \le \phi\gamma + \sum_{i=1}^{N} \eta_i\beta_i + \mu\alpha. \quad (10.9)$$

If $\phi = 0$, then they are called *distributed consensus controllers*.

Note that it does not mean that the bound $\phi\gamma + \sum_{i=1}^{N} \eta_i\beta_i + \mu\alpha$ increases as the size N increases, where the chosen parameters $\phi > 0$, $\eta_i > 0$ and $\mu > 0$ may be different for the increasing size N, even for the same system.

Definition 10.5 The designed filters (10.3) or (10.4) are said to be *distributed bounded consensus filters* (*distributed consensus filters*) if the controllers in (10.3) or (10.4) are *distributed bounded consensus controllers* (*distributed consensus controllers*).

In the following, the convergence of the bound for $\frac{1}{N}\mathbf{E}\left(\sum_{i=1}^{N} \| x_i(t) - s(t)\|^2\right)$ will be analyzed.

10.2 Distributed Consensus Filters Design for Sensor Networks with Fully-Pinned Controllers

For simplicity, consider using linear state-feedback controllers

$$u_i(t) = -ck_i(x_i(t) - y_i(t)), i = 1, 2, \ldots, N, \tag{10.10}$$

where $k_i > 0$ is the feedback control gain [151]. Let $K = \text{diag}(k_1, k_2, \ldots, k_N)$, $B = (b_{ij})_{N\times N} = A - K$, and $\lambda_{\max}(B)$ denote the largest eigenvalue of matrix B.

Subtracting (10.1) from (10.4) with controllers (10.10) yields the following error dynamical network:

$$\begin{aligned} de_i(t) = {} & [f(x_i(t), t) - f(s(t), t) + c\sum_{j=1}^{N} a_{ij}e_j(t) - ck_ie_i(t)]dt - v(t)dv \\ & + ck_i\sigma_i(t)d\omega_i, \end{aligned} \tag{10.11}$$

where $e_i(t) = x_i(t) - s(t)$, $i = 1, 2, \ldots, N$.

Theorem 10.6 *Suppose that Assumptions 10.1 and 10.2 hold. The designed controllers (10.10) are distributed consensus controllers if*

$$\theta + c\lambda_{\max}(B) < 0. \tag{10.12}$$

The estimated bound is given by

$$\lim_{t\to\infty} \frac{1}{N}\mathbf{E}\left(\sum_{i=1}^{N} \| x_i(t) - s(t)\|^2\right) \le \frac{\alpha + \frac{c^2}{N}\sum_{i=1}^{N} k_i^2\beta_i}{-2(\theta + c\lambda_{\max}(B))}. \tag{10.13}$$

Proof. Consider the following Lyapunov function candidate:

$$V(t) = \frac{1}{2}\sum_{i=1}^{N} e_i^T(t)e_i(t). \tag{10.14}$$

From the Itô formula [105], one obtains the following stochastic differential:

$$dV(t) = \mathscr{L}V(t)dt + \sum_{i=1}^{N} e_i^T(t)[-v(t)dv + ck_i\sigma_i(t)d\omega_i]. \tag{10.15}$$

By Assumption 10.2, the weak infinitesimal operator $\mathscr{L}$ of the stochastic process gives

$$\mathscr{L}V(t) = \sum_{i=1}^{N}\left\{ e_i^T(t)\left[f(x_i(t), t) - f(s(t), t) + c\sum_{j=1}^{N} a_{ij}e_j(t) - ck_ie_i(t)\right]\right.$$

$$
\begin{aligned}
&+ \frac{1}{2}v^T(t)v(t) + \frac{c^2k_i^2}{2}\sigma_i^T(t)\sigma_i(t) \Bigg\} \\
&\leq \theta \sum_{i=1}^{N} e_i^T(t)e_i(t) + c\sum_{i=1}^{N}\sum_{j=1}^{N} b_{ij}e_i^T(t)e_j(t) + \frac{1}{2}N\alpha + \frac{c^2}{2}\sum_{i=1}^{N} k_i^2\beta_i \\
&\leq (\theta + c\lambda_{\max}(B))\sum_{i=1}^{N} e_i^T(t)e_i(t) + \frac{1}{2}N\alpha + \frac{c^2}{2}\sum_{i=1}^{N} k_i^2\beta_i. \qquad (10.16)
\end{aligned}
$$

In view of Lemma 2.32, $\lambda_{\max}(B) < 0$. From condition (10.12), it follows that

$$
\begin{aligned}
\mathscr{L}V(t) &\leq (\theta + c\lambda_{\max}(B))\sum_{i=1}^{N} e_i^T(t)e_i(t) + \frac{1}{2}N\alpha + \frac{c^2}{2}\sum_{i=1}^{N} k_i^2\beta_i \\
&= N(\theta + c\lambda_{\max}(B))\left[\frac{1}{N}\sum_{i=1}^{N} e_i^T(t)e_i(t) - \frac{\alpha + \frac{c^2}{N}\sum_{i=1}^{N} k_i^2\beta_i}{-2(\theta + c\lambda_{\max}(B))}\right]. \qquad (10.17)
\end{aligned}
$$

From the Itô formula, it follows that

$$
\begin{aligned}
&\mathbf{E}V(t) - \mathbf{E}V(0) \\
&= \mathbf{E}\int_0^t \mathscr{L}V(s)ds \\
&\leq -N(\theta + c\lambda_{\max}(B))\int_0^t \left[\frac{1}{N}\mathbf{E}\sum_{i=1}^{N} e_i^T(t)e_i(t) - \frac{\alpha + \frac{c^2}{N}\sum_{i=1}^{N} k_i^2\beta_i}{-2(\theta + c\lambda_{\max}(B))}\right]ds.
\end{aligned}
$$

Under condition (10.13), if $\frac{1}{N}\mathbf{E}\left(\sum_{i=1}^{N} \| e_i(t)\|^2\right) > \frac{\alpha+\frac{c^2}{N}\sum_{i=1}^{N} k_i^2\beta_i}{-2(\theta+c\lambda_{\max}(B))}$, then $\mathbf{E}V(t) - \mathbf{E}V(0) < 0$. This completes the proof. □

By letting $k_i = k > 0, i = 1, 2, \ldots, N$, in (10.10), one has the following corollary.

Corollary 10.7 *Suppose that Assumptions 10.1 and 10.2 hold. The designed controllers (10.10) are distributed consensus controllers if*

$$\theta - ck < 0. \qquad (10.18)$$

The estimated bound is given by

$$\lim_{t\to\infty}\frac{1}{N}\mathbf{E}\left(\sum_{i=1}^{N} \| x_i(t) - s(t)\|^2\right) \leq \frac{\alpha + \frac{c^2}{N}\sum_{i=1}^{N} k^2\beta_i}{2(ck - \theta)}. \qquad (10.19)$$

Proof. By Lemma 2.6 and Theorem 10.6, the proof can be easily completed. □

Denote $g(k) = \frac{\alpha + \frac{c^2}{N}\sum_{i=1}^{N} k^2\beta_i}{2(ck-\theta)}$. Then,

$$g'(k) = \frac{\frac{c^3}{N}\sum_{i=1}^{N}\beta_i k^2 - 2\frac{c^2}{N}\sum_{i=1}^{N}\beta_i\theta k - c\alpha}{2(ck-\theta)^2}.$$

Let $g'(k) = 0$. Then, it follows that $\overline{k}_{1,2} = \frac{\theta \pm \sqrt{\theta^2 + \frac{\alpha}{\frac{1}{N}\sum_{i=1}^{N}\beta_i}}}{c}$. Since $\overline{k}_1 > k > \theta/c > \overline{k}_2$, if $\overline{k}_1 > k > 0$, then $g'(k) < 0$; if $k > \overline{k}_1$, then $g'(k) > 0$. So, there is a minimum value of $g(k)$ at $k = \overline{k}_1$.

If $N = 1$, i.e., only one sensor is used to track the target, one has the following corollary. Without loss of generality, let the first node be this sensor.

Corollary 10.8 *Suppose that Assumptions 10.1 and 10.2 hold. The designed controllers (10.10) are distributed consensus controllers if*

$$\theta - ck_1 < 0. \tag{10.20}$$

The estimated bound is given by

$$\lim_{t\to\infty} \mathbf{E}(\| x_1(t) - s(t)\|^2) \leq \frac{\alpha + c^2k_1^2\beta_1}{2(ck_1 - \theta)}. \tag{10.21}$$

Corollary 10.9 *Suppose that Assumptions 10.2 and 10.3 hold. The designed controllers (10.10) are distributed bounded consensus controllers, and the estimated bound is given by*

$$\lim_{t\to\infty} \frac{1}{N}\mathbf{E}\left(\sum_{i=1}^{N} \| x_i(t) - s(t)\|^2\right) \leq \left(\frac{\gamma + \sqrt{\gamma^2 - c\lambda_{\max}(B)(\alpha + \frac{c^2}{N}\sum_{i=1}^{N} k_i^2\beta_i)/2}}{-c\lambda_{\max}(B)}\right)^2. \tag{10.22}$$

Proof. Choose (10.14) as the Lyapunov function candidate. By Assumption 10.3, the weak infinitesimal operator $\mathscr{L}$ of the stochastic process yields

$$\mathscr{L}V(t) \leq \sum_{i=1}^{N} e_i^T(t)[f(x_i(t),t) - f(s(t),t)] + c\lambda_{\max}(B)\sum_{i=1}^{N} e_i^T(t)e_i(t)$$

$$+\frac{1}{2}N\alpha + \frac{c^2}{2}\sum_{i=1}^{N} k_i^2 \beta_i$$

$$\leq 2\gamma \sum_{i=1}^{N} \| e_i(t) \| + c\lambda_{\max}(B) \sum_{i=1}^{N} \| e_i(t)\|^2 + \frac{1}{2}N\alpha + \frac{c^2}{2}\sum_{i=1}^{N} k_i^2 \beta_i. \quad (10.23)$$

From Lemma 2.32, it follows that $\lambda_{\max}(B) < 0$. Note that

$$\left(\sum_{i=1}^{N} \| e_i(t) \|\right)^2 = \sum_{i=1}^{N}\sum_{j=1}^{N} \| e_i(t) \|\| e_j(t) \|$$

$$\leq \frac{1}{2}\sum_{i=1}^{N}\sum_{j=1}^{N} (\| e_i(t)\|^2 + \| e_j(t)\|^2) = N\sum_{i=1}^{N} \| e_i(t)\|^2.$$

Then, it follows that

$$\mathscr{L}V(t) \leq c\lambda_{\max}(B) \sum_{i=1}^{N} \| e_i(t)\|^2 + 2\gamma\sqrt{N}\sqrt{\sum_{i=1}^{N} \| e_i(t)\|^2} + \frac{1}{2}N\alpha + \frac{c^2}{2}\sum_{i=1}^{N} k_i^2 \beta_i$$

$$= cN\lambda_{\max}(B)\left[\frac{1}{N}\sum_{i=1}^{N} \| e_i(t)\|^2 + \frac{2\gamma}{c\lambda_{\max}(B)} \times \sqrt{\frac{1}{N}\sum_{i=1}^{N} \| e_i(t)\|^2} - \frac{\alpha + \frac{c^2}{N}\sum_{i=1}^{N} k_i^2\beta_i}{-2c\lambda_{\max}(B)}\right]. \quad (10.24)$$

Let $z = \sqrt{\frac{1}{N}\sum_{i=1}^{N} \| e_i(t)\|^2}$ and $g(z) = z^2 + \frac{2\gamma}{c\lambda_{\max}(B)}z - \frac{\alpha + \frac{c^2}{N}\sum_{i=1}^{N} k_i^2\beta_i}{-2c\lambda_{\max}(B)}$. It is easy to see that $g(z) = 0$ has two solutions:

$$z_{1,2} = \frac{-\gamma \pm \sqrt{\gamma^2 - c\lambda_{\max}(B)(\alpha + \frac{c^2}{N}\sum_{i=1}^{N} k_i^2\beta_i)/2}}{c\lambda_{\max}(B)}, \quad (10.25)$$

where $z_1 < 0$ and $z_2 > 0$. If $z(t) \geq z_2$, then $g(z) \geq 0$ and $\mathbf{E}V(t) - \mathbf{E}V(0) < 0$. The proof is completed. □

Remark 10.10 One might wonder, if only one sensor can achieve the filtering performance as shown by Corollary 10.8, why so many (N) sensors are used in Theorem 10.6 and Corollary 10.7. This can be explained as follows. First, in a sensor network consisting of a large number of sensor nodes in a wide spatial region, it is impossible to have only one centralized processor that can collect the measurements from all the sensors, especially in a remote area. Usually, each sensor may be able to use only local information and communicate with neighbors so that the estimation can be achieved in a distributed way. Second, in a sensor network, using a large number of sensors to observe the target can usually obtain a more accurate estimation than using a single sensor node. Third, in (10.2), every state of the target can be observed, which is not always the case in applications. The next few subsections will deal with the case where a sensor can only measure some states of the target. In that case, using only one sensor cannot estimate the target accurately.

Remark 10.11 From $\overline{k}_1 = \frac{1}{c}\left(\theta + \sqrt{\theta^2 + \frac{\alpha}{\frac{1}{N}\sum_{i=1}^{N}\beta_i}}\right)$, it follows that the minimal bound increases as α or θ increases, while it decreases when c or $\frac{1}{N}\sum_{i=1}^{N}\beta_i$ increase. Therefore, it is unreasonable to use a very large control gain if the noise intensity β_i is relatively high.

10.3 Distributed Consensus Filters Design for Sensor Networks with Pinning Controllers

In many cases, it is literally impossible to add controllers to all sensors.

Here, the pinning strategy is applied to a small fraction δ $(0 < \delta < 1)$ of the sensors in network (10.3). Without loss of generality, let the first $l = \lfloor \delta N \rfloor$ nodes be controlled, where $\lfloor \cdot \rfloor$ is the integer part of a real number. Thus, the designed pinning controllers can be described by

$$\begin{aligned} u_i(t) &= -ck_i(x_i(t) - y_i(t)), i = 1, 2, \dots, l, \\ u_i(t) &= 0, i = l+1, l+2, \dots, N, \end{aligned} \tag{10.26}$$

where $k_i > 0$ is the feedback control gain. Let $\overline{K} = \mathrm{diag}(\underbrace{k_1, \dots, k_l}_{l}, \underbrace{0, \dots, 0}_{N-l})$, $\overline{B} = A - \overline{K} = (\overline{b}_{ij})_{N\times N}$, and $\lambda_{\max}(\overline{B})$ denote the largest eigenvalue of matrix $\overline{B}$.

Subtracting (10.1) from (10.4) with controllers (10.26) yields the following error dynamical network:

$$\begin{aligned} de_i(t) &= \left[f(x_i(t), t) - f(s(t), t) + c\sum_{j=1}^{N} \overline{b}_{ij} e_j(t)\right] dt - v(t)dv + ck_i\sigma_i(t)d\omega_i, \\ & i = 1, 2, \dots, l, \\ de_i(t) &= \left[f(x_i(t), t) - f(s(t), t) + c\sum_{j=1}^{N} \overline{b}_{ij} e_j(t)\right] dt - v(t)dv, \\ & i = l+1, l+2, \dots, N, \end{aligned} \tag{10.27}$$

where $e_i(t) = x_i(t) - s(t)$, $i = 1, 2, \dots, N$.

Theorem 10.12 *Suppose that Assumptions 10.1 and 10.2 hold. Then the designed controllers (10.26) are distributed consensus controllers if*

$$\theta + c\lambda_{\max}(\overline{B}) < 0. \tag{10.28}$$

The estimated bound is given by

$$\lim_{t\to\infty} \frac{1}{N}\mathbf{E}\left(\sum_{i=1}^{N} \| x_i(t) - s(t)\|^2\right) \le \frac{\alpha + \frac{c^2}{N}\sum_{i=1}^{l} k_i^2\beta_i}{-2(\theta + c\lambda_{\max}(\overline{B}))}. \tag{10.29}$$

Proof. By Lemma 2.32, $\overline{B}$ is negative definite. Thus, using Theorem 1 completes the proof. □

Corollary 10.13 *Suppose that Assumptions 10.2 and 10.3 hold. The designed controllers (10.10) are distributed and bounded consensus controllers, and the estimated bound is given by*

$$\lim_{t\to\infty} \frac{1}{N}\mathbf{E}\left(\sum_{i=1}^{N} \| x_i(t) - s(t)\|^2\right)$$
$$\le \frac{\gamma + \sqrt{\gamma^2 - c\lambda_{\max}(\overline{B})(\alpha + \frac{c^2}{N}\sum_{i=1}^{l} k_i^2\beta_i)/2}}{-c\lambda_{\max}(\overline{B})}. \tag{10.30}$$

Rewrite $\overline{B}$ as $\overline{B} = \begin{pmatrix} A_1 - \widetilde{K} & A_2 \\ A_2^T & \widetilde{B} \end{pmatrix}$, where $\widetilde{K} = \text{diag}(k_1, \ldots, k_l)$, A_1 and A_2 are matrices with appropriate dimensions, and $\widetilde{B}$ is obtained by removing the $1, 2, \ldots, l$ row-column pairs from matrix A.

Corollary 10.14 *Suppose that Assumptions 10.1 and 10.2 hold. If the control gains k_i are sufficiently large, then the condition (10.28) is equivalent to*

$$\theta + c\lambda_{\max}(\widetilde{B}) < 0. \tag{10.31}$$

The estimated bound is given by

$$\lim_{t\to\infty} \frac{1}{N}\mathbf{E}\left(\sum_{i=1}^{N} \| x_i(t) - s(t)\|^2\right) \le \frac{\alpha + \frac{c^2}{N}\sum_{i=1}^{l} \beta_i}{-2(\theta + c\lambda_{\max}(\widetilde{B}))}. \tag{10.32}$$

For simplicity, one may select $\widetilde{K} > \lambda_{\max}(A_1 - A_2\widetilde{B}^{-1}A_2^T)I_l$ as in [151].

Proof. In order to derive the result $\lambda_{\max}(\widetilde{B}) = \lambda_{\max}(\overline{B})$, it only needs to prove that $\widetilde{B} - \lambda I_{N-l} < 0$ is equivalent to $\overline{B} - \lambda I_N < 0$ for any positive $\lambda > 0$.

It is easy to see that if $\overline{B} - \lambda I_N < 0$ then $\widetilde{B} - \lambda I_{N-l} < 0$. So, it suffices to prove that if $\widetilde{B} - \lambda I_{N-l} < 0$ then $\overline{B} - \lambda I_N < 0$. Choose $\widetilde{K} > A_1 - A_2\widetilde{B}^{-1}A_2^T$. Then, by Lemma 3.12, $\overline{B} - \lambda I_N < 0$. The proof is completed. □

Remark 10.15 In this subsection, only a small fraction of sensors are used to measure the target, which is more practical in real applications. Note that the sensors with measurements can communicate with the neighboring sensors, some of which cannot observe the target. The whole process may be considered as the leader-follower behavior, where the target is the true leader, the measured sensors are virtual leaders, and the other sensors are the followers.

Remark 10.16 Although $-\theta - c\lambda_{\max}(\overline{B}) < -\theta - c\lambda_{\max}(B)$ in Theorems 10.6 and 10.12, the sum of the bounds (from 1 to l) in Theorem 10.12 (10.14) is lower than that in Theorem 10.6 (from 1 to N). Therefore, the bound in Theorem 2 is not necessarily lower than that in Theorem 10.6. In fact, a higher-dimensional condition (10.29) can be equivalent to a lower-dimensional condition (10.31).

10.4 Distributed Consensus Filters Design for Sensor Networks with Pinning Observers

In many cases, it is impractical to assume that a sensor can observe all the states of the target in (10.2). Therefore, assume that the first $l = \lfloor \delta N \rfloor$ sensors can measure the signals by

$$\widetilde{y}_i(t) = H_i s(t) + \sigma_i(t) d\omega_i/dt, i = 1, 2, \dots, l, \tag{10.33}$$

where $\widetilde{y}_i(t) \in R^m$ is the measurement of sensor i by observing the target $s(t)$, and $H_i \in R^{m\times n}$, $i = 1, 2, \dots, l$.

The designed distributed consensus filter controller is described by

$$\begin{aligned} u_i(t) &= -cD_i(H_i x_i(t) - \widetilde{y}_i(t)), i = 1, 2, \dots, l, \\ u_i(t) &= 0, i = l+1, l+2, \dots, N, \end{aligned} \tag{10.34}$$

where $D_i \in R^{n\times m}$ is the feedback control gain matrix.

Subtracting (10.1) from (10.4) with observers (10.33) and controllers (10.34) yields the following error dynamics:

$$\begin{aligned} de_i(t) &= \left[f(x_i(t), t) - f(s(t), t) + c\sum_{j=1}^{N} a_{ij} e_j(t) - cD_i H_i e_i \right] dt \\ &\quad - v(t)dv + cD_i H_i \sigma_i(t) d\omega_i, i = 1, 2, \dots, l, \\ de_i(t) &= \left[f(x_i(t), t) - f(s(t), t) + c\sum_{j=1}^{N} a_{ij} e_j(t) \right] dt - v(t)dv, \\ &\quad i = l+1, l+2, \dots, N, \end{aligned} \tag{10.35}$$

where $e_i(t) = x_i(t) - s(t)$, $i = 1, 2, \dots, N$.

Definition 10.17 The state j $(1 \leq j \leq n)$ of the target is said to be observable by sensor i $(1 \leq k \leq l)$, if there is a gain matrix D_i and a positive constant ϵ_j, such that

$$-s^T D_i H_i s \leq -\epsilon_j s_j^2$$

for any $s = (s_1, s_2, \ldots, s_n)^T \in R^n$.

Definition 10.18 The state j $(1 \leq j \leq n)$ of the target is said to be *observable* by the sensor network if it is observable by a sensor k, where $1 \leq k \leq l$.

Definition 10.19 The target is said to be *observable* by the sensor network, if each of its state is observable by the sensor network.

Assumption 10.20 *Suppose that the target is observable and, without loss of generality, suppose that the target is observable by the first $\widetilde{l}$ $(\widetilde{l} \leq l)$ sensors. Then, there exist matrices D_i, $i = 1, 2, \ldots, \widetilde{l}$, such that*

$$-\sum_{i=1}^{\widetilde{l}} e_i^T(t) D_i H_i e_i \leq -\sum_{j=1}^{n} \sum_{k \in \mathscr{M}_j} \epsilon_{kj} e_{kj}^2, \tag{10.36}$$

where the state j is observable by sensors k $(1 \leq k \leq \widetilde{l})$, ϵ_{kj} are positive constants, and $\mathscr{M}_j$ are the sets of all sensors that can observe the target state j, $j = 1, \ldots, n$.

Theorem 10.21 *Suppose that Assumptions 10.1, 10.2 and 10.20 hold. The designed controllers (10.34) are distributed consensus controllers if*

$$\theta + c\lambda_{\max}(A - \Xi) < 0, \tag{10.37}$$

where $\lambda_{\max}(A - \Xi) = \max_j(\lambda_{\max}(A - \Xi_j))$ *and* $\Xi_j = \mathrm{diag}(0, \ldots, \underbrace{\epsilon_{j_1 j}}_{j_1}, \ldots, \underbrace{\epsilon_{j_p j}}_{j_p}, \ldots, 0) \in R^{N \times N}$, *i.e., the j_ith diagonal element is $\epsilon_{j_i j}$, and $j_i \in \mathscr{M}_j$, $j = 1, 2, \ldots, n$, $i = 1, \ldots, p$.*

The estimated bound is given by

$$\lim_{t \to \infty} \frac{1}{N} \mathbf{E} \left(\sum_{i=1}^{N} \| x_i(t) - s(t) \|^2 \right) \leq \frac{\alpha + \frac{c^2}{N} \sum_{i=1}^{\widetilde{l}} \lambda_{\max}(H_i^T D_i^T D_i H_i) \beta_i}{-2(\theta + c\lambda_{\max}(A - \Xi))}. \tag{10.38}$$

Proof. Consider the Lyapunov function candidate (10.14). The weak infinitesimal operator $\mathscr{L}$ of the stochastic process gives

$$\mathscr{L} V(t) = \sum_{i=1}^{N} \{ e_i^T(t) [f(x_i(t), t) - f(s(t), t) + c \sum_{j=1}^{N} a_{ij} e_j(t) + \frac{1}{2} v^T(t) v(t) \Bigg]$$

$$+\sum_{i=1}^{l}\left\{e_i^T(t)\left[-cD_iH_ie_i+\frac{c^2}{2}\sigma_i^T(t)(D_iH_i)^T(D_iH_i)\sigma_i(t)\right]\right\}$$

$$\leq\theta\sum_{i=1}^{N}e_i^T(t)e_i(t)+c\sum_{i=1}^{N}\sum_{j=1}^{N}a_{ij}e_i^T(t)e_j(t)-c\sum_{i=1}^{l}e_i^T(t)D_iH_ie_i$$

$$+\frac{1}{2}N\alpha+\frac{c^2}{2}\sum_{i=1}^{l}\lambda_{\max}(H_i^TD_i^TD_iH_i)\beta_i. \tag{10.39}$$

Then, by Assumption 10.20,

$$\sum_{i=1}^{N}\sum_{j=1}^{N}a_{ij}e_i^T(t)e_j(t)-\sum_{i=1}^{l}e_i^T(t)D_iH_ie_i$$

$$\leq\sum_{i=1}^{N}\sum_{j=1}^{N}a_{ij}e_i^T(t)e_j(t)-\sum_{j=1}^{n}\sum_{k\in\mathscr{M}_j}\epsilon_{kj}e_{kj}^2-\sum_{i=\widetilde{l}+1}^{l}e_i^T(t)D_iH_ie_i$$

$$=\sum_{j=1}^{n}\widetilde{e}_j^T(t)A\widetilde{e}_j(t)-\sum_{j=1}^{n}\widetilde{e}_j^T(t)\Xi_j\widetilde{e}_j(t)-\sum_{i=\widetilde{l}+1}^{l}e_i^T(t)D_iH_ie_i, \tag{10.40}$$

where $\widetilde{e}_j=(e_{1j},e_{2j},\dots,e_{Nj})^T, j=1,2,\dots,n$.

For simplicity, choose $D_i=0$ for $i=\widetilde{l}+1,\dots,l$. Under condition (10.37), it follows that

$$\mathscr{L}V(t)\leq(\theta+c\lambda_{\max}(A-\Xi))\sum_{i=1}^{N}e_i^T(t)e_i(t)+\frac{1}{2}N\alpha$$

$$+\frac{c^2}{2}\sum_{i=1}^{\widetilde{l}}\lambda_{\max}(H_i^TD_i^TD_iH_i)\beta_i \tag{10.41}$$

$$=N(\theta+c\lambda_{\max}(A-\Xi))\left[\frac{1}{N}\sum_{i=1}^{\widetilde{l}}e_i^T(t)e_i(t)-\frac{\alpha+\frac{c^2}{N}\sum_{i=1}^{\widetilde{l}}\lambda_{\max}(H_i^TD_i^TD_iH_i)\beta_i}{-2(\theta+c\lambda_{\max}(A-\Xi))}\right].$$

Under condition (10.38), if $\frac{1}{N}\mathbf{E}\left(\sum_{i=1}^{N}\|e_i(t)\|^2\right)>\frac{\alpha+\frac{c^2}{N}\sum_{i=1}^{\widetilde{l}}\lambda_{\max}(H_i^TD_i^TD_iH_i)\beta_i}{-2(\theta+c\lambda_{\max}(A-\Xi))}$, then $\mathbf{E}V(t)-\mathbf{E}V(0)<0$. This completes the proof. □

Corollary 10.22 *Suppose that Assumptions 10.1, 10.3, and 10.20 hold. Then, the designed controllers (10.34) are distributed bounded consensus controllers, and the estimated bound is given by*

$$\lim_{t\to\infty}\frac{1}{N}\mathbf{E}\left(\sum_{i=1}^{N}\| x_i(t)-s(t)\|^2\right)\leq\frac{\gamma+\sqrt{\gamma^2-\widetilde{\alpha}}}{-c\lambda_{\max}(A-\Xi)},$$

where $\widetilde{\alpha}=c\lambda_{\max}(A-\Xi)(\alpha+\frac{c^2}{N}\sum_{i=1}^{\widetilde{l}}\lambda_{\max}(H_i^T D_i^T D_i H_i)\beta_i)/2$, $\lambda_{\max}(A-\Xi)=\max_j(\lambda_{\max}(A-\Xi_j))$ *and* $\Xi_j=\mathrm{diag}(0,\ldots,\underbrace{\epsilon_{j_1 j}}_{j_1},\ldots,\underbrace{\epsilon_{j_p j}}_{j_p},\ldots,0)\in R^{N\times N}$, *i.e., the* j_i*th diagonal element is* $\epsilon_{j_i j}$, *and* $j_i\in\mathcal{M}_j$, $j=1,2,\ldots,n$, $i=1,\ldots,p$.

Remark 10.23 In (10.2), each sensor can observe the full state of the target. However, this is not always the case in practice. In this subsection, only partial state information about the target is assumed to be observed by the sensors in (10.34), and that the target is observable by the sensor network (Definition 10.19), which is very reasonable for real sensor networks.

Remark 10.24 If the coupling matrix A is not irreducible, which means that the network is not connected, then the whole network can be composed of many connected components. The filtering design in this section can be extended to the disconnected network by studying its connected components.

10.5 Simulation Examples

In this section, some simulation examples are provided to verify the designed distributed consensus filters.

Distributed Consensus Filters in Sensor Networks with Pinning Controllers

Consider an ER random network [35], and suppose that there are $N=100$ nodes with connection probability $p=0.1$. This random network has about $pN(N-1)/2\approx 500$ connections. If there is a connection between node i and j, then $a_{ij}=a_{ji}=1$ $(i\neq j)$, $i,j=1,2,\ldots,100$.

The target and the measurement models are described by (10.1) and (10.2), respectively, where $f(s(t))=\cos(t)$ with initial condition $s(0)=0$, $v(t)=0.5$, and $\sigma_i(t)=0.4$. The designed distributed consensus filter is (10.3) with controllers (10.26), where $c=1$, $k_i=5$, and $l=20$. Here, a simulation-based analysis on the controlled random network is performed by using random pinning scheme. In the random pinning scheme, one randomly selects $\lfloor\delta N\rfloor$ nodes with a small fraction $\delta=0.2$ to pin. The orbits of the measured data y_i in sensors and the filtering data x_j are shown in Fig. 10.1 for $i=1,2,\ldots,20$ and $j=1,2,\ldots,100$. By Corollary 10.13, the designed pinning controllers are distributed bounded consensus controllers. The orbits of the errors e_i are illustrated in Fig. 10.2.

Distributed Consensus Filters in Sensor Networks with Pinning Observers

Here, a simulation-based analysis on the controlled scale-free network is performed by using the high-degree pinning scheme. In the simulated scale-free network [8],

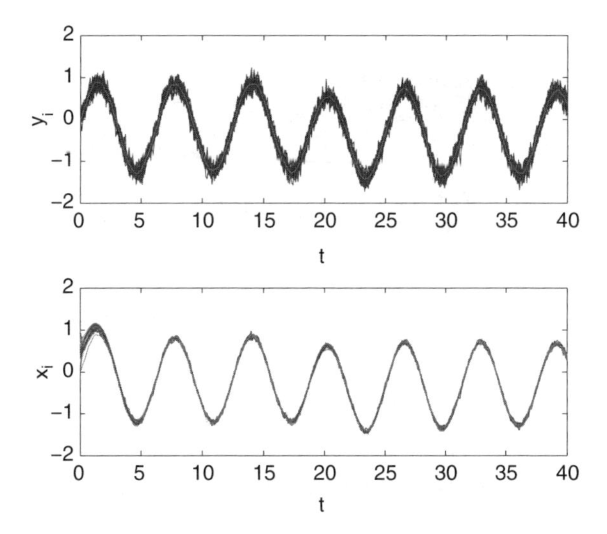

Figure 10.1 Orbits of the measured data y_i in sensors and filtering data x_j, $i = 1, 2, \ldots, 20$ and $j = 1, 2, \ldots, 100$

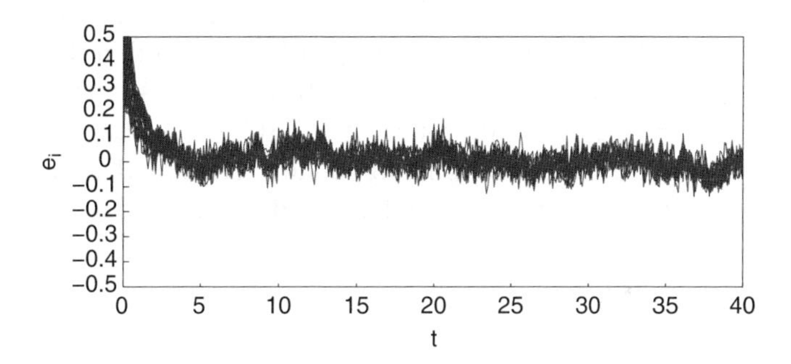

Figure 10.2 Orbits of the errors e_i, $i = 1, 2, \ldots, 100$

$N = 100$, $m_0 = m = 3$, which contains about 3000 connections. The high-degree pinning scheme first pins the node with the highest degree, and then continues to choose and pin the other nodes in the monotonically decreasing order of node degrees.

The target model is described by Chua's circuit [24]

$$\begin{cases} ds_1 = [\eta(-s_1 + s_2 - l(s_1))]dt + v_1(t)dv, \\ ds_2 = [s_1 - s_2 + s_3]dt + v_2(t)dv, \\ ds_3 = [-\beta s_2]dt + v_3(t)dv, \end{cases} \tag{10.42}$$

where $l(x_1) = bx_1 + 0.5(a - b)(|x_1 + 1| - |x_1 - 1|)$ and $v_i(t) = 0.5$, $i = 1, 2, 3$. The system (10.42) is chaotic without noise when $\eta = 10$, $\beta = 18$, $a = -4/3$, and $b = -3/4$, as shown in Fig. 10.3. In view of Assumption 10.1, by computation, one obtains $\theta = 5.1623$.

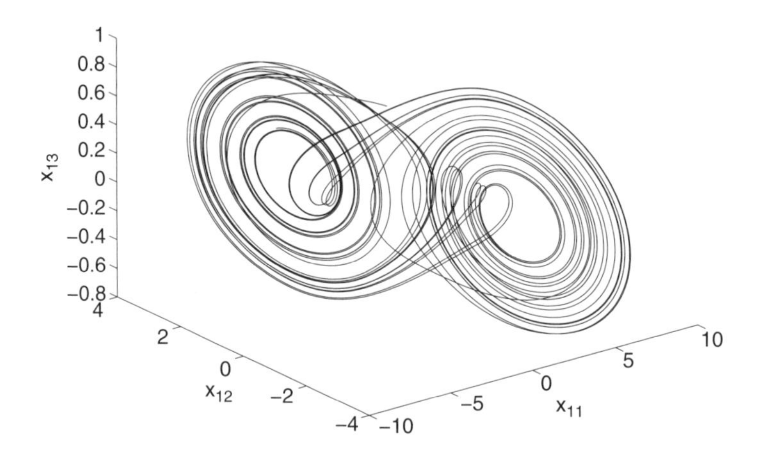

Figure 10.3 Chaotic orbits of Chua's circuit

The measurement model is described by (10.33), where H_i is chosen as $(1,0,0)$, $(0,1,0)$, or $(0,0,1)$, with $\sigma_i(t) = 0.5$. The designed distributed consensus filter is (10.3) with controllers (10.34), where $c = 15$, $l = 20$, and the corresponding D_i is chosen as $(50,0,0)^T$, $(0,50,0)^T$, or $(0,0,50)^T$. Assumption 10.20 is satisfied for $\epsilon_{kj} = 50$, $k = 1,\dots,l$ and $j = 1,2,3$. The orbits of the estimation x_i are shown in Fig. 10.4 for $i = 1,2,\dots,100$. Since $\theta = 5.1623 < -c\lambda_{\max}(A - \Xi) = 6.8565$, by Corollary 10.22, the designed pinning controllers are distributed consensus controllers. The orbits of the errors e_i are illustrated in Fig. 10.5.

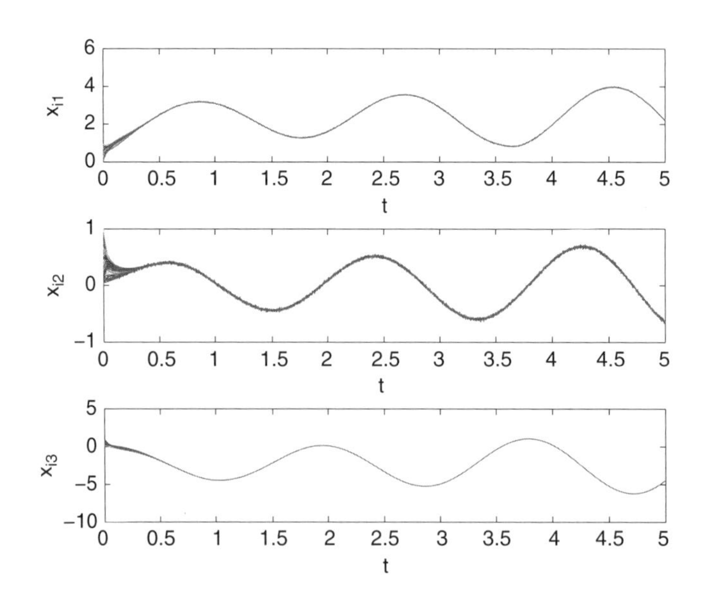

Figure 10.4 Orbits of estimations x_i, $i = 1,2,\dots,100$

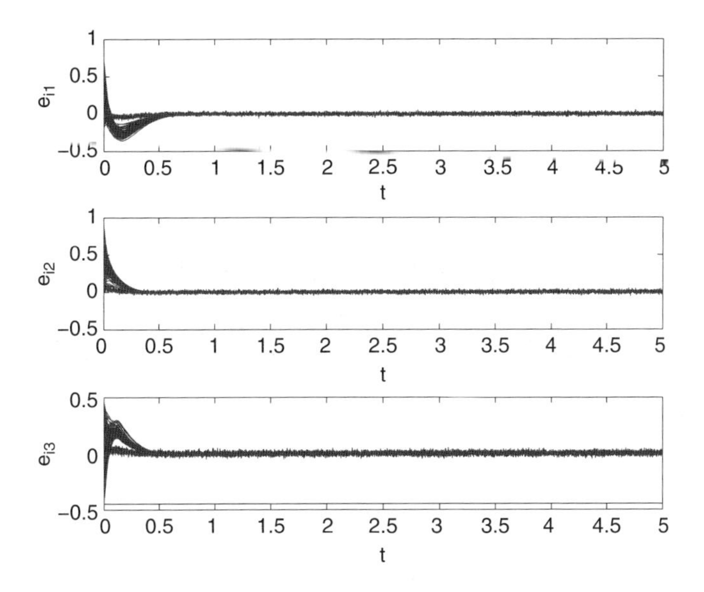

Figure 10.5 Orbits of the errors e_i, $i = 1, 2, \dots, 100$

10.6 Notes

In this chapter, some distributed consensus filters in sensor networks have been designed, analyzed, and simulated. Three scenarios have been considered. Under the condition that each sensor can measure the target, the distributed controllers could be added to all sensor nodes. By using the pinning control approach, assuming that only a small fraction of sensors can measure the target, the network still could be controlled. Moreover, pinning observers have been designed under the condition that a sensor can only observe partial states of the target. All these designs have been analyzed and validated by simulations.

Distributed consensus filter are every effective, with many advantages for its easier implementation, low cost, fast speed, etc. A general framework for distributed consensus filter design has been presented in this section, which is a promising approach therefore deserving further investigation in the near future.

11

Delay-Induced Consensus and Quasi-Consensus in Multi-agent Systems

This chapter studies consensus and quasi-consensus in multi-agent dynamical systems [152]. A linear consensus protocol for systems with second-order dynamics is designed, where both the current and delayed position information are utilized. Time delay, in a common perspective, can induce periodic oscillations or even chaos in dynamical systems. However, it is found that consensus and quasi-consensus in a multi-agent system cannot be reached without using delayed position information under the given protocol while they can be achieved with a relatively small time delay by appropriately choosing the coupling strengths. A necessary and sufficient condition is established for reaching consensus in multi-agent dynamical systems. It is shown that consensus and quasi-consensus can be achieved if and only if the time delay is bounded by some critical values that depend on the coupling strengths and the largest eigenvalue of the Laplacian matrix of the network. The motivation for studying quasi-consensus is revealed, where the potential relationship between the second-order multi-agent system with delayed positive feedback and the first-order system with distributed-delay control input is discussed.

11.1 Problem Statement

Recently, second-order dynamics [50, 51, 95, 93, 109, 146, 149, 158] have received increasing attention due to many real-world applications where agents are governed by both position and velocity dynamics. In [146], in particular, some necessary and sufficient conditions for second-order consensus in multi-agent dynamical systems with directed topologies were established. It was found that both the real and imaginary

Distributed Cooperative Control of Multi-agent Systems, First Edition.
Wenwu Yu, Guanghui Wen, Guanrong Chen, and Jinde Cao.

parts of the eigenvalues of the Laplacian matrix associated with the corresponding network topology play key roles in reaching consensus. However, as shown in [50, 51, 95], the velocity states of agents are often unavailable, therefore some observers were designed with some additional variables involved, which leads to the study of higher-order dynamical systems.

It is well known that time delay, a destructive character in dynamics, may result in oscillatory behaviors [116], network instability (periodic oscillation and even chaos) [140], or network desynchronization with a general coupling function [40, 56]. On the other hand, consensus can be reached for any finite time delay on the neighboring agents [70]. However, in [85], it was shown that time delay can induce system stability in linear time-invariant systems, where both the stability regions called pockets in the domain of time delay and the number of unstable characteristic roots at any given pocket were theoretically analyzed. In [2], delayed positive feedback was designed to stabilize the systems with second-order oscillations. Differing from the results in [2, 85], quasi-consensus behavior is considered and the systems are coupled in this chapter. In particular, the motivation for studying quasi-consensus is revealed where the potential relationship between the second-order multi-agent system with delayed positive feedback and the first-order system with distributed-delay control input is discussed. In [19], consensus in first-order multi-agent systems with current and outdated position states was discussed, showing that the delay-involved algorithm converges faster than the standard consensus protocol without time delays.

In many real-world applications, the relative velocities of neighboring agents are more difficult to measure than relative positions [50, 51]. For example, a camera can be used for relative position measurements. In general, relative velocity measurements require more expensive sensors. In some experimental work, each mobile robot, equipped with range sensors, obtains the position information of its own and its neighbors through some localization algorithms. In the settings of such formation control problems with range-only sensing, the velocity information is difficult to obtain directly. By using delayed position information in the memory and without knowing the velocity information of agents in second-order dynamics as in [50, 51, 93], it is found in this chapter that consensus can be reached by appropriately choosing network parameters while consensus may not be achieved without time delay. This implies that, similar to the delay-induced stability in linear time-invariant systems [85], time delay can induce consensus in multi-agent dynamical systems, which is the primary motivation of the present work.

It is assumed that there are no physical communication delays in the network. This context essentially explores the combination of the current relative position with outdated relative position data (stored in memory) to help achieve consensus. As a result, there is no need to measure relative velocities. In the designed system, the delays are not the real communication delays existing in the network but are outdated data stored in the memory.

A new type of consensus called quasi-consensus is defined in this chapter, where the velocity states of agents asymptotically converge to a common value but there are relative position differences among agents depending on the initial conditions, which is different from flocking [82] and formation control [7, 33, 83] in multi-agent systems.

In [7], a behavior-based decentralized control for formation control architecture was proposed. Formation stabilization of a group of autonomous agents with linear dynamics was investigated by using structural potential functions in [83]. Then, a leader-follower problem for maintaining a desired formation was considered in [33]. For formation control and flocking in multi-agent systems, a geometrically desirable formation will be designed first, while for quasi-consensus to be studied in this chapter, the final position configuration changes with different initial states.

Definition 11.1 The multi-agent system is said to achieve *quasi-consensus* if, for any initial conditions,

$$\lim_{t\to\infty} \| x_i(t) - x_j(t) \| = c_{ij}, \ \lim_{t\to\infty} \| v_i(t) - v_j(t) \| = 0,$$
$$\forall i,j = 1,2,\cdots,N,$$

where c_{ij} are constants. Particularly, if $c_{ij} = 0, \forall i,j = 1,2,\cdots,N$, then the quasi-consensus becomes *consensus*.

Since $\dot{x}_i(t) - \dot{x}_j(t) = v_i(t) - v_j(t)$ in (3.2), one only needs to check if the final velocity states of all the agents are the same for quasi-consensus.

In [50, 51, 95], distributed observers were designed for dynamics of multi-agent systems, where the velocity states were assumed to be unavailable, and some slack variables were introduced and a higher-order controller was designed. In this chapter, by using delayed position information, it will be shown that consensus and quasi-consensus can be reached in the multi-agent systems. To do so, the following consensus protocol with both current and delayed position information is considered:

$$\begin{aligned}
&\dot{x}_i(t) = v_i,\\
&\dot{v}_i(t) = \alpha \sum_{j=1, j\neq i}^{N} G_{ij}(x_j(t) - x_i(t)) - \beta \sum_{j=1, j\neq i}^{N} G_{ij}(x_j(t-\tau) - x_i(t-\tau)),\\
&\quad i = 1,2,\cdots,N,
\end{aligned} \tag{11.1}$$

where $\tau \geq 0$ is a time delay, and α and β are the coupling strengths. Then, this system can be equivalently rewritten as

$$\begin{aligned}
&\dot{x}_i(t) = v_i,\\
&\dot{v}_i(t) = -\alpha \sum_{j=1}^{N} L_{ij}x_j(t) + \beta \sum_{j=1}^{N} L_{ij}x_j(t-\tau), i = 1,2,\cdots,N.
\end{aligned} \tag{11.2}$$

11.2 Delay-Induced Consensus and Quasi-Consensus in Multi-agent Dynamical Systems

Let $\eta_i = (x_i, v_i)^T$, $C = \begin{pmatrix} 0 & 1 \\ 0 & 0 \end{pmatrix}$, and $D = \begin{pmatrix} 0 & 0 \\ 1 & 0 \end{pmatrix}$. Then, network (11.2) can be rewritten as

$$\dot{\eta}_i(t) = C\eta_i(t) - \alpha \sum_{j=1}^{N} L_{ij} D\eta_j(t) + \beta \sum_{j=1}^{N} L_{ij} D\eta_j(t-\tau),$$
$$i = 1, 2, \dots, N. \tag{11.3}$$

Note that a solution of an isolated node satisfies

$$\dot{s}(t) = Cs(t), \tag{11.4}$$

where $s(t) = (s_1, s_2)^T$ is the state vector. Let $\eta = (\eta_1^T, \dots, \eta_N^T)^T$ and rewrite system (11.3) into a matrix form:

$$\dot{\eta}(t) = [(I_N \otimes C) - \alpha(L \otimes D)]\eta(t) + \beta(L \otimes D)\eta(t-\tau), \tag{11.5}$$

where $\otimes$ is the Kronecker product [53]. Let Λ be the diagonal form associated with matrix L, i.e., there exists an unitary matrix P such that $P^T L P = \Lambda$, where $\Lambda = \mathrm{diag}(\lambda_1, \lambda_2, \dots, \lambda_N)$. Then,

$$\begin{aligned}
&(P^T \otimes I_2)\dot{\eta}(t) \\
&= [(P^T \otimes I_2)(I_N \otimes C) - \alpha(\Lambda \otimes D)(P^T \otimes I_2)]\eta(t) \\
&\quad + \beta(\Lambda \otimes D)(P^T \otimes I_2)\eta(t-\tau) \\
&= [(P^T \otimes C) - \alpha(\Lambda \otimes D)(P^T \otimes I_2)]\eta(t) + \beta(\Lambda \otimes D)(P^T \otimes I_2)\eta(t-\tau) \\
&= [(I_N \otimes C) - \alpha(\Lambda \otimes D)](P^T \otimes I_2)\eta(t) + \beta(\Lambda \otimes D)(P^T \otimes I_2)\eta(t-\tau).
\end{aligned}$$

Let $\xi(t) = (P^T \otimes I_2)\eta(t) = (\xi_1^T, \dots, \xi_N^T)^T$, $\xi_i = (\xi_{i1}, \xi_{i2})^T$, $x = (x_1, \dots, x_N)^T$, $v = (v_1, \dots, v_N)^T$, $\widetilde{\xi}_1 = (\xi_{11}, \dots, \xi_{N1})^T$, and $\widetilde{\xi}_2 = (\xi_{12}, \dots, \xi_{N2})^T$. Then, the above multi-agent system can be transformed to

$$\dot{\xi}(t) = [(I_N \otimes C) - \alpha(\Lambda \otimes D)]\xi(t) + \beta(\Lambda \otimes D)\xi(t-\tau), \tag{11.6}$$

or

$$\dot{\xi}_i(t) = (C - \alpha\lambda_i D)\xi_i(t) + \beta\lambda_i D\xi_i(t-\tau), i = 1, \dots, N. \tag{11.7}$$

Theorem 11.2 *Suppose that the network $\mathcal{G}$ is connected. Quasi-consensus in the multi-agent system (11.1) can be reached if and only if, in (11.6) or (11.7),*

$$\lim_{t\to\infty} \| \xi_{i2} \| = 0, i = 2, \dots, N. \tag{11.8}$$

Proof. (Sufficiency). Since the network is connected, $p_1 = 1_N/\sqrt{N}$ is the unit eigenvector of the Laplacian matrix L associated with the simple zero eigenvalue $\lambda_1 = 0$, where $LP = P\Lambda$ and $P = (p_1, \ldots, p_N)$. Since $\lim_{t\to\infty} \parallel \xi_{i2} \parallel = 0$ for $i = 2, \ldots, N$ and $v(t) = P\widetilde{\xi}_2(t) = \sum_{j=1}^{N} p_i\xi_{i2} \to p_1\xi_{i2}$, one has

$$\lim_{t\to\infty} \left\| v(t) - \frac{1}{\sqrt{N}}(\xi_{12}(t)^T, \ldots, \xi_{N2}(t)^T)^T \right\| = 0,$$

where $\dot{\xi}_1(t) = C\xi_1(t)$.

(Necessity). If quasi-consensus in the multi-agent system (11.1) can be reached, then there exists a value $v^*(t) \in R$ such that $\lim_{t\to\infty} \parallel v(t) - 1_N \otimes v^*(t) \parallel = 0$. Since $0_N = P^T L 1_N = \Lambda P^T 1_N = (\lambda_1 p_1^T 1_N, \ldots, \lambda_N p_N^T 1_N)^T$, we have $p_i^T 1_N = 0$ for $i = 2, \ldots, N$. Therefore, $\parallel \xi_{i2}(t) \parallel = \parallel p_i^T v(t) \parallel \to \parallel (p_i^T 1_N) \otimes v^*(t) \parallel \to 0$, as $t \to \infty$ for all $i = 2, \ldots, N$. □

Corollary 11.3 *Suppose that the network $\mathcal{G}$ is connected. Quasi-consensus in the multi-agent system (11.1) can be reached if and only if each of the following $N - 1$ equations,*

$$\lambda^2 + \lambda_i(\alpha - \beta e^{-\lambda\tau}) = 0, i = 2, \ldots, N, \tag{11.9}$$

has a simple zero root and the real parts of all the other roots are negative.

Proof. It suffices to prove that $\lim_{t\to\infty} \parallel \xi_{i2} \parallel = 0, i = 2, \ldots, N$, if and only if each of the equations in (11.9) has a simple zero root and the real parts of all the other roots are negative.

The characteristic equation of the multi-agent system (11.7) is

$$\begin{aligned} &\det(\lambda I_2 - C + \alpha\lambda_i D - \beta\lambda_i e^{-\lambda\tau} D) \\ &= \det \begin{pmatrix} \lambda I_n & -I_n \\ (\alpha - \beta e^{-\lambda\tau})\lambda_i I_n & \lambda I_n \end{pmatrix} \\ &= \lambda^2 + (\alpha - \beta e^{-\lambda\tau})\lambda_i, i = 2, \ldots, N. \end{aligned} \tag{11.10}$$

(Sufficiency). If each of the equations in (11.9) has a simple zero root and the real parts of all the other roots are negative, then the states in (11.7) converge to some constants. Suppose that $\lim_{t\to\infty} \parallel \xi_{i2}(t) \parallel = \widetilde{c}_i \neq 0$. Then, it follows that $\lim_{t\to\infty} \parallel \xi_{i1}(t) \parallel = \infty$, which is a contradiction.

(Necessity). From $\lim_{t\to\infty} \parallel \xi_{i2}(t) \parallel = 0$ and $\dot{\xi}_{i1}(t) = \xi_{i2}(t)$, one has $\lim_{t\to\infty} \xi_{i1}(t) = c_i, i = 2, \ldots, N$, where c_i are constants. If each of the equations in (11.9) has at least one nonzero root with nonnegative real part, then $\xi_{i1}(t)$ or $\xi_{i2}(t)$ cannot converge; or if one of the equations in (11.9) has more than one zero root, then $\alpha = \beta = 0$ or $\alpha = \beta$ and $\tau = 0$. In both cases, $\xi_{i1}(t)$ or $\xi_{i2}(t)$ cannot converge. □

Corollary 11.4 *Suppose that the network $\mathcal{G}$ is connected. Consensus in the multi-agent system (11.1) can be reached if and only if, in (11.6) or (11.7),*

$$\lim_{t\to\infty} \| \xi_i \| = 0, i = 2, \dots, N,$$

or equivalently, if and only if the real parts of all the roots in (11.9) are negative.

Proof. The result can be proved through examining the state $x(t)$ following the same process as in the proofs of Theorem 11.2 and Corollary 11.3. □

Some necessary and sufficient conditions for reaching consensus or quasi-consensus in the multi-agent system (11.1) have been obtained in Corollaries 11.3 and 11.4 above. Next, it will be shown that consensus and quasi-consensus in multi-agent system (11.1) cannot be achieved when $\tau = 0$; however, they can be reached by appropriately choosing the time delay $\tau > 0$ and the coupling strengths α and β.

Lemma 11.5 *Suppose that the network $\mathcal{G}$ is connected. Consensus and quasi-consensus in the multi-agent systems (11.1) cannot be reached when $\tau = 0$. However, for a sufficiently small $\tau > 0$ and given fixed control gains α and β, consensus (resp. quasi-consensus) can be reached if and only if $\alpha > \beta > 0$ (resp. $\alpha = \beta > 0$).*

Proof. From (11.9), $\lambda = \pm\sqrt{\lambda_i(\beta - \alpha)}$ when $\tau = 0$. If $\alpha = \beta$, each of the equations (11.9) has two zero roots; if $\alpha \neq \beta$, there exists at least one nonzero root with nonnegative real part. Therefore, consensus and quasi-consensus in the multi-agent systems (11.1) cannot be reached if $\tau = 0$.

From (11.9), $\lambda^2 = \lambda_i(\beta e^{-\lambda\tau} - \alpha)$, $i = 2, \dots, N$. If $\mathcal{R}(\lambda) \geq 0$, then $|e^{-\lambda\tau}| \leq 1$. Thus, it follows that $|\lambda|^2 \leq \lambda_i(|\beta| + \alpha)$, which indicates that $|\lambda|$ is bounded. If $\mathcal{R}(\lambda) < 0$, the orders of λ^2 and $e^{-\lambda\tau}$ with regard to λ are different when $\tau > 0$, and thus $|\lambda|$ is bounded. For a sufficiently small τ,

$$\begin{aligned}\lambda^2 &= \lambda_i \left[\beta(1 - \lambda\tau) + \beta\frac{(\lambda\tau)^2}{2} - \alpha + o(\tau^2)\right] \\ &= \lambda_i \left[\frac{\beta\tau^2}{2}\lambda^2 - \beta\tau\lambda + (\beta - \alpha) + o(\tau^2)\right].\end{aligned}$$

It follows that

$$\left(1 - \frac{\lambda_i\beta\tau^2}{2}\right)\lambda^2 + \lambda_i\beta\tau\lambda + \lambda_i(\alpha - \beta) + o(\tau^2) = 0. \tag{11.11}$$

By Lemma 2.7, $\lambda_i > 0$, $i = 2, \dots, N$. From (11.9), it is easy to verify that zero is a simple root if and only if $\alpha = \beta \neq 0$ since $\lambda^2 + \lambda_i\alpha(1 - e^{-\lambda\tau}) = 0$ when $\lambda = 0$. If $\alpha = \beta > 0$, then $\lambda \approx 0$ or $\lambda \approx -\frac{\lambda_i\beta\tau}{1-\frac{\lambda_i\beta\tau^2}{2}} < 0$. Therefore, quasi-consensus can be reached for a sufficiently small $\tau > 0$ if and only if $\alpha = \beta > 0$.

From (11.11), one obtains

$$\lambda \approx \frac{-\lambda_i \beta\tau \pm \sqrt{(\lambda_i\beta\tau)^2 - 4(1-\frac{\lambda_i\beta\tau^2}{2})\lambda_i(\alpha-\beta)}}{2\left(1-\frac{\lambda_i\beta\tau^2}{2}\right)}.$$

The real parts of all the roots in (11.9) are negative if and only if $\alpha > \beta > 0$, where $(\lambda_i\beta\tau)^2 - 4(1-\frac{\lambda_i\beta\tau^2}{2})\lambda_i(\alpha-\beta) < 0$ for a sufficiently small $\tau > 0$. □

Remark 11.6 It is easy to verify from Lemma 11.5 that consensus (resp. quasi-consensus) in the multi-agent systems (11.1) cannot be reached without delay, i.e., $\tau = 0$, but interestingly they can be reached even for a sufficiently small $\tau > 0$ by choosing some appropriate coupling strengths $\alpha > \beta > 0$ (resp. $\alpha = \beta > 0$). It is well known that the time delay may result in oscillatory behaviors or network instability (periodic oscillation and chaos) [140]. However, as shown by Lemma 11.5, time delay here can induce consensus in the multi-agent system (11.1). Moreover, in order to reach consensus in the multi-agent system (11.1), the coupling strength of the current states should be larger than that of the outdated states, i.e., $\alpha > \beta$, at all the nodes of the network.

Lemma 11.7 *Suppose that the network $\mathcal{G}$ is connected. Each of the equations in (11.9) has a purely imaginary root if and only if*

$$\tau = \frac{k\pi}{\sqrt{\lambda_i(\alpha \pm \beta)}}, \text{ when } \alpha > \beta > 0; k = 1, 2, \ldots;\ i = 2, \ldots, N, \qquad (11.12)$$

or if and only if

$$\tau = \frac{k\pi}{\sqrt{\lambda_i(\alpha + \beta)}}, \text{ when } \alpha = \beta > 0; k = 1, 2, \ldots;\ i = 2, \ldots, N. \qquad (11.13)$$

Proof. Let $\lambda = \mathbf{i}\omega$. Without loss of generality, suppose that $\omega > 0$. From (11.9),

$$\omega^2 = \lambda_i(\alpha - \beta e^{-\mathbf{i}\omega\tau}). \qquad (11.14)$$

Separating the real and imaginary parts of (11.14) yields

$$\omega\tau = k\pi, k = 1, 2, \ldots$$

It follows that $\cos(\omega\tau) = \pm 1$. If $\alpha > \beta > 0$, then $\omega^2 = \lambda_i(\alpha \pm \beta)$. Since $\omega > 0$, one has $\omega^2 = \lambda_i(\alpha + \beta)$ when $\alpha = \beta > 0$. □

Lemma 11.8 *[79] Consider the exponential polynomial*

$$\begin{aligned} &P(\lambda, e^{-\lambda\tau_1}, \cdots, e^{-\lambda\tau_m}) \\ &= \lambda^n + p_1^{(0)}\lambda^{n-1} + \cdots + p_{n-1}^{(0)}\lambda + p_n^{(0)} + [p_1^{(1)}\lambda^{n-1} + \cdots + p_{n-1}^{(1)}\lambda + p_n^{(1)}]e^{-\lambda\tau_1} \\ &\quad + \cdots + [p_1^{(m)}\lambda^{n-1} + \cdots + p_{n-1}^{(m)}\lambda + p_n^{(m)}]e^{-\lambda\tau_m}, \end{aligned}$$

where $\tau_i \geq 0$ $(i = 1, 2, \cdots, m)$ *and* $p_j^{(i)}$ $(i = 0, 1, \cdots, m; j = 1, 2, \cdots, n)$ *are constants. As* $(\tau_1, \tau_2, \cdots, \tau_m)$ *are varied, the sum of the orders of the zeros of* $P(\lambda, e^{-\lambda\tau_1}, \cdots, e^{-\lambda\tau_m})$ *on the open right-half plane can change only if a zero appears on or across the imaginary axis.*

Lemma 11.9 *Suppose that the network* $\mathscr{G}$ *is connected. Let* λ *be a solution in equation (11.9). Then,*

$$\mathscr{R}\left(\frac{d\lambda}{d\tau}\right)\bigg|_{\tau=\frac{k\pi}{\sqrt{\lambda_i(\alpha+\beta)}}} > 0, \alpha \geq \beta > 0; k = 1, 2, \ldots; i = 2, \ldots, N. \tag{11.15}$$

Proof. Let $f_i(\lambda, \tau) = \lambda^2 + \alpha\lambda_i - \beta\lambda_i e^{-\lambda\tau}$, $\tau_{ki} = \frac{k\pi}{\sqrt{\lambda_i(\alpha+\beta)}}$, and $\omega_i = \sqrt{\lambda_i(\alpha+\beta)}$. Then, from Lemma 11.8, $f_i(\mathbf{i}\omega_i, \tau_{ki}) = 0$. Since $f_i(\lambda, \tau)$ is continuous around the point $(\mathbf{i}\omega_i, \tau_{ki})$, $\frac{\partial f_i}{\partial \lambda}$ and $\frac{\partial f_i}{\partial \tau}$ are continuous, and $\frac{\partial f_i}{\partial \lambda}|_{(\mathbf{i}\omega_i, \tau_{ki})} \neq 0$, and λ is differentiable with respect to τ around the point $(\mathbf{i}\omega_i, \tau_{ki})$ according to the implicit function theorem.

Taking the derivative of λ with respect to τ in $f_i(\lambda, \tau) = 0$, one obtains

$$2\lambda\frac{d\lambda}{d\tau} + \beta\lambda_i\left(\lambda + \tau\frac{d\lambda}{d\tau}\right)e^{-\lambda\tau} = 0. \tag{11.16}$$

It follows that

$$\begin{aligned}
&\frac{d\lambda}{d\tau} \\
&= -\frac{\beta\lambda_i\lambda e^{-\lambda\tau}}{2\lambda + \beta\lambda_i\tau e^{-\lambda\tau}} \\
&= -\beta\lambda_i\frac{\lambda}{2\lambda e^{\lambda\tau} + \beta\lambda_i\tau} \\
&= -\frac{\beta\lambda_i(\mathscr{R}(\lambda) + \mathbf{i}\mathscr{I}(\lambda))}{\theta(\lambda, \tau)},
\end{aligned}$$

where $\theta(\lambda, \tau) = 2e^{\mathscr{R}(\lambda)\tau}\{[\mathscr{R}(\lambda)\cos(\mathscr{I}(\lambda)\tau) - \mathscr{I}(\lambda)\sin(\mathscr{I}(\lambda)\tau)] + \mathbf{i}[\mathscr{R}(\lambda)\sin(\mathscr{I}(\lambda)\tau) + \mathscr{I}(\lambda)\cos(\mathscr{I}(\lambda)\tau)]\} + \beta\lambda_i\tau$.

Let

$$\begin{aligned}
q(\lambda, \tau) = &(2e^{\mathscr{R}(\lambda)\tau}[\mathscr{R}(\lambda)\cos(\mathscr{I}(\lambda)\tau) - \mathscr{I}(\lambda)\sin(\mathscr{I}(\lambda)\tau)] + \beta\lambda_i\tau)^2 \\
&+(2e^{\mathscr{R}(\lambda)\tau}[\mathscr{R}(\lambda)\sin(\mathscr{I}(\lambda)\tau) + \mathscr{I}(\lambda)\cos(\mathscr{I}(\lambda)\tau)])^2 \geq 0.
\end{aligned}$$

By simple calculations,

$$\begin{aligned}
&-\frac{q(\lambda, \tau)}{\beta\lambda_i}\mathscr{R}\left(\frac{d\lambda}{d\tau}\right) \\
&= \mathscr{R}(\lambda)(2e^{\mathscr{R}(\lambda)\tau}[\mathscr{R}(\lambda)\cos(\mathscr{I}(\lambda)\tau) - \mathscr{I}(\lambda)\sin(\mathscr{I}(\lambda)\tau)] + \beta\lambda_i\tau)
\end{aligned}$$

$$+\mathcal{I}(\lambda)(2e^{\mathcal{R}(\lambda)\tau}[\mathcal{R}(\lambda)\sin(\mathcal{I}(\lambda)\tau)+\mathcal{I}(\lambda)\cos(\mathcal{I}(\lambda)\tau)])$$
$$=\mathcal{R}(\lambda)\beta\lambda_i\tau+2e^{\mathcal{R}(\lambda)\tau}|\lambda|^2\cos(\mathcal{I}(\lambda)\tau).$$

If $\tau = \tau_{ik}$, then $\lambda = \mathbf{i}\omega_i$ and $\cos(\mathcal{I}(\lambda)\tau) = -1$. Finally,

$$\mathcal{R}\left(\frac{d\lambda}{d\tau}\right)\bigg|_{\tau=\frac{k\pi}{\sqrt{\lambda_i(\alpha+\beta)}}} = \frac{2\beta\lambda_i^2(\alpha+\beta)}{\beta^2\lambda_i^2\tau^2+4(\alpha+\beta)} > 0.$$

□

Theorem 11.10 *Suppose that the network $\mathcal{G}$ is connected.*

1. *Consensus can be reached in the multi-agent system (11.1) if and only if*

$$\frac{2k\pi}{\sqrt{\lambda_i(\alpha-\beta)}} < \tau < \frac{(2k+1)\pi}{\sqrt{\lambda_i(\alpha+\beta)}}, \alpha > \beta > 0, \tag{11.17}$$

   for all $i = 2, \dots, N$, where $k = 0, 1, \dots$.

2. *Quasi-consensus can be reached in the multi-agent system (11.1) if and only if*

$$0 < \tau < \frac{\pi}{\sqrt{\lambda_N(\alpha+\beta)}}, \alpha = \beta > 0. \tag{11.18}$$

Proof.

1. The proof can be completed by the Nyquist criterion in the frequency domain. Eq. (11.9) can be written as [2]

$$1 - G(s) = 1 - \frac{\lambda_i\beta e^{-s\tau}}{s^2+\lambda_i\alpha} = 0, i = 2, \dots, N. \tag{11.19}$$

   Note that if $\tau = 0$, the Nyquist plot always encircles the point $(-1, 0)$. Therefore, a necessary condition for the stability of (11.20) is that $\alpha > \beta > 0$. Otherwise, there is at least one clockwise encirclement about $(-1, 0)$ in the Nyquist plot.

   One only needs to consider the number of encirclements of $(-1, 0)$ by the Nyquist plot:

$$-G(\mathbf{i}\omega) = \frac{-\lambda_i\beta e^{-\mathbf{i}\omega\tau}}{\lambda_i\alpha-\omega^2} \; i = 2, \dots, N. \tag{11.20}$$

   Consider all the intersections of the polar plot with the negative real axis. Then, there are no encirclements about $(-1, 0)$ of the Nyquist plot if and only if [2]

$$\frac{2k\pi}{\sqrt{\lambda_i(\alpha-\beta)}} < \tau < \frac{(2k+1)\pi}{\sqrt{\lambda_i(\alpha+\beta)}}, i = 2, \dots, N, \tag{11.21}$$

   where $k = 0, 1, \dots$.

2. Since the network $\mathcal{G}$ is connected and from Lemma 11.5, quasi-consensus can be reached for a sufficiently small $\tau > 0$ if and only if $\alpha = \beta > 0$. Consider τ as a parameter varying from 0 to $+\infty$. By Lemma 11.7, a purely imaginary root first emerges when $\tau = \frac{\pi}{\sqrt{\lambda_N(\alpha+\beta)}}$. In view of Lemmas 11.8 and 11.9, Eq. (11.9) has a simple zero root and the real parts of all the other roots are negative if $0 < \tau < \frac{\pi}{\sqrt{\lambda_N(\alpha+\beta)}}$ and $\alpha = \beta > 0$, and there is at least one nonzero root with non-negative real part if $\tau \geq \frac{\pi}{\sqrt{\lambda_N(\alpha+\beta)}}$. Therefore, quasi-consensus can be reached in the multi-agent system (11.1) if and only if $0 < \tau < \frac{\pi}{\sqrt{\lambda_N(\alpha+\beta)}}$ and $\alpha = \beta > 0$. □

For a fixed network topology, consensus can be reached in the multi-agent system (11.1) if and only if τ is bounded by some critical values via choosing appropriate coupling strengths $\alpha > \beta > 0$. On the other hand, when the time delay τ is fixed, an interesting problem is how to design the coupling strengths such that consensus can be reached. This issue is addressed by the following result.

Corollary 11.11 *Suppose that the network $\mathcal{G}$ is connected. Consensus (quasi-consensus) can be reached in the multi-agent system (11.1) if*

$$\alpha + \beta < \frac{\pi^2}{\lambda_N \tau^2}, \alpha > \beta > 0 \ (\alpha = \beta > 0). \tag{11.22}$$

Remark 11.12 In the multi-agent system (11.1), the velocity states of the agents are updated based on the current and delayed position states of their neighboring agents. If the control gain $\alpha + \beta$ is very large, then from the conditions in Theorem 11.10 and Corollary 11.11, the allowable time delay should be very small such that the delayed information can follow the states of neighboring agents in real time. However, if the time delay is large, this delayed position information may be outdated and therefore unable to reflect the real-time states of neighboring agents. Thus, the larger the coupling strength $\alpha + \beta$ is, the smaller the time delay τ should be.

11.3 Motivation for Quasi-Consensus Analysis

In the above section, quasi-consensus in multi-agent system (11.1) was introduced. In order to motivate the idea for defining the new concept of quasi-consensus in multi-agent systems with second-order dynamics, it is interesting to see that the studied model is the exact first-order multi-agent system with the control input involving the distributed delay. Actually, in the multi-agent system (11.1), each agent needs some memory to store the outdated information of its neighboring agents.

Next, a typical multi-agent system with memory of distributed delay is considered:

$$\dot{y}_i(t) = -\gamma \int_{t-\tau}^{t} \sum_{j=1}^{N} L_{ij} y_j(z) dz, i = 1, 2, \cdots, N, \tag{11.23}$$

where $y_i \in R^n$ is the state of agent i, L_{ij} is defined as above, and $\gamma > 0$ is the coupling strength. If the initial condition for (11.23) is well defined such that $\dot{y}_i(t)$ is differentiable, then

$$\dot{y}_i(t) = u_i(t)$$
$$\dot{u}_i(t) = -\gamma \sum_{j=1}^{N} L_{ij} y_j(t) + \gamma \sum_{j=1}^{N} L_{ij} y_j(t-\tau), i = 1, 2, \cdots, N, \tag{11.24}$$

which is exactly the system (11.1) or (11.2) with $\alpha = \beta = \gamma$.

Corollary 11.13 *Suppose that the network $\mathcal{G}$ is connected. Quasi-consensus can be reached in the multi-agent system (11.23) if and only if*

$$0 < \tau < \frac{\pi}{\sqrt{2\gamma\lambda_N}}. \tag{11.25}$$

Proof. Choose $y_i = x_i$ and $\gamma = \alpha = \beta$ in (11.1). Then, the result in (11.24) can be easily obtained by Theorem 11.10. □

Remark 11.14 In the multi-agent system (11.23), only quasi-consensus can be reached if the time delay τ is less than a critical value $\frac{\pi}{\sqrt{2\gamma\lambda_N}}$, and it should be noted that consensus in (11.23) cannot be reached for any time delay $\tau > 0$ and any coupling strength γ. To satisfy the condition $\alpha > \beta$ for reaching consensus as in Theorem 11.10, a modified system of (11.23) can be considered:

$$\dot{y}_i(t) = -\gamma \int_{t-\tau}^{t} \sum_{j=1}^{N} L_{ij} g(z,t) y_j(z) dz, i = 1, 2, \cdots, N, \tag{11.26}$$

where $g(\cdot)$ is a weighting function. For example, one can choose $g(s,t) = \frac{1}{1-e^{-\tau}} e^{-(t-z)}$ satisfying $\int_{t-\tau}^{t} g(z,t) = 1$ [141].

11.4 Simulation Examples

In this section, some simulation examples are given to verify the theoretical analysis.

Consensus and quasi-consensus in a scale-free complex network

A scale-free network is generated in simulation, where the number of initial nodes is five, and at each time step a new node is introduced and connected to five existing nodes in the network with degree preferential attachment, until the total number of nodes $N = 100$ [8]. By computation, $\lambda_N = 32.5389$. Let $\alpha = 1$ and $\beta = 0.5$. From Theorem 11.10, consensus can be reached in the multi-agent system (11.1) if and only if $\tau \in (0, \frac{\pi}{\sqrt{\lambda_N(\alpha+\beta)}}) = (0, 0.4497)$. The position and velocity states of all the

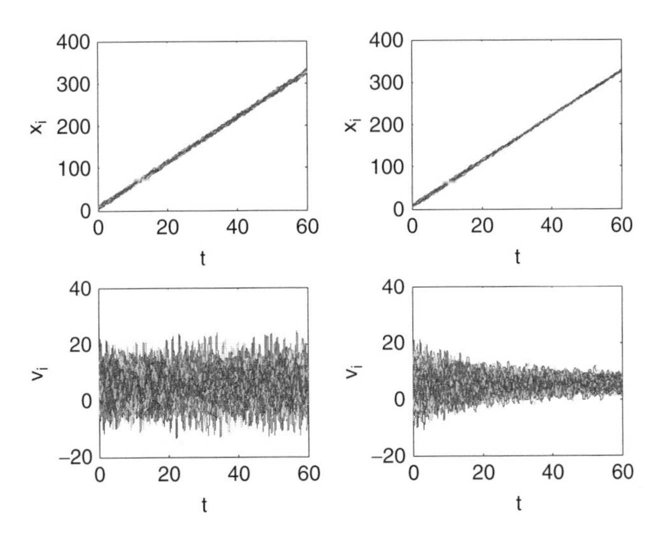

Figure 11.1 Position and velocity states of agents in a multi-agent dynamical system with a scale-free network topology, where $\tau = 0$ (a: left) and $\tau = 0.01$ (b: right)

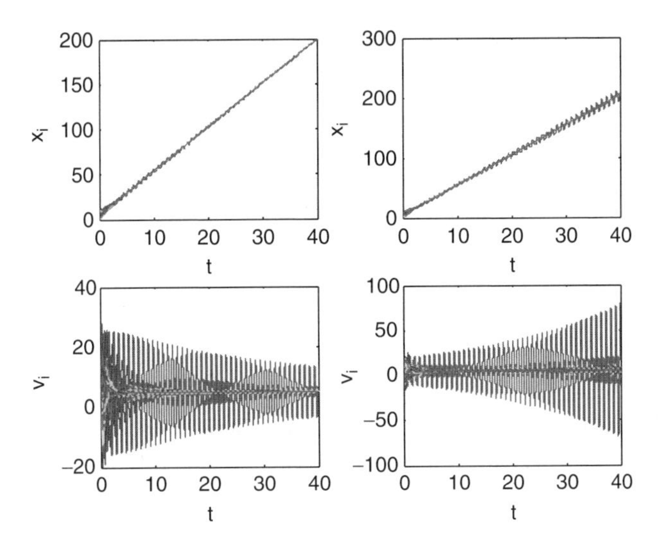

Figure 11.2 Position and velocity states of agents in a multi-agent dynamical system with a scale-free network topology, where $\tau = 0.444$ (a: left) and $\tau = 0.454$ (b: right)

agents are shown in Figs. 11.1 and 11.2, where consensus cannot be achieved when $\tau = 0$ (Fig. 11.1(a)) and $\tau = 0.454$ (Fig. 11.2(b)) but it can be reached if $\tau = 0.01$ (Fig. 11.1(b)) and $\tau = 0.444$ (Fig. 11.2(a)). It is easy to see that the numerical simulations clearly confirm the theoretical analysis.

Actually, the real parts of all the roots in Eq. (11.9) are negative for $\tau \in (0, \frac{\pi}{\sqrt{\lambda_N(\alpha+\beta)}}) = (0, 0.4497)$ in this example. It is quite easy to see that the convergence

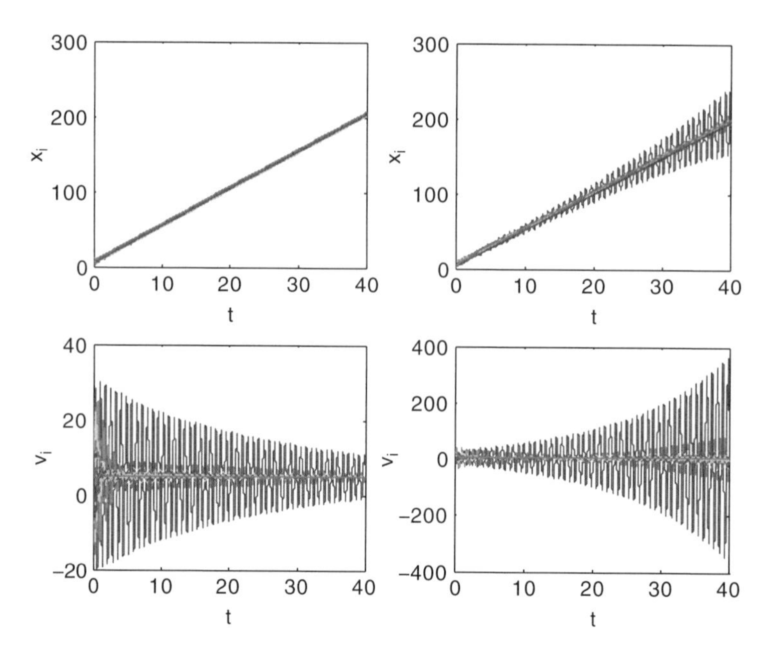

Figure 11.3 Position and velocity states of agents in a multi-agent dynamical system with a scale-free network topology, where $\tau = 0.384$ (a: left) and $\tau = 0.394$ (b: right)

rate can be facilitated by choosing an appropriate time delay τ. By simple calculation, the time delay τ_m for reaching a fast convergence rate satisfies $\tau_m = \max_\tau \min_i(\mathcal{R}(\lambda))$, where $\lambda^2 + \lambda_i(\alpha - \beta e^{-\lambda\tau}) = 0$.

Consider the multi-agent system with the same network structure as above. Let $\alpha = 1$ and $\beta = 1$. From Theorem 11.10, quasi-consensus can be reached in the multi-agent system (11.1) if and only if $\tau \in (0, \frac{\pi}{\sqrt{\lambda_N(\alpha+\beta)}}) = (0, 0.3894)$. The position and velocity states of all the agents are shown in Fig. 11.3, where consensus cannot be achieved when $\tau = 0.394$ (Fig. 11.3(b)) but it can be reached if $\tau = 0.384$ (Fig. 11.3(a)).

Consensus and quasi-consensus in a random network

A random network is also performed in the simulation, where each pair of nodes is connected with the probability $p = 0.1$ and the total number of nodes $N = 100$. By simple calculation, $\lambda_N = 26.1441$. Let $\alpha = 1$ and $\beta = 0.5$. From Theorem 11.10, consensus can be reached in multi-agent system (11.1) if and only if $\tau \in (0, \frac{\pi}{\sqrt{\lambda_N(\alpha+\beta)}}) = (0, 0.5017)$. The position and velocity states of all the agents are shown in Figs. 11.4. From Lemma 11.5, consensus cannot be achieved when $\tau = 0$ (Fig. 11.4(a)) while it can be reached if $\tau = 0.01$ (Fig. 11.4(b)), which indicates that a small time delay can induce consensus in multi-agent system (11.1).

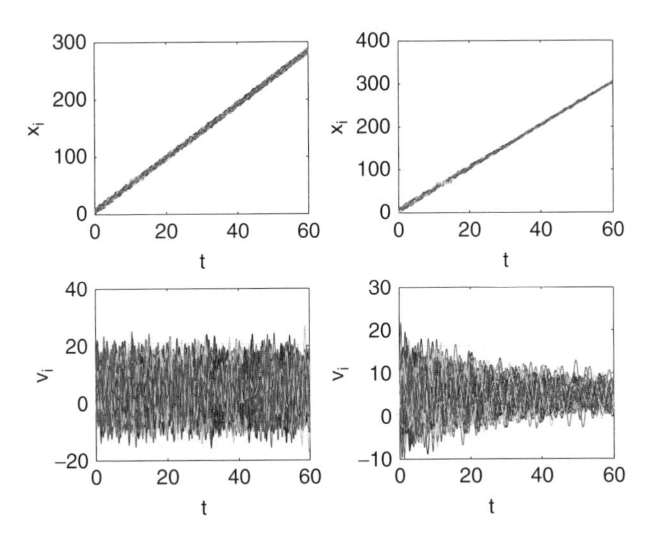

Figure 11.4 Position and velocity states of agents in a multi-agent dynamical system with a random network topology, where $\tau = 0$ (a: left) and $\tau = 0.01$ (b: right)

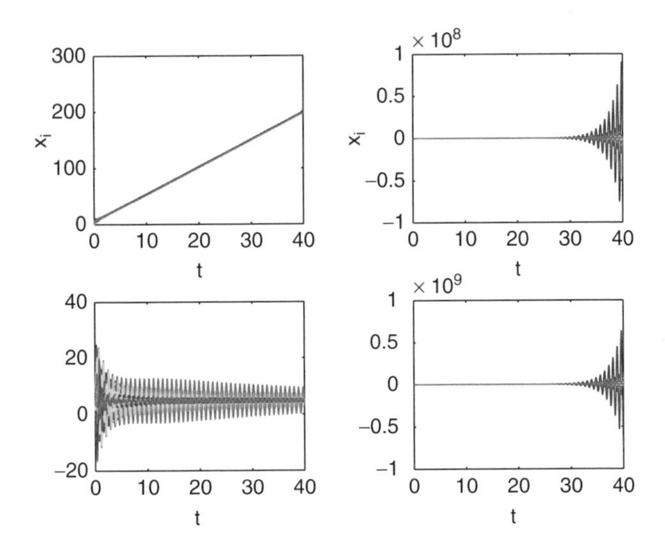

Figure 11.5 Position and velocity states of agents in a multi-agent dynamical system with a random network topology, where $\tau = 0.43$ (a: left) and $\tau = 0.44$ (b: right)

In order to verify the quasi-consensus in multi-agent system (11.1), the same parameters $\alpha = 1$ and $\beta = 1$ are used. From Theorem 11.10, quasi-consensus can be reached in multi-agent system (11.1) if and only if $\tau \in (0, \frac{\pi}{\sqrt{\lambda_N(\alpha+\beta)}}) = (0, 0.4345)$. The position and velocity states of all the agents are shown in Fig. 11.5, where consensus cannot be achieved when $\tau = 0.43$ (Fig. 11.5(b)) but it can be reached if $\tau = 0.44$ (Fig. 11.5(a)).

11.5 Notes

In this chapter, a linear consensus protocol with second-order dynamics has been designed based on both current and delayed position information of agents. Time delay, usually a destructive character in dynamics, can induce periodic oscillations and even chaos in dynamical systems. However, it has been found here that consensus and quasi-consensus in a multi-agent system cannot be reached without time delay under the given protocol while they can be achieved with a relatively small time delay by appropriately choosing the network coupling strengths. A necessary and sufficient condition for reaching consensus has been derived, which shows that consensus and quasi-consensus can be achieved in a multi-agent system if and only if the time delay is bounded by some critical values depending on the coupling strengths and the largest eigenvalue of the Laplacian matrix in the network. The designed consensus protocol with both current and delayed position information is very useful, especially when the velocity information of the neighboring agents is unavailable. The allowable maximum communication delay for reaching consensus has been theoretically analyzed, which is helpful for the design and implementation of collective behaviors in multi-agent systems.

There are still many related interesting problems deserving further investigations. For example, it is of interest to study multi-agent systems with nonuniform time delays and general directed topologies, the critical time delays for reaching the fastest convergence, and more general protocols with negative weights in (11.26), and so on.

12

Conclusions and Future Work

12.1 Conclusions

This book has studied distributed cooperative control of multi-agent systems in a broad discipline, which includes complex dynamics, hybrid control, adaptive control, distributed filtering, etc. In summary, the book has addressed several important notions as follows.

1. Three types of consensus, i.e., first-order, second-order, and higher-order consensus, have been studied in Chapters 2–4. In Chapter 2, first-order consensus for cooperative agents with nonlinear dynamics in a directed network was discussed. Several sufficient conditions for reaching both local and global consensus were derived by using tools from complex analysis, local consensus manifold approach, Lyapunov methods, algebraic graph theory, and matrix theory. In Chapter 3, some necessary and sufficient conditions concerning the control gains and the spectra of the networks for second-order consensus in linear multi-agent dynamical systems with or without time delays were established. Then, some sufficient conditions were derived for reaching second-order consensus in multi-agent systems with nonlinear dynamics based on algebraic graph theory, matrix theory, and Lyapunov control methods. Finally, consensus regions were characterized to design higher-order distributed consensus protocols for linear multi-agent dynamical systems in Chapter 4.
2. In Chapters 5 and 6, swarming and flocking behaviors in multi-agent systems were investigated. In Chapter 5, some conditions were derived in terms of the system parameters for the stability of a continuous-time swarm model with stochastic noise, nonlinear profiles, changing communication topologies, and unbounded repulsive functions by using nonsmooth analysis. In Chapter 6, a distributed leader-follower flocking algorithm for multi-agent dynamical systems with time-varying velocities was developed using tools from algebraic graph theory, pinning observer-based control, and nonsmooth analysis in combination with

Distributed Cooperative Control of Multi-agent Systems, First Edition.
Wenwu Yu, Guanghui Wen, Guanrong Chen, and Jinde Cao.

techniques of collective potential functions, velocity consensus and navigation feedback.

3. In Chapters 7 and 8, hybrid control of multi-agent systems was discussed. In particular, based on sampled full or partial information, several distributed consensus protocols with second-order dynamics were designed, where agents can communicate with neighboring agents at some time instants instead of all-time communications. Some criteria for choosing sampling periods were established. Furthermore, consensus of second-order multi-agent systems with intermittent communication was discussed, where agents only communicate with their neighboring agents on some disconnected time intervals. Several distributed consensus protocols for multi-agent systems with intermittent communication were designed and analyzed.
4. In Chapter 9, distributed adaptive control for multi-agent systems was studied, which can overcome the disadvantage of almost all published schemes that need to check global information (Laplacian matrix or its eigenvalues) for reaching cooperative control, which is inconsistent with the distributed protocol. First, some distributed adaptive algorithms were designed to tune the coupling weights of a network based on local information of nodes' dynamics for reaching synchronization. Then, the design of distributed control gains for consensus in multi-agent systems with second-order nonlinear dynamics was discussed.
5. In Chapter 10, some applications of collective behaviors in multi-agent systems for sensor networks were studied. A new type of distributed consensus filters was designed to track the leader in a sensor network by using pinning observers and local information of neighboring agents.
6. In Chapter 11, delay-induced consensus and quasi-consensus in multi-agent dynamical systems were discussed. A linear consensus protocol for second-order dynamics was designed, where both the current and delayed position information were utilized. Furthermore, the motivation for studying quasi-consensus was discussed, where the potential relationship between the second-order multi-agent system with delayed positive feedback and the first-order system with distributed-delay control input was revealed.

12.2 Future Work

Some interesting topics for future research are suggested as follows.

1. Pinning control schemes for collective behaviors in complex networks and systems utilizing key agents for better and faster reaching of goals, with good understanding of the role of the pinning-informed agents in the network structures and the network dynamics.

2. Studies in collective behavioral models with hybrid events, such as observers design, distributed adaptive control protocols, and associated issues about time delays, sampled data, communication noise, and intermittent measurements.
3. Applications to practical systems with distributed cooperative control, for example, robotic systems, UAVs, power systems, smart grids, sensor networks, brain neural networks, biological genetic regulatory networks, social networks, etc.

Bibliography

[1] A. Abdessameud, I.G. Polushin, and A. Tayebi, "Motion coordination of thrust-propelled underactuated vehicles with intermittent and delayed communications," *Syst. Control Lett.*, vol. 79, pp. 15–22, 2015.

[2] C. Abdallah, P. Dorato, J. Benites-Read, and R. Byrne, "Delayed positive feedback can stabilize oscillatory systems," *American Control Conference*, pp. 3106–3107, 1993.

[3] M. Arcak, "Passivity as a design tool for group coordination," *IEEE Trans. Auto. Contr.*, vol. 52, no. 8, pp. 1380–1389, 2007.

[4] A. Arenas, A. Díaz-Guilera, J. Kurths, Y. Moreno, and C. Zhou, "Synchronization in complex networks," *Physics Reports*, vol. 468, no. 3, pp. 93–153, 2009.

[5] K.J. Aström and B. Wittenmark, *Adaptive Control*. New York: Addison-Wesley Publishing Company, 1995.

[6] J.P. Aubin and H. Frankowska, *Set Valued Analysis*. Boston: Birkhauser, 1990.

[7] T. Balch and R.C. Arkin, "Behavior-based formation control for multirobot teams," *IEEE Trans. Auto. Contr.*, vol. 14, no. 6, pp. 926–939, 1982.

[8] A.L. Barabási and R. Albert, "Emergence of scaling in random networks," *Science*, vol. 286, no. 5439, pp. 509–512, 1999.

[9] B.R. Barmish, *New Tools for Robustness of Linear Systems*. New York: Macmillan Publishing Company, 1994.

[10] J.A. Benediktsson and P.H. Swain, "Consensus theoretic classification methods," *IEEE Trans. Systems, Man, and Cybernetics*, vol. 22, no. 4, pp. 688–704, 1992.

[11] P. Bliman and G. Ferrari-Trecate, "Average consensus problems in networks of agents with delayed communications," *Automatica*, vol. 44, no. 8, pp. 1985–1995, 2008.

[12] S. Boyd, L.E. Ghaoui, E. Feron, and V. Balakrishnan, *Linear Matrix Inequalities in System and Control Theory*. Philadelphia, PA: SIAM, 1994.

[13] R.A. Brualdi and H.J. Ryser, *Combinatorial Matrix Theory*. Cambridge: Cambridge University Press, 1991.

[14] J. Cao, P. Li, and W. Wang, "Global synchronization in arrays of delayed neural networks with constant and delayed coupling," *Phys. Lett. A*, vol. 353, no. 4, pp. 318–325, 2006.

Distributed Cooperative Control of Multi-agent Systems, First Edition.
Wenwu Yu, Guanghui Wen, Guanrong Chen, and Jinde Cao.

[15] M. Cao, A.S. Morse, and B.D.O. Anderson, "Reaching a consensus in a dynamically changing environment: a graphical approach," *SIAM J. Control and Optimization*, vol. 47, no. 2, pp. 575–600, 2008.
[16] M. Cao, A.S. Morse, and B.D.O. Anderson, "Reaching a consensus in a dynamically changing environment: convergence rates, measurement delays and asynchronous events," *SIAM J. Control and Optimization*, vol. 47, no. 2, pp. 601–623, 2008.
[17] Y. Cao, W. Yu, W. Ren, and G. Chen, "An overview of recent progress in the study of distributed multi-agent coordination," *IEEE Trans. Indus. Infor.*, vol. 9, no. 1, pp. 427–438, 2013.
[18] Y. Cao and W. Ren, "Multi-vehicle coordination for double-integrator dynamics under fixed undirected/directed interaction in a sampled-data setting," *Int. J. Robust and Nonlinear Control*, vol. 20, no. 9, pp. 987–1000, 2010.
[19] Y. Cao, W. Ren, and Y. Chen, "Multi-agent consensus using both current and outdated states," in *Proc. of the 17th World Congress IFAC*, vol. 41, no. 2, pp. 2874–2879, 2008.
[20] G. Chen and F. L. Lewis, "Distributed adaptive tracking control for synchronization of unknown networked Lagrangian systems," *IEEE Trans. Systems, Man, and Cybernetics-Part B*, vol. 41, no. 3, pp. 805–816, 2011.
[21] G. Chen and T. Ueta, "Yet another chaotic attractor," *Int. J. Bifur. Chaos*, vol. 9, no. 7, pp. 1465–1466, 1999.
[22] L. Chen and K. Aihara, "Stability of genetic regulatory networks with time delay," *IEEE Trans. Circuits Syst. I*, vol. 49, no. 5, pp. 602–608, 2002.
[23] T. Chen, X. Liu, and W. Lu, "Pinning complex networks by a single controller," *IEEE Trans. Circuits Syst. I*, vol. 54, no. 6, pp. 1317–1326, 2007.
[24] L.O. Chua, "The genesis of Chua's circuit," *Arch Elektron Ubertragung*, vol. 46, no. 3, pp. 250–257, 1992.
[25] F.H. Clarke, *Optimization and Nonsmooth Analysis*, New York: Wiely, 1983.
[26] J. Cortés, "Distributed algorithms for reaching consensus on general functions," *Automatica*, vol. 44, no. 3, pp. 726–737, 2007.
[27] A. Das and F.L. Lewis, "Distributed adaptive control for networked multi-robot systems," *Multi-Robot Systems, Trends and Development*, pp. 33–52, 2011.
[28] M.H. DeGroot, "Reaching a consensus," *J. Am. Statist. Assoc.*, vol. 69, no. 345, pp. 118–121, 1974.
[29] P. DeLellis, M. diBernardo, and F. Garofalo, "Novel decentralized adaptive strategies for the synchronization of complex networks," *Automatica*, vol. 45, no. 5, pp. 1312–1318, 2009.
[30] P. DeLellis, M. diBernardo, and F. Garofalo, "Synchronization of complex networks through local adaptive coupling," *Chaos*, vol. 18, no. 3, pp. 037110-1–037110-8, 2008.
[31] Z. Duan, G. Chen, and L. Huang, "Disconnected synchronized regions of complex dynamical networks," *IEEE Trans. Auto. Contr.*, vol. 54, no. 4, pp. 845–849, 2009.
[32] Z. Duan, C. Liu, and G. Chen, "Network synchronizability analysis: The theory of subgraphs and complementary graphs," *Physica D*, vol. 237, no. 7, pp. 1006–1012, 2008.
[33] M. Egerstedt and X. Hu, "Formation constrained multi-agent control," *IEEE Trans. Robotics and Automation*, vol. 17, no. 6, pp. 947–951, 2001.
[34] M.B. Elowitz and S. Leibler, "A synthetic oscillatory network of transcriptional regulators," *Nature*, vol. 403, no. 6767, pp. 335–338, 2000.
[35] P. Erdös and A. Rényi, "On random graphs," *Pub. Math.*, vol. 6, pp. 290–297, 1959.

[36] J.A. Fax and R.M. Murray, "Information flow and cooperative control of vehicle formations," *IEEE Trans. Auto. Contr.*, vol. 49, no. 9, pp. 1465–1476, 2004.

[37] M. Fiedler, "Algebraic connectivity of graphs," *Czechoslovak Math. J.*, vol. 23, no. 2, pp. 298–305, 1973.

[38] A.F. Filippov, *Differential Equations with Discontinuous Righthand Sides*. Mathematics and its applications. Boston: Kluwer Academic Publishers, 1988.

[39] M. C. de Oliveira and R. E. Skelton, "Stability tests for constrained linear systems," *Perspectives in Robust Control*. New York: Springer–Verlag, 2001.

[40] V. Flunkert, S. Yanchuk, T. Dahms, and E. Schöll, "Synchronizing distant nodes: a universal classification of networks," *Phys. Rev. Lett.*, vol. 105, no. 25, pp. 254101-1–254101-4, 2010.

[41] Y. Gao, L. Wang, G. Xie, and B. Wu, "Consensus of multi-agent systems based on sampled-data control," *Int. J. Control*, vol. 82, no. 12, pp. 2193–2205, 2009.

[42] Y. Gao and L. Wang, "Sampled-data based consensus of continuous-time multi-agent systems with time-varying topology," *IEEE Trans. Auto. Contr.*, vol. 56, no. 5, pp. 1226–1231, 2011.

[43] Y. Gao and L. Wang, "Consensus of multiple dynamic agents with sampled information," *IET Control Theory and Applications*, vol. 4, no. 6, pp. 945–956, 2010.

[44] V. Gazi and K.M. Passino, "Stability analysis of swarms," *IEEE Trans. Auto. Contr.*, vol. 48, no. 4, pp. 692–697, 2003.

[45] V. Gazi and K.M. Passino, "Stability analysis of social foraging swarms," *IEEE Trans. Systems, Man, and Cybernetics-Part B*, vol. 34, no. 1, pp. 539–557, 2004.

[46] C. Godsil and G. Royle, *Algebraic Graph Theory*. New York: Springer-Verlag, 2001.

[47] R.O. Grigoriev, M.C. Cross, and H.G. Schuster, "Pinning control of spatiotemporal chaos," *Phys. Rev. Lett.*, vol. 79, no. 15, pp. 2795–2798, 1997.

[48] K. Gu, "An integral inequality in the stability problem of time-delay systems," in *Proc. 39th IEEE Conference Decision and Control,* pp. 2805–2810, 2000.

[49] J. Hale and S.V. Lunel, *Introduction to Functional Differential Equations*. New York: Springer, 1993.

[50] Y. Hong, G. Chen, and L. Bushnell, "Distributed observers design for leader-following control of multi-agent networks," *Automatica*, vol. 44, no. 3, pp. 846–850, 2008.

[51] Y. Hong, J. Hu, and L. Gao, "Tracking control for multi-agent consensus with an active leader and variable topology," *Automatica*, vol. 42, no. 7, pp. 1177–1182, 2006.

[52] R.A. Horn and C.R. Johnson, *Matrix Analysis*. Cambridge, U.K.: Cambridge Univ. Press, 1985.

[53] R.A. Horn and C.R. Johnson, *Topics in Matrix Analysis*. Cambridge, U.K.: Cambridge Univ. Press, 1991.

[54] L. Huang, *Linear Algebra in System and Control Theory*. Beijing, China: Science Press, 1984.

[55] S. Huang, "Gene expression profilling, genetic networks, and cellular states: an integrating concept for tumorigenesis and drug discovery," *J. Molecular Medicine*, vol. 77, no. 6, pp. 469–480, 1999.

[56] D. Hunt, G. Korniss, and B.K. Szymanski, "Network synchronization in a noisy environment with time delays: fundamental limits and trade-offs," *Phys. Rev. Lett.*, vol. 105, no. 6, pp. 068701-1–068701-4, 2010.

[57] A. Jadbabaie, J. Lin, and A.S. Morse, "Coordination of groups of mobile autonomous agents using nearest neighbor rules," *IEEE Trans. Auto. Contr.*, vol. 48, no. 6, pp. 985–1001, 2003.

[58] J. Jin, P. Liang, and G. Beni, "Stability of synchronized distributed control of discrete swarm structures," *in Proc. IEEE Int. Conf. Robotics Automation*, San Diego, CA, pp. 1033–1038, May 1994.

[59] L. Kocarev and P. Amato, "Synchronization in power-law networks," *Chaos*, vol. 15, no. 2, 024101, 2005.

[60] C. Li, L. Chen, and K. Aihara, "Stability of genetic networks with SUM regulatory logic: Lur'e System and LMI approach," *IEEE Trans. Circuits Syst. I*, vol. 53, no. 11, pp. 2451–2458, 2006.

[61] C. Li, L. Chen, and K. Aihara, "Stochastic synchronization of genetic oscillator networks," *BMC Syst. Biol.*, vol. 1, article no. 6, 2007.

[62] X. Li, X. Wang, and G. Chen, "Pinning a complex dynamical network to its equilibrium," *IEEE Trans. Circuits Syst. I*, vol. 51, no. 10, pp. 2074–2087, 2004.

[63] W. Lin and W. He, "Complete synchronization of the noise-perturbed Chua's circuits," *Chaos*, vol. 15, no. 2, pp. 023705-1–023705-9, 2005.

[64] Z. Lin, B. Francis, and M. Maggiore, "State agreement for continuous-time coupled nonlinear systems," *SIAM J. Control and Optimization*, vol. 46, no. 1, pp. 288–307, 2007.

[65] C. Liu, Z. Duan, G. Chen, and L. Huang, "Analyzing and controlling the network synchronization regions," *Physica A*, vol. 386, no. 1, pp. 531–542, 2007.

[66] H. Liu, G. Xie, and L. Wang, "Necessary and sufficient conditions for solving consensus problems of double-integrator dynamics via sampled control," *Int. J. Robust and Nonlinear Control*, vol. 20, no. 15, pp. 1706–1722, 2010.

[67] X. Liu and T. Chen, "Boundedness and synchronization of y-coupled lorenz systems with or without controller," *Physica D*, vol. 237, no. 5, pp. 630–639, 2008.

[68] E.N. Lorenz, "Deterministic nonperiodic flow," *J. Atmos. Sci.*, vol. 20, no. 2, pp. 130–141, 1963.

[69] W. Lu and T. Chen, "Dynamical behaviors of Cohen–Grossberg neural networks with discontinuous activation functions," *Neural Netw.*, vol. 18, no. 3, pp. 231–242, 2005.

[70] W. Lu and T. Chen, "New approach to synchronization analysis of linearly coupled ordinary differential systems," *Physica D*, vol. 213, no. 2, pp. 214–230, 2006.

[71] J. Lü and G. Chen, "A new chaotic attractor coined," *Int. J. Bifur. Chaos*, vol. 12, no. 3, pp. 659–661, 2002.

[72] J. Lü and G. Chen, "A time-varying complex dynamical network models and its controlled synchronization criteria," *IEEE Trans. Auto. Contr.*, vol. 50, no. 6, pp. 841–846, 2005.

[73] H. Lütkepohl, *Handbook of Matrices*. New York: Wiley, 1996.

[74] N.A. Lynch, *Distributed Algorithms*. San Francisco, CA: Morgan Kaufmann, 1997.

[75] B. Mohar, "Eigenvalues, diameter, and mean distance in graphs," *Graphs and Combinatorics*, vol. 7, no. 1, pp. 53–64, 1991.

[76] B. Mohar, "The Laplacian spectrum of graphs," *Graph Theory, Combin. Applicat.*, Y.A.G. Chartrand, O.R. Ollerman, and A.J. Schwenk, Eds., vol. 2, pp. 871–898, 1991.

[77] L. Moreau, "Stability of multiagent systems with time-dependent communication links," *IEEE Trans. Auto. Contr.*, vol. 50, no. 2, pp. 169–182, 2005.

[78] I.P. Nathanson, *Theory of Functions of a Real Variable*. New York: Unger Publishing., 1961.

[79] S.I. Niculescu, *Delay Effects on Stability: A Robust Control Approach*. London: Springer–Verlag, 2001.

[80] M.E.J. Newman and D.J. Watts, "Renormalization group analysis of the small-world network model," *Phys. Lett. A*, vol. 263, no. 4, pp. 341–346, 1999.

[81] R. Olfati-Saber and R.M. Murray, "Consensus problems in networks of agents with switching topology and time-delays," *IEEE Trans. Auto. Contr.*, vol. 49, no. 9, pp. 1520–1533, 2004.

[82] R. Olfati-Saber, "Flocking for multi-agent dynamic systems: algorithms and theory," *IEEE Trans. Auto. Contr.*, vol. 51, no. 3, pp. 401–420, 2006.

[83] R. Olfati-Saber, "Distributed cooperative control of multiple vehicle formations using structural potential functions," in *Proc. 15th IFAC World Congress IFAC*, Barcelona, Spain, pp. 495–500, Jul. 2002.

[84] R. Olfati-Saber, J. Fax, and R.M. Murray, "Consensus and cooperation in networked multi-agent systems," *Proc. IEEE*, vol. 95, no. 1, pp. 215–233, 2007.

[85] N. Olgac and R. Sipahi, "An exact method for the stability analysis of time-delayed linear time-invariant (LTI) systems," *IEEE Trans. Auto. Contr.*, vol. 47, no. 5, pp. 793–797, 2002.

[86] A. Olshevsky and J.N. Tsitsiklis, "On the nonexistence of quadratic Lyapunov functions for consensus algorithms," *IEEE Trans. Auto. Contr.*, vol. 53, no. 11, pp. 2642–2645, 2008.

[87] B. Paden and S. Sastry, "A calculus for computing Filippov's differential inclusion with application to the variable structure control of robot manipulators," *IEEE Trans. Circuits Syst.*, vol. 34, no. 1, pp. 73–82, 1987.

[88] P.C. Parks and V. Hahn, *Stability Theory*. Upper Saddle River, NJ: Prentice Hall, 1993.

[89] N. Parekh, S. Parthasarathy, and S. Sinha, "Global and local control of spatiotemporal chaos in coupled map lattices," *Phys. Rev. Lett.*, vol. 81, no. 7, pp. 1401–1404, 1998.

[90] L.M. Pecora and T.L. Carroll, "Master stability functions for synchronized coupled systems," *Phys. Rev. Lett.*, vol. 80, no. 10, pp. 2109–2112, 1998.

[91] Z. Qu, J. Wang, and R.A. Hull, "Cooperative control of dynamical systems with application to autonomous vehicles," *IEEE Trans. Auto. Contr.*, vol. 53, no. 4, pp. 894–911, 2008.

[92] F. Ren and J. Cao, "Asymptotic and robust stability of genetic regulatory networks with time-varying delays," *Neurocomputing*, vol. 71, no. 4, pp. 834–842, 2008.

[93] W. Ren, "Second-order consensus algorithm with extensions to switching topologies and reference models," in *Proc. 2007 American Control Conference*, pp. 1431–1436, Jul. 2007.

[94] W. Ren, "Multi-vehicle consensus with a time-varying reference state," *Syst. Control Lett.*, vol. 56, no. 7, pp. 474–483, 2007.

[95] W. Ren, "On consensus algorithms for double-integrator dynamics," *IEEE Trans. Auto. Contr.*, vol. 58, no. 6, pp. 1503–1509, 2008.

[96] W. Ren, "Synchronization of coupled harmonic oscillators with local interaction," *Automatica*, vol. 44, no. 12, pp. 3195–3200, 2008.

[97] W. Ren and E. Atkins, "Second-order consensus protocols in multiple vehicle systems with local interactions," *AIAA Guidance, Navigation, and Control Conference and Exhibit*, San Francisco, California, pp. 15–18, Aug. 2005.

[98] W. Ren and R.W. Beard, "Consensus seeking in multiagent systems under dynamically changing interaction topologies," *IEEE Trans. Auto. Contr.*, vol. 50, no. 5, pp. 655–661, 2005.

[99] W. Ren and R.W. Beard, *Distributed Consensus in Multi-vehicle Cooperative Control*. London: Springer–Verlag, 2008.

[100] W. Ren, K.L. Moore, and Y. Chen, "High-order and model reference consensus algorithms in cooperative control of multi-vehicle systems," *J. Dyn. Syst-T ASME*, vol. 129, no. 5, pp. 678–688, 2007

[101] C.W. Reynolds, "Flocks, herds, and schools: a distributed behavior model," *Computer Graphics*, vol. 21, no. 4, pp. 25–34, 1987.

[102] W. Rudin, *Principles of Mathematical Analysis*. New York-Auckland-Dusseldorf: McGraw-Hill Book Co., 1976.

[103] A.V. Savkin, "Coordination collective motion of groups of autonomous mobile robots: analysis of Vicsek's model," *IEEE Trans. Auto. Contr.*, vol. 49, no. 6, pp. 981–983, 2004.

[104] L. Scardovi and R. Sepulchre, "Synchronization in networks of identical linear systems," *Automatica*, vol. 45, no. 11, pp. 2557–2562, 2009.

[105] Z. Schuss, *Theory and Applications of Stochastic Differential Equations*. New York: Wiley, 1980.

[106] D. Shevitz and B. Paden, "Lyapunov stability theory of nonsmooth systems," *IEEE Trans. Auto. Contr.*, vol. 39, no. 9, pp. 1910–1914, 1994.

[107] J.J.E. Slotine and W. Li, *Applied Nonlinear Control*. Upper Saddle River, NJ: Prentice Hall, 1991.

[108] H. Su, X. Wang, and Z. Lin, "Flocking of multi-agents with a virtual leader," *IEEE Trans. Auto. Contr.*, vol. 54, no. 2, pp. 293–307, 2009.

[109] Y. Sun and J. Cao, "Adaptive synchronization between two different noise-perturbed chaotic systems with fully unknown parameters," *Physica A*, vol. 376, pp. 253–265, 2007.

[110] H.G. Tanner, A. Jadbabaie, and G.J. Pappas, "Stable flocking of mobile agents, part I: fixed topology," in *Proc. 42nd IEEE Conf. Decision and Control*, pp. 2010–2015, Dec. 2003.

[111] H.G. Tanner, A. Jadbabaie, and G.J. Pappas, "Stable flocking of mobile agents, part II: dynamic topology," in *Proc. 42nd IEEE Conf. Decision and Control*, pp. 2016–2021, Dec. 2003.

[112] H.G. Tanner, A. Jadbabaie, and G.J. Pappas, "Flocking in fixed and switching networks," *IEEE Trans. Auto. Contr.*, vol. 52, no. 5, pp. 863–868, 2007.

[113] F. Taousser, M. Defoort, and M. Djemai, "Consensus for linear multi-agent system with intermittent information transmissions using the time-scale theory," *Int. J. Contr.*, vol. 89, no. 1, pp. 210–220, 2016.

[114] Y. Tian and C. Liu, "Consensus of multi-agent systems with diverse input and communication delays," *IEEE Trans. Auto. Contr.*, vol. 53, no. 9, pp. 2122–2128, 2008.

[115] T. Vicsek, A. Cziok, E.B. Jacob, I. Cohen, and O. Shochet, "Novel type of phase transition in a system of self-driven particles," *Phys. Rev. Lett.*, vol. 75, no. 6, pp. 1226–1229, 1995.

[116] Q. Wang, M. Perc, Z. Duan, and G. Chen, "Delay-induced multiple stochastic resonances on scale-free neuronal networks," *Chaos*, vol. 19, no. 2, pp. 023112-1–023112-7, 2009.
[117] X. Wang and G. Chen, "Synchronization in scale-free dynamical networks: robustness and fragility," *IEEE Trans. Circuits Syst. I*, vol. 49, no. 1, pp. 54–62, 2002.
[118] X. Wang and G. Chen, "Synchronization in small-world dynamical networks," *Int. J. Bifur. Chaos*, vol. 12, no. 1, pp. 187–192, 2002.
[119] X. Wang and G. Chen, "Pinning control of scale-free dynamical networks," *Physica A*, vol. 310, no. 3, pp. 521–531, 2002.
[120] X. Wang and G. Chen, "Complex networks: small-world, scale-free, and beyond," *IEEE Control Syst. Mag.*, vol. 3, no. 1, pp. 6–20, 2003.
[121] X. Wang and X. Wang, "Semi-global consensus of multi-agent systems with intermittent communications and low-gain feedback," *IET Control Theory Appl.*, vol. 9, no. 5, pp. 766–774, 2015.
[122] Z. Wang, F. Yang, D.W.C. Ho, and X. Liu, "Robust finite-horizon filtering for stochastic systems with missing measurements," *IEEE Signal Process. Lett.*, vol. 12, no. 6, pp. 437–440, 2005.
[123] D.J. Watts and S.H. Strogatz, "Collective dynamics of 'small-world' networks," *Nature*, vol. 393, pp. 440–442, 1998.
[124] G. Wen, Z. Duan, G. Chen, and W. Yu, "Second-order consensus for nonlinear multi-agent systems with intermittent measurements," in *Proc. 2011 Chinese Control and Decision Conference*, pp. 3710–3714, May 2011.
[125] G. Wen, Z. Duan, W. Ren, and G. Chen, "Distributed consensus of multi-agent systems with general linear node dynamics and intermittent communications," *Int. J. Robust and Nonlinear Control*, vol. 24, no. 16, pp. 2438–2457, 2014.
[126] G. Wen, Z. Duan, W. Yu, and G. Chen, "Consensus in multi-agent systems with communication constraints," *Int. J. Robust and Nonlinear Control*, vol. 22, no. 2, pp. 170–182, 2012.
[127] G. Wen, Z. Duan, W. Yu, and G. Chen, "Consensus of multi-agent systems with nonlinear dynamics and sampled-data information: A delayed-input approach," *Int. J. Robust and Nonlinear Control*, vol. 23, no. 6, pp. 602–619, 2013.
[128] G. Wen, Z. Duan, W. Yu, and G. Chen, "Consensus of second-order multi-agent systems with delayed nonlinear dynamics and intermittent communications," *Int. J. Control*, vol. 86, no. 2, pp. 322–331, 2013.
[129] G. Wen, G. Hu, W. Yu, J. Cao, and G. Chen, "Consensus tracking for higher-order multi-agents systems with switching directed topologies and occasionally missing control inputs," *Syst. Control Lett.*, vol. 62, no. 12, pp. 1151–1158, 2013.
[130] G. Wen, W. Yu, G. Hu, J. Cao, and X. Yu, "Pinning synchronization of directed networks with switching topologies: A multiple Lyapunov functions approach," *IEEE Trans. Neural Netw. Learn. Syst.*, vol. 26, no. 6, pp. 3239–3250, 2015.
[131] P. Wieland, J.S. Kim, H. Scheu, and F. Allgöwer, "On consensus in multi-agent systems with linear higher-order agents," in *Proc. 17th Word Congress IFAC*, pp. 1541–1546, Jul. 2008.
[132] C. Wu, "Synchronization in arrays of coupled nonlinear systems with delay and nonreciprocal time-varying coupling," *IEEE Trans. Circuits Syst. II*, vol. 52, no. 5, pp. 282–286, 2005.

[133] C. Wu, "Synchronization in networks of nonlinear dynamical systems coupled via a directed graph," *Nonlinearity*, vol. 18, no. 3, pp. 1057–1064, 2005.

[134] C. Wu and L.O. Chua, "Synchronization in an array of linearly coupled dynamical systems," *IEEE Trans. Circuits Syst. I*, vol. 42, no. 8, pp. 430–447, 1995.

[135] J. Xiang and G. Chen, "On the V-stability of complex dynamical networks," *Automatica*, vol. 43, no. 6, pp. 1049–1057, 2007.

[136] F. Yang, Z. Wang, and Y.S. Hung, "Robust Kalman filtering for discrete time-varying uncertain systems with multiplicative noises," *IEEE Trans. Auto. Contr.*, vol. 47, no. 7, pp. 1179–1183, 2002.

[137] W. Yu and J. Cao, "Stability and Hopf bifurcation analysis on a four-neuron BAM neural network with time delays," *Phys. Lett. A*, vol. 351, no. 1-2, pp. 64–78, 2006.

[138] W. Yu and J. Cao, "Stability and Hopf bifurcation on a two-neuron system with time delay in the frequency domain," *Int. J. Bifur. Chaos*, vol. 17, no. 4, pp. 1355–1366, 2007.

[139] W. Yu, J. Cao, and G. Chen, "Robust adaptive control of unknown modified Cohen-Grossberg neural networks with delay," *IEEE Trans. Circuits Syst. II*, vol. 54, no. 6, pp. 502–506, 2007.

[140] W. Yu, J. Cao, and G. Chen, "Stability and Hopf bifurcation of a general delayed recurrent neural network," *IEEE Trans. Neural Networks*, vol. 19, no. 5, pp. 845–854, 2008.

[141] W. Yu, J. Cao, G. Chen, J. Lü, J. Han, and W. Wei, "Local synchronization of a complex network model," *IEEE Trans. Systems, Man, and Cybernetics-Part B*, vol. 39, no. 1, pp. 230–241, 2009.

[142] W. Yu, J. Cao, and J. Lü, "Global synchronization of linearly hybrid coupled networks with time-varying delay," *SIAM J. Applied Dynamical Systems*, vol. 7, no. 1, pp. 108–133, 2008.

[143] W. Yu, J. Cao, K.W. Wong, and J. Lü, "New communication schemes based on adaptive synchronization," *Chaos*, vol. 17, no. 3, pp. 033114-1–033114-13, 2007.

[144] W. Yu, J. Cao, and K. Yuan, "Synchronization of switched system and application in communication," *Phys. Lett. A*, vol. 372, no. 24, pp. 4438–4445, 2008.

[145] W. Yu, G. Chen, J. Cao, J. Lü, and U. Parlitz, "Parameter identification of dynamical systems from time series," *Phy. Rev. E*, vol. 75, no. 6, 067201, 2007.

[146] W. Yu, G. Chen, and M. Cao, "Some necessary and sufficient conditions for second-order consensus in multi-agent dynamical systems," *Automatica*, vol. 46, no. 6, pp. 1089–1095, 2010.

[147] W. Yu, G. Chen, and M. Cao, "Consensus in directed networks of agents with nonlinear dynamics," *IEEE Trans. Auto. Contr.*, vol. 56, no. 6, pp. 1436–1441, 2011.

[148] W. Yu, G. Chen, M. Cao, J. Lü, and H. Zhang, "Swarming behaviors in multi-agent systems with nonlinear dynamics," *Chaos*, vol. 23, no. 4, pp. 043118-1–043118-12, 2013.

[149] W. Yu, G. Chen, M. Cao, and J. Kurths, "Second-order consensus for multi-agent systems with directed topologies and nonlinear dynamics," *IEEE Trans. Systems, Man, and Cybernetics-Part B*, vol. 40, no. 3, pp. 881–891, 2010.

[150] W. Yu, G. Chen, and M. Cao, "Distributed leader-follower flocking control for multi-agent dynamical systems with time-varying velocities," *Syst. Control Lett.*, vol. 59, no. 9, pp. 543–552, 2010.

[151] W. Yu, G. Chen, and J. Lü, "On pinning synchronization of complex dynamical networks," *Automatica*, vol. 45, no. 2, pp. 429–435, 2009.
[152] W. Yu, G. Chen, W. Ren, and M. Cao, "Delay-induced consensus and quasi-consensus in multi-agent dynamical systems," *IEEE Trans. Circuits Syst. I*, vol. 60, no. 10, pp. 2679–2687, 2013.
[153] W. Yu, G. Chen, W. Ren, J. Kurths, and W. Zheng, "Distributed higher order consensus protocols in multiagent dynamical systems," *IEEE Trans. Circuits Syst. I*, vol. 58, no. 8, pp. 1924–1932, 2011.
[154] W. Yu, G. Chen, Z. Wang, and W. Yang, "Distributed consensus filtering in sensor networks," *IEEE Trans. Systems, Man, and Cybernetics-Part B*, vol. 39, no. 6, pp. 1568–1577, 2009.
[155] W. Yu, J. Lü, G. Chen, Z. Duan, and Q. Zhou, "Estimating uncertain delayed genetic regulatory networks: an adaptive filtering approach," *IEEE Trans. Auto. Contr.*, vol. 54, no. 4, pp. 892–897, 2009.
[156] W. Yu, P. De Lellis, G. Chen, M. di Bernardo, and J. Kurths, "Distributed adaptive control of synchronization in complex networks," *IEEE Trans. Auto. Contr.*, vol. 57, no. 8, pp. 2153–2158, 2012.
[157] W. Yu, W. Ren, W. Zheng, G. Chen, and J. Lü, "Distributed control gains design for consensus in multi-agent systems with second-order nonlinear dynamics," *Automatica*, vol. 49, no. 7, pp. 2107–2115, 2013.
[158] W. Yu, W. Zheng, G. Chen, W. Ren, and J. Cao, "Second-order consensus in multi-agent dynamical systems with sampled position data," *Automatica*, vol. 47, no. 7, pp. 1496–1503, 2011.
[159] W. Yu, L. Zhou, J. Lü, X. Yu, and R. Lu, "Consensus in multi-agent systems with second-order dynamics and sampled data," *IEEE Trans. Indus. Infor.*, vol. 9, no. 4, pp. 2137–2146, 2013.
[160] C.H. Yuh, H. Bolouri, and E.H. Davidson, "Genomic cis-regulatory logic: experimental and computational analysis of a sea urchin gene," *Science*, vol. 279, no. 5358, pp. 1896–1902, 1998.
[161] Y. Zhang and Y. Tian, "Consentability and protocol design of multi-agent systems with stochastic switching topology," *Automatica*, vol. 45, no. 5, pp. 1195–1201, 2009.
[162] Y. Zhang and Y. Tian, "Consensus of data-sampled multi-agent systems with random communication delay and packet loss," *IEEE Trans. Auto. Contr.*, vol. 55, no. 4, pp. 939–943, 2010.
[163] H. Zhao, S. Xu, and D. Yuan, "Consensus of data-sampled multi-agent systems with Markovian switching topologies," *Asian J. Control*, vol. 14, no. 5, pp. 1366–1373, 2012.
[164] C. Zhou and J. Kurths, "Dynamical weights and enhanced synchronization in adaptive complex networks," *Phys. Rev. Lett.*, vol. 96, 164102, 2006.
[165] J. Zhou, J. Lu, and J. Lü, "Adaptive synchronization of an uncertain complex dynamical network," *IEEE Trans. Auto. Contr.*, vol. 51, no. 4, pp. 652–656, 2006.
[166] J. Zhou, J. Lu, and J. Lü, "Pinning adaptive synchronization of a general complex dynamical network," *Automatica*, vol. 44, no. 4, pp. 996–1003, 2008.

Index

algebraic connectivity, 19
α-lattice, 113
attraction and repulsion function, 74

balanced graph, 19
Barabási and Albert (BA), 1
Brownian motion, 199

collective potential function, 97
complex networks, 2
connected, 12
consensus, 14
  manifold, 15
  region, 63
coupling configuration matrix, 14
coupling strength, 15
Courant-Fischer min-max theorem, 22

delay-induced consensus, 214
delay-input approach, 145
directed
  path, 12
  rooted tree, 12
  spanning tree, 12
  tree, 12
distributed adaptive control, 176
distributed bounded consensus
  controllers, 200
  filters, 199, 200
distributed delay, 223

Erdös and Rényi (ER), 1

Filippov solution, 85
flocking control, 96
Frobenius normal form, 23

general algebraic connectivity, 19
generalized directional derivative, 100
generalized eigenvector, 34
generalized gradient, 100
global consensus, 16

higher-order consensus, 57

inner compling strength, 57
interaction force, 74
intermittent communication, 159

Kharitonov's theorem, 61
Kronecker product, 13

Distributed Cooperative Control of Multi-agent Systems, First Edition.
Wenwu Yu, Guanghui Wen, Guanrong Chen, and Jinde Cao.

Laplacian matrix, 12
leader-follower control, 64
local consensus, 15

multi-agent systems, 2

network spectrum, 125
Newmann and Watts (NW), 1
noisy environment, 80
nonlinear multi-agent systems, 15
nonsmooth analysis, 99

observers, 67
outer coupling strength, 57

pinning control, 98
pinning edges control, 178
positive vector, 21

quasi-consensus, 216

reducible, 12
regular function, 100

sampled data, 118
sampling periods, 122
Schur complement, 44
second-order consensus, 33
set-value map, 85
spectral radius, 21
strongly connected, 12
swarming behaviors, 73
switched topologies, 82

undirected, 12
unmanned air vehicles (UAVs), 1

velocity consensus, 97
virtual networks, 28

Watts and Strogatz (WS), 1